GEOGRAPHIES OF THE BOOK

Geographies of the Book

Edited by

MILES OGBORN
Queen Mary University of London, UK

CHARLES W.J. WITHERS
University of Edinburgh, UK

ASHGATE

Published by
Ashgate Publishing Limited
Wey Court East
Union Road
Farnham
Surrey, GU9 7PT
England

Ashgate Publishing Company
Suite 420
101 Cherry Street
Burlington
VT 05401-4405
USA

www.ashgate.com

British Library Cataloguing in Publication Data
Geographies of the book.
1. Books--History. 2. Transmission of texts--History.
3. Literature and globalization. 4. Publishers and
publishing--History.
I. Ogborn, Miles. II. Withers, Charles W. J.
002'.09-dc22

Library of Congress Cataloging-in-Publication Data
Geographies of the book / edited by Miles Ogborn and Charles W.J. Withers.
    p. cm.
  Includes bibliographical references and index.
  ISBN 978-0-7546-7850-2 (hardback) -- ISBN 978-0-7546-9675-9 (ebook)
1. Book industries and trade--Political aspects. 2. Book industries and trade--
Social aspects. 3. Books and reading--Political aspects. 4. Books and reading--Social
aspects. 5. Publishers and publishing--Political aspects. 6. Publishers and publishing--
Social aspects. 7. Literature and state. I. Ogborn, Miles. II. Withers, Charles W. J.

  Z278.G45 2010
  302.23'2--dc22

                                                                              2009045898

ISBN 9780754678502 (hbk)
ISBN 9780754696759 (ebk)

Printed and bound in Great Britain by
MPG Books Group, UK

Contents

PART III   GEOGRAPHIES OF RECEPTION

List of Figures and Tables

Tables

Notes on Contributors

Fiona A. Black is Director of the School of Information Management, Dalhousie University. Her research interests include work on public libraries' interactions with their communities and the application of geographic information systems (GIS) technology to the study of book history and print culture. She is (with Yvan Lamonde and Patricia L. Fleming) co-editor of *History of the Book in Canada, Volume 2, 1840–1918* (Toronto 2005).

Michael L. Dorn is Research Assistant Professor, Center for Medical Humanities, Compassionate Care, and Bioethics, Department of Preventive Medicine, Stony Brook University School of Medicine, Stony Brook, New York. His current research examines the place of medical geographic ideas in the formation of American regional identities, linking early conjectures and projections of Ohio Valley as a region for European settlement and civilization into the larger body of scholarship on the Atlantic World. He completed his PhD dissertation, 'Climate, alcohol and the American body politic: the medical and moral geographies of Daniel Drake (1785–1852),' at the University of Kentucky in 2003.

Aileen Fyfe is a Lecturer in the Department of History at the National University of Ireland in Galway. She is currently completing a monograph on the impact of steam technologies on the availability of cheap print. She is the author of *Science and Salvation: Evangelicals and Popular Science Publishing in Victorian Britain* (Chicago 2004) and, with Bernard Lightman, co-editor of *Science in the Marketplace: Nineteenth-Century Sites and Experiences* (Chicago 2007).

Deryck W. Holdsworth is Professor of Geography in the Department of Geography, Pennsylvania State University. In addition to his work linking the words and worlds of trade, he has research interests in the historical geographies of office districts, and in the roles of speculators and landowners in the emergence of the Philadelphia landscape. He is (with Peter Ennals) co-author of *Homeplace: The Making of the Canadian Dwelling over Three Centuries* (Toronto 1998).

Daniel Hopkins is an Associate Professor of Geography at the University of Missouri-Kansas City. His research interests centre upon the historical geography of Denmark's colonies in the Caribbean islands and on the west coast of Africa in the eighteenth and nineteenth centuries.

Innes M. Keighren is a Postdoctoral Research Associate in the Institute of Geography at the University of Edinburgh. He received his PhD from that University in 2008. He has published on the making and reception of geographical knowledge in the nineteenth century, and his book *Bringing Geography to Book: Ellen Semple and the Reception of Geographical Knowledge* will be published by I.B. Tauris in 2010.

Robert J. Mayhew is Professor of Historical Geography and Intellectual History at the University of Bristol. He has published extensively on geography, print culture and the humanities in the British geographical tradition. His books include *Enlightenment Geography: The Political Languages of British Geography, c.1650–1850* (London 2000) and *Landscape, Literature and English Religious Culture, 1660–1800* (London 2004).

Miles Ogborn is Professor of Geography at Queen Mary University of London. He is (with Charles Withers) co-editor of *Georgian Geographies: Essays on Space, Place and Landscape in the Eighteenth Century* (Manchester 2004) and, most recently, author of *Indian Ink: Script and Print in the Making of the English East India Company* (Chicago 2007) and *Global Lives: Britain and the World, 1550–1800* (Cambridge 2008).

Rosa Salzberg is the UK Society for Renaissance Studies Ruth and Nicolai Rubinstein Postdoctoral Fellow. She received her PhD from Queen Mary University of London in 2009, and is continuing to research the peddling of print and other items in Renaissance Italy.

Charles W.J. Withers is Professor of Historical Geography at the University of Edinburgh and a Fellow of the British Academy. Recent publications include *Geography and Revolution* (Chicago 2005), co-edited with David Livingstone, and *Placing the Enlightenment: Thinking Geographically about the Age of Reason* (Chicago 2007). His *Geography and Science in Britain 1831–1939: A Study of the British Association for the Advancement of Science* is due to be published in 2010.

Preface

If this book has a single defining originating time and place, it is over fifty years ago and in another country. The publication in 1958 in Paris of *L'Apparition du Livre* by Lucien Febvre and Henri-Jean Martin, translated into English as *The Coming of the Book*, marked a moment in the geographical examination of the importance of print and print culture to European civilisation. As perhaps no other work, their book helped set in train that lively scholarly enterprise we know now as the history of the book. It was also an exercise in scholarly collaboration across disciplinary boundaries. Where Henri-Jean Martin brought his expertise as one of the world's leading authorities of the history of the book to bear on the problem of the movement over space of print and print culture, Lucien Febvre brought the craft of the historical geographer and recognition of the importance of place and culture to the making of meaning in print. So, too, it is our concern in this collaborative and inter-disciplinary volume to bring together scholars in different fields – in geography, in history, in book history, intellectual history and the history of science especially and in the humanities generally – in order to continue conversations and to develop research agendas around the geographical study of print in historical context.

More particularly, this collection has its origins in conference sessions organised under the title 'Geographies of the Book', held at the annual meeting of the Association of American Geographers in Philadelphia in 2004, and at the Material Cultures meeting in the University of Edinburgh in July 2005. The papers presented in these two cities, well known for the part they played in the history of print culture, provided the germ of the idea that the time might indeed be right for a more sustained enquiry into geography and the book from different scholarly perspectives, and from younger scholars as well as from more established figures. Through invitation, discussion and revision the collection has taken the shape that it has, and we are delighted with the shape it has taken. Our contributors make their own acknowledgements in their own chapters to those who have assisted them in their work: here, it is our joint great pleasure to acknowledge not just the quality of their individual chapters, but also their collective forbearance and tolerance as we have sought to bring this collection forward for publication. It is with pleasure too that we acknowledge the support we have received from Valerie Rose and the publishing staff at Ashgate: she and they have been all we could have wished for. We would also like to thank Edward Oliver of the Department of Geography, Queen Mary University of London who drew Figures 2.1, 2.2 and 4.1 with such great skill. Others, even if they have not always known it, have likewise been there when we needed them: colleagues, archivists and librarians and, importantly, Catherine Nash and Anne Withers. We owe you all a great deal.

Miles Ogborn (London)
Charles W.J. Withers (Edinburgh)

Introduction: Book Geography, Book History

Miles Ogborn and Charles W.J. Withers

The Coming of the Geography of the Book

The geography of the book is as old as the history of the book. Yet it can also be said to still be defining what it might become.[1] In *The Coming of the Book* (1958), Lucien Febvre and Henri-Jean Martin's seminal examination of the impact of printing since its development in Europe in the middle of the fifteenth century, the authors devoted a chapter to the geography of the book. The chapter comes after their account of the introduction of paper in Europe; after their discussion of the solution of the technical problems of printing with moveable type; after their delineation of the forms that the printed book took and its relationship to, and differences from, the manuscripts that preceded it; after their accounting of the economics of the book as a commodity; and after their depiction of the relationships between journeymen, masters and authors in the print shop's 'little world of the book'. In charting what they describe as the geography of the book their chapter considers the wider world of the printed codex. It begins when printing has been invented, when the form of the book has been settled, and when this new invention, its producers and its products are ready to enter the world. Recognizing that this invention is already too well known and too important to be kept secret, the apprentices of Gutenberg and Schoeffer, and the workmen who had learned from those apprentices, begin an epic series of journeys. For Febvre and Martin:

> They were willing to leave their master's shop and travel the length and breadth of Europe, like many other journeymen of the period, carrying their equipment with them and practising and instructing in the new art. They must have led nomadic lives, their chief asset experience, their equipment of the most elementary sort; they would stop in a town, hope for orders locally, probably suffering poverty very often. What they sought was someone to provide capital so that they could

---

1   "If the history of the book is now an established discipline, the geography of the book is still making up its rules", Leah Price quoted in Innes Keighren, "Bringing Geography to the Book: Charting the Reception of *Influences of Geographic Environment*," *Transactions of the Institute of British Geographers* 31 (2006), 528.

establish themselves permanently, and a town which met the conditions for
the establishment of a successful printing shop. Nothing stopped them: did not
Jerome Münzer, a doctor of Nuremberg, meet three German printers already
established in Grenada in 1494, only two years after the city's liberation from
the Arabs? Two others, from Strasbourg and Nordlingen, settled in far away São
Tomé, an unhealthy island off the African coast in the Gulf of Guinea.[2]

So, too, Febvre and Martin laid out the road they and others have travelled
in thinking about the geography of the book. By investigating the interests of
influential figures and institutions in making texts accessible, and the economic
attractions that certain places offered for printing, Febvre and Martin set out a
three-century-long history of how printing 'spread slowly across the whole of
Western Europe'. The church played a significant role, with its demand for missals
and breviaries. The Archbishop of Mainz described printing as 'a divine art' which
spread the word of God far and wide. Yet piety alone was not enough. Printing
only flourished where there was a wider market for its products. Presses thrived
in university towns, printing the works of Aristotle, Duns Scotus and Galen for
scholars to teach and argue over, as well as devotional works. Printers flourished
where there were lawyers too, selling law books and works of literature and
current affairs. But the esoteric world of scholarship and law were not sufficient
either. Printers needed a broader demand to 'put their business on a firm footing
and make profits'.[3] The geography of printing was always shaped by the benefits
of supplying a popular market with cheap little books that could be frequently
reprinted.

The geography of the book was from the outset associated with the cartography
of print production and mobility. With the aid of distribution maps, showing where
and when printing was established, Febvre and Martin demonstrated the adoption
of printing in Europe in the decades before 1500. As they put it, 'If we consider
all the difficulties involved in setting up a new industry formed out of a number
of different skills and processes, and in building up the commercial contacts to
provide outlets for mass-produced books, then we must admit that printing in fact
spread surprisingly rapidly, and that the men of the 15th century were particularly
keen to innovate, as a few dates and a glance at a map will show'.[4] Their maps
show clusters in northern Italy, the German Rhinelands and the low countries.
They also showed, by about 1500, a patterning of printers across Europe in 236
towns from Faro in the south-west to Danzig in the north-east, with important

---

2   Lucien Febvre and Henri-Jean Martin, *The Coming of the Book: The Impact of
Printing, 1450–1800* (London: Verso, 1976, orig. 1958), 167–8. For earlier uses of the term
see, A. Emile Egger, *Histoire du Livre Depuis ses Origins Jusqu'a Nos Jours* (Paris: Hetzel,
1880) and François de Dainville, "La Géographie du Livre en France de 1764 à 1945," *Le
Courier Graphique* (1951): Jan–Feb, Mar–Apr.

3   Febvre and Martin, *The Coming of the Book*, 170–1.

4   Ibid., 181.

centres in Venice, Paris, Lyons and Leipzig. This geography of the book was not one that simply conformed to emerging nation-state boundaries. France was 'typographically two zones'.[5] Paris dominated the north, in close contact with Cologne and Basel, and produced books for the English market too. Lyons ruled the south, also dealing with the printers of Basel and the Rhineland, and with Italian and Spanish booksellers. During the sixteenth century this European typographic landscape was reshaped by the demands and prohibitions of Reformation and Counter-Reformation: publishing suffered in Leipzig, for example, but expanded in Wittenburg.

There were also counter tendencies, reshaping the map in accordance with the hardening borders of nation-states. English book production was initially heavily dependent upon expertise from the continent. From 1476 to 1536, two-thirds of the printers, booksellers and bookbinders resident in England were from elsewhere. Along with other Reformation measures limiting the export of books from the continent, legislation passed in 1523 restricted the numbers of foreign apprentices and journeymen employed by English print shops. These tendencies were reinforced in the seventeenth century with the rise of vernacular printing, which meant that 'the major part of the book trade ceased to be an international European affair'.[6]

What of beyond Europe? In a section of their formative chapter entitled 'Printing Conquers the World', Febvre and Martin, with the help of Mme A. Basanoff (on the Slav countries) and the Reverend Father Henri Bernard-Maitre (on the Far East) tried in a few pages to assess how the new technology fared in the countries of Northern Europe 'with their less dense populations and their greater distance from the original centres of the trade'; in the Slav countries which used a different alphabet; in the New World 'where it was necessary to master vast expanses of territory which long remained almost uninhabited'; and in Asia which already had comparable techniques of reproduction. In dealing with the Americas and Asia they sought to examine the role of the printing press in an epoch in which 'Western civilization has acted to transform the rest of the world'.[7] In South America, this included the Spanish printing of chivalric romances which fired the imaginations of the Conquistadors, and the printing – first in Madrid and Antwerp, later in Mexico City (1539) and Lima (1584) – of the volumes necessary for missionary work among the indigenous people and 'the indispensable works of religious piety'.[8] In seventeenth-century North America, printers survived only if they were appointed official printer to one of the colonies, and were paid to publish its laws and regulations, along with a less profitable diet of almanacs, ABCs, sermons and prayer books. By the eighteenth century, with increased European settlement, printers discovered the profits that could be made from newspapers

5   Ibid., 189.
6   Ibid., 195.
7   Ibid., 207.
8   Ibid., 208.

and became that 'essentially American phenomenon' the printer-journalist, and sometimes even printer-journalist-postmaster to ensure distribution as well as content and production.[9]

Looking to the east, there were presses operated by the Portuguese in Goa in 1557, Macao in 1588 and Nagasaki in 1590. In China, as the authors have it, Jesuit missionaries encountered, initially used, and then quickly moved to replace the already existing method of reproducing writing by printing with wood-cuts: xylography. As the chapter concludes, racing through the centuries, the printed book played an important role in the process by which 'Europe achieved almost total dominance in the East'.[10] In tracing this historical geography, the authors moved a long way from Gutenberg's little print shop in mid fifteenth-century Mainz.

As a 'Geography of the Book' this account has much to recommend it, and much that will be echoed in the chapters that follow. Febvre and Martin's account is concerned with a range of geographies at a series of scales from the local factors of supply and demand that affect the location and success of print shops to the grand dramas of the Reformation and the making of empires. They attempted to weave these scales together in explaining where printing was profitable and useful, and where its spread was more or less rapid. They paid attention to what the reconstruction of the changing geographical distribution of printers and their presses can tell us about the important historical process which goes under the name of the 'impact' of printing. Questions of economic, political, intellectual and cultural history are tied together in their explanation of these geographies. Such concerns continue to animate many historians of the book, and now include the use of new tools such as Geographical Information Systems (GIS) which can produce maps far more sensitive to variety, pattern and pace of change than Febvre and Martin's cartographers were able to.[11]

Yet there are limitations here too. Febvre and Martin's 'Geography of the Book' fits into *The Coming of the Book* as only one particular part of the jigsaw. It accounts for the spread of an already formed and fixed technology, the printing-press with moveable type; its product, the printed codex; and its producers, the printers, apprentices and journeymen who worked the press. It is, in essence, a story of the diffusion of something (or some things) already made. As a result it has little to say about how we might construct an account of the technologies of making books (printed or otherwise), of those involved in producing those books, of the books themselves and, crucially, of their readers which would show how their diverse geographies can illuminate how those people and things actually come to

---

9   Ibid., 211.

10   Ibid., 215.

11   Alexis Weedon, "The Uses of Quantification," in *A Companion to the History of the Book*, ed. Simon Eliot and Jonathan Rose (Oxford: Blackwell Publishing, 2007), 33–49; Fiona A. Black, Bertrum H. MacDonald, and J. Malcolm W. Black, "Geographic Information Systems: A New Research Method for Book History," *Book History* 1 (1998): 11–31.

be as they are. How questions of geography are central to the very constitution of 'the book' itself. Such an alternative account could also show that what is taken by the diffusionist vision of the impact of printing to be the same (a book, a scribe, a reader, or a printing press) might be constituted quite differently in different places. And what is taken to be different (for example, wood-block printing versus moveable type) might be more similar than expected.[12]

Our aim in this introduction, therefore, is to use a broader body of work on books, and on the histories and geographies of knowledge, to sketch out what this sort of geography of the book might look like. We start with a significant dispute about the nature of books, of reading and of 'print culture' that turns out to be an argument about geography. It also shows how deeply geography is involved in the production, distribution and consumption of books, and how that makes a difference to the ways in which books and their histories should be understood.

Geographies of the Book

In her influential and complex work *The Printing Press as an Agent of Change* (1979) Elizabeth Eisenstein argued that printing changed the conditions under which texts were produced and read, and, therefore, changed the sort of knowledge that was produced. She took issue with what she saw as Febvre and Martin's gradualism and stressed the advent of printing as a revolutionary event rather than an evolutionary change. Yet that historical 'event' was in fact the very geographical distribution identified in *The Coming of the Book*. As Eisenstein says, 'My point of departure ... is not one device invented in one Mainz shop but the establishment of print shops in many urban centers throughout Europe over the course of two decades or so' in the fifteenth century.[13] Concentrating on Europe's Latin-reading intellectual elite, she argued that these printing presses, in contrast to the scribal practices that preceded them, brought a relative fixity and standardization to the texts that readers encountered, and a broader dissemination of those texts. What became revolutionary, then, was not the effect on any single text of producing it with moveable type. Instead, it was the distribution across Europe of many texts that could be read by many people in many places with at least the assumption that other people in other places were looking at the same thing. Printed words alone might be relatively incidental here. Images produced by woodblock printing, and combined with type, played a crucial role in her argument. For Eisenstein, 'The

---

12   Kai-Wing Chow, "Reinventing Gutenberg: Woodblock and Moveable-Type Printing in Europe and China," in *Agent of Change: Print Culture Studies after Elizabeth L. Eisenstein*, ed. Sabrina Alcorn Baron, Eric N. Lindquist and Eleanor F. Shevlin (Amherst and Boston: University of Massachusetts Press, 2007): 169–92.

13   Elizabeth L. Eisenstein, *The Printing Press as an Agent of Change: Communications and Cultural Transformations in Early-Modern Europe* (Cambridge: Cambridge University Press, 1979), xv.

fact that identical images, maps and diagrams could be viewed simultaneously by scattered readers constituted a kind of communications revolution in itself'.[14] In natural philosophy this had decisive effects, since 'The advantages of issuing identical images bearing identical labels to scattered observers who could feed back information to publishers enabled astronomers, geographers, botanists and zoologists to expand data pools far beyond all previous limits'.[15] Printing, and the products of the press, allowed, in Eisenstein's view, a different orientation to knowledge because it made possible the preservation of information, its accumulation, comparison and correction, and its wide dissemination. A new geography of (printed) books is what makes a new sort of knowledge possible.

The account of printing in *The Printing Press as an Agent of Change* has been the subject of much criticism and debate.[16] In a significant questioning of Eisenstein's emphasis on the revolutionary impact of printing and its effects on natural philosophy, Adrian Johns focused on the attribution of fixity and the standardization of texts to the process of printing, and to the printing press itself.[17] While Eisenstein admitted that early printed texts could never be absolutely standardized due to the nature of the work methods of the print shop, she argued that quibbling over the details of a few errors or emendations here or there would obscure the crucial fact that print had meant that these texts 'were sufficiently uniform for scholars in different regions to correspond with each other about the same citations and for the same emendations and errors to be spotted by many eyes'.[18] They were standardized and fixed enough. Johns, in contrast, argued that the very existence of the possibility of non-standard, non-fixed texts as a result of hand-press printing, meant that authors, and more particularly the authors of works of natural philosophy, had to worry over the production of their books. They had to worry about how to make sure that they were standard and fixed enough. They had to be concerned with the processes of the production of books. So while Eisenstein celebrates a new meeting of diverse interests in the places of print production as print 'brought bookworms and mechanics together in person as collaborators within the same workshops', Johns focuses attention on the ugly battles between stationers, printers and authors over the production of texts in the very same places where each side tried to fight for its own interests.[19] For Johns, it is the tense social relationships of print production, and of the print shop itself, that become crucial in determining the nature of the books that were produced,

---

14   Ibid., 53.

15   Ibid., 687.

16   For a range of views, see Sabrina Alcorn Baron, Eric N. Lindquist and Eleanor F. Shevlin, eds, *Agent of Change: Print Culture Studies after Elizabeth L. Eisenstein* (Amherst and Boston: University of Massachusetts Press, 2007).

17   Adrian Johns, *The Nature of the Book: Print and Knowledge in the Making* (Chicago: University of Chicago Press, 1998).

18   Eisenstein, *The Printing Press as an Agent of Change*, 81.

19   Ibid., 250.

including their standardization and fixity, and the credibility that they might then claim from their readers.

What is important for us here is that Johns's account of these social relationships is also strikingly geographical even as it presents quite a different geography from that offered by Eisenstein. In *The Nature of the Book*, Johns presents a detailed 'social geography' of the spaces of print production in early modern London.[20] His account works through a series of 'domains' or 'social spaces' that were the 'abodes of the book trade', 'examples of which include the printing house and the bookshop, but also the city square, courtroom, and coffee-house – [which] were dynamic localities defined by physical environment, work, and sociability. Discrete but interlocking, they both exhibited and were constituted by particular clusters of representations, practices and skills'.[21] This account centres on the work of stationers in their printing houses and bookshops, but sets that within a constellation of connected spaces that reaches out into the urban environment to provide 'a topography of the places of printed material in early modern London'.[22] Moreover, the intention is not simply to detail the processes of printing, and the distribution and use of printed materials, in one place and at one time. Instead, Johns seeks to make an argument about how these spaces and places, this micro-geography of the book, shaped the knowledge that was available at the time of the "Scientific Revolution" of the late seventeenth century.

This historical geography of print in England's capital city details the people, institutions and localities associated with printing. This is a geography of bookshops around Amen Corner, St Paul's Churchyard and Paternoster Row and along Cheapside, Fleet Street and over London Bridge. There were, in addition, the many coffee-shops where newspapers were available to all who could pay for their cup, and that might be read aloud to those who could not read them themselves. Outside bookshops and coffee-shops there were chapmen, higglers and 'Mercury-women' who sold printed material in the streets and to a wider market beyond the city. Print shops themselves were more widely dispersed than the bookshops. They were to be found in Blackfriars and Smithfield, and close to St Paul's Cathedral.

Where those making and selling books were found, and what these places were themselves like, mattered. Johns's argument is that the character of these localities, and of the individual printing houses themselves, what we might think of as their 'cultural geography', affected the 'perceived epistemic status' of the works they produced.[23] These strange mixtures of 'library, scriptorium, study, home, and workshop' were complex spaces.[24] They organized particular technologies, skills

---

20   Johns, *The Nature of the Book*, 62. Interestingly he gets this term from Felicity Heal, *Hospitality in Early Modern England* (Oxford: Oxford University Press, 1990) rather than from the long tradition of social geography itself.

21   Johns, *The Nature of the Book*, 59–60.

22   Ibid., 61.

23   Ibid., 73.

24   Ibid., 75.

and workers in specific ways. They were artisanal spaces in which those who worked with print – pressmen, compositors, correctors, 'devils', 'spirits' and 'flies' – shaped the way in which a text came into the world as a material object. The many processes, decisions and practices involved meant that the places of production had a significant effect on what 'books' themselves actually were.

This geography had other important effects. Significantly, these were homes as well as places of work, with domestic rooms above the printing presses, and often a shop on the ground floor. What was important was the good order of these spaces, something that mixed their social order, the political and religious views of the stationer, and their printing practices: from the quality of their work to their proclivity for piracy. These were spaces that were supposed to be shaped by the moral principles of domesticity and patriarchy, such that 'one of the costs of failing to conform to the domestic ideal could therefore be a questioning not just of the printer, but of the texts he or she produced'.[25] Concerned, along with other historians of science, with matters of credit, credibility and civility in the making of new forms of knowledge, Johns argues that one way in which early modern Londoners read books and decided whether to trust them was through their knowledge of *where* they had been produced.[26] As he puts it, 'Whether a book contained safe, reliable knowledge could be questioned by asking whether it had been produced in conditions of propriety, or affirmed by asserting that it had'.[27] Johns' aim is to show that such geographical considerations played a crucial role in the making and crediting of natural philosophical knowledge in print, such as in the epic battle over the nature of the heavens between Sir Isaac Newton and John Flamsteed, the Astronomer Royal. In his account of this conflict, familiarity with printing houses, bookshops, and coffee-shops, with what went on within them and how it might be turned to advantage, is as important as the ideas that were being fought over. What is significant in the geography of the book for Johns is not the new patterns of distribution of adequately standardized texts and images identified by Eisenstein, but the intricate and small-scale geographies of where those texts were produced in the first place.

The differences between these positions have themselves been fought over in print. Elizabeth Eisenstein herself points out that these approaches to print culture 'diverge on the "geography of the book"', but she restricts that divergence to a matter of scale – the difference between Johns's study of English printing practices and her own 'cosmopolitan' approach to the history of printing in Western

---

25   Ibid., 77.

26   These concerns are developed in Steven Shapin and Simon Shaffer, *Leviathan and the Air-Pump: Hobbes, Boyle, and the Experimental Life* (Princeton: Princeton University Press, 1985) and Steven Shapin, *A Social History of Truth: Civility and Science in Seventeenth-Century England* (Chicago: University of Chicago Press, 1994).

27   Johns, *The Nature of the Book*, 128.

Europe.[28] While significant, this divergence is less important than the more basic differences of approach between an account of the 'far-flung book trade' and one that adopts 'a local perspective' on the practices of print.[29] Each takes the geographies of books as vital to the historical changes that they were part of, but in different ways. The former, Eisenstein argues, meant that 'atlases, ephemeredes, logarithm tables, and other paper tools that were used throughout a cosmopolitan commonwealth of learning did transcend locale'. As she reiterates, 'printed copies *were* sufficiently alike to change conditions within the learned world, to make it possible, for example, for scholars to correspond about a common text – such as Copernicus's *De Revolutionibus* – while referring to the same passages'.[30] The latter, Johns replies, would provide a cultural history of print that recognizes that 'the decisions and actions that sustained the world of the book not only occurred in specific situations but took their character from their particularities'. Moreover, 'if one's concern is to understand processes, then a concentration on the places where they occurred is nevertheless appropriate'.[31] Questions of the 'cosmopolitan' and the 'local' seem to present two contrasting ways in which the geography of books is essential to the history of the book.

If seen in this way – as a difference of geographical interpretation and of geographical scale rather than through the historical lens as either acknowledging a 'print revolution' or flatly denying it – the possibility of some reconciliation between these perspectives seems to open up. Roger Chartier points out that what he calls the 'dissemination' and 'constructivist' models need to be brought together in order to understand printing within a global history of trade and empire which has to account for how Western printing techniques could come to be invested with such divergent meanings, and be part of such diverse practices, in different parts of the world. He also notes that beyond the arena of the history of science – where notions of comparability, standardization, credibility and fact are particularly highly charged – how readers read books might be quite different.[32] With that in mind we are perhaps left with the sense that printed materials (and there is no reason why the same cannot be said of scribal ones) and their place in history should be understood in terms of both their geographical distribution (how far were they flung, and who and where were they flung to?) and the local conditions of their production, movement and consumption.

Within the history of science, James Secord has made similar arguments. Taking issue with the 'localist' turn in the histories and geographies of science he has

---

28   Elizabeth L. Eisenstein, "An Unacknowledged Revolution Revisited," *American Historical Review* 107:1 (2002): 90.

29   Ibid., 95 and Adrian Johns, "How to Acknowledge a Revolution," *American Historical Review* 107:1 (2002): 125.

30   Eisenstein, "An Unacknowledged Revolution Revisited," 94–5.

31   Johns, "How to Acknowledge a Revolution," 117.

32   Roger Chartier, "The Printing Revolution: A Reappraisal," in *Agent of Change*, ed. Baron, Lindquist and Shevlin, 397–408.

argued that the making of scientific knowledge is always a matter of communicative action. The 'construction' of knowledge and its 'dissemination' cannot be treated as separable processes. Developing an idea of 'knowledge in transit', he argues for research that will understand the ways in which knowledge is not just local but is made mobile between different places. As he argues, 'It is not so much a question of seeing how knowledge transcends the local circumstances of its production but instead of seeing how every local situation has within it connections with and possibilities for interaction with other settings'.[33]

Overall, therefore, taking the geography of the book seriously in parallel with the history of the book means that such geographies must be about more than just mapping the distribution of printers, printing presses and printed words. As we have argued, the geography of the book enters into the very nature of the book itself. Geographical questions, and they can be of many different sorts, address fundamental issues of the production, distribution and consumption of books. We must, at one and the same time, explore the local places in which written materials were produced, and their impact on the nature of books as material and signifying objects; the patterns of dispersal and modalities of movement through which books travelled, and the implications of those for forms of knowledge; and the geographical positioning of readers whose located reading practices shape how books were consumed.[34] In what follows we explore some of the ways in which that has been done. Our account is organized in terms of the key conceptual developments which have led the resurgence of interest in the history of the book: the book as a material object, understood in terms of production and distribution; and questions of reading, the consumption of books.

Making and Moving Books: Geographies of Production and Circulation

Key works in the history of the book have emphasized the material nature of books.[35] Robert Darnton has focused on the practicalities of gathering paper, print and the workers to combine them in the making of the *Encyclopédie*. His 'communications circuit' is built upon following the 'book' as it moves between the various 'book people' that are part of the 'different stages of its life cycle'.[36] Donald McKenzie has delved into the practices of the printing house to demonstrate

---

33   James A. Secord, "Knowledge in Transit," *Isis* 95 (2004): 664.

34   We have attempted this in different ways in Miles Ogborn, *Indian Ink: Script and Print in the Making of the English East India Company* (Chicago: University of Chicago Press, 2007) and Charles W. J. Withers, *Placing the Enlightenment: Thinking Geographically about the Age of Reason* (Chicago: University of Chicago Press, 2007).

35   For example, the main section of *Companion to the History of the Book*, ed. Eliot and Rose, is entitled "The History of the Material Text."

36   Robert Darnton, *The Business of Enlightenment: A Publishing History of the* Encyclopédie, *1775–1800* (Cambridge: The Belknap Press of Harvard University Press,

the impact of processes such as concurrent printing on the final form that a book takes. More expansively, he has developed a notion of the sociology of the text in order to investigate the ways a text comes into the world as a particular sort of material object – as manuscript, printed book, broadsheet or digital file – and the impact that that has on its future uses.[37] Similarly, Roger Chartier has tried to elucidate the complex processes whereby the materiality of the book is part of the meanings made by readers, neither fully determining them, nor open to any and all interpretations.[38]

These concerns with materiality are ones that mark the history of the book off from other versions of the historical study of literature, ideas or systems of thought. They are shared by Febvre and Martin, Eisenstein, Johns and Secord as they chart in their different ways the history of the effects of printing technologies on knowledge. There are also endeavours in literary studies and science studies which seek to understand printed words on paper as an 'intellectual tool' for managing and producing knowledge. For example, Neil Rhodes and Jonathan Sawday in the *Renaissance Computer*, their exploration of knowledge technology in the first age of print, and Bruno Latour in *Science in Action*, both develop the idea of a 'paper world'.[39] For Rhodes and Sawday this term signals a new world of extensive print production and circulation which helped to create a European network of scholars for whom the book was the technology – with its 'tables of contents and indexes, running heads, chapters and partitions' – which enabled the searching, compilation and comparison of information from many different sources. As they put it, 'print taught readers what to expect of a book, and enabled them to pass with ease from one book to another'.[40] For Latour, paper, and printing on paper, are also part of practices that permit new ways of dealing with information and making knowledge. Various forms of inscription permit the displacements of the things inscribed from one site to another 'without

---

1979); Idem., "What is the History of Books?" *Daedalus* 111:3 (1982): 65–83; Idem., "'What is the History of Books?' Revisited," *Modern Intellectual History* 4:3 (2007): 504.

37 Donald F. McKenzie, *Bibliography and the Sociology of Texts* (London: British Library, 1985); Idem., *Making Meaning: "Printers of the Mind" and Other Essays*, ed. Peter McDonald and Michael F. Suarez (Amherst and Boston: University of Massachusetts Press, 2002).

38 Roger Chartier, *The Order of Books: Readers, Authors, and Libraries in Europe between the Fourteenth and Eighteenth Centuries* (Stanford: Stanford University Press, 1994); Idem., *Forms and Meanings: Texts, Performances, and Audiences from Codex to Computer* (Philadelphia: University of Pennsylvania Press, 1995); Idem., "*The Order of Books* Revisited," *Modern Intellectual History* 4:3 (2007): 509–19.

39 Neil Rhodes and Jonathan Sawday, "Introduction. Paperworlds: Imagining the Renaissance Computer," in *The Renaissance Computer: Knowledge Technology in the First Age of Print*, ed. Neil Rhodes and Jonathan Sawday (London: Routledge, 2000), 1; Bruno Latour, *Science in Action: How to Follow Scientists and Engineers Through Society* (Cambridge: Harvard University Press, 1987), 226.

40 Rhodes and Sawday, "Introduction. Paperworld," 7–8.

deformation through massive transformations'.[41] Using examples such as the late eighteenth-century transcribing of a map of Sakhalin Island drawn in the sand of a Pacific beach onto paper and its subsequent transportation back to France and Tycho Brahe's collation of observations of the heavens from astronomers across Europe on his own pre-printed forms, Latour developed the notion of 'immutable and combinable mobiles'. These vehicles allow things such as a coastline or the stars to be displaced to another site – to Versailles or to Brahe's observatory on Hven island – through their transformation into another material (paper) form. Once there they can be combined, compared and worked on in new ways, with other maps and other observations, permitting the world or the skies to be seen anew. As Latour puts it, emphasizing the difference that paper makes, 'they all end up at such scale that a few men or women can dominate them by sight; at one point or another, they all take the shape of a flat surface or paper that can be archived, pinned on a wall and combined with others'. This is, therefore, a matter of material rather than simply intellectual transformation since, as Latour stresses, 'Many things can be done with this paper world that cannot be done with the world'.[42]

All of these considerations of materiality ground the history of books in the stuff of the world. They seek to understand books as part of a range of material cultures within which their meanings are produced through particular work with pens, paper, ink and printing presses that make specific sorts of signifying objects. Their production is understood as a material process that shapes what books are, what they mean and how they are used.[43] The materiality of books – their size, shape, quality, portability, and the marks of their significance or insignificance – shapes what is done with them: everything from being revered to being rejected.[44] Most significantly here, the recognition of the materiality of the book means that it is an object that must have a geography. Its making can be located and its movements can be mapped. Its history as an object is shaped by where it is made and where it can subsequently be found. This makes for many possible geographies. The material conditions of the production and circulation of books take place on every scale from that of the printed page itself to the global networks of trade and empire.

---

41   Bruno Latour, *Reassembling the Social: An Introduction to Actor-Network-Theory* (Oxford: Clarendon Press, 2005), 223.

42   Latour, *Science in Action,* 226–7. Latour's interpretations of these particular examples have been questioned in Michael T. Bravo, "Ethnographic Navigation and the Geographical Gift," in *Geography and Enlightenment*, ed. David N. Livingstone and Charles W.J. Withers (Chicago: University of Chicago Press, 1999), 199–235 and Johns, "How to Acknowledge a Revolution."

43   On the particular meanings of varieties of paper within an office, although not usually within books, see Andrea Pellegram, "The Meaning in Paper," in *Material Cultures: Why Some Things Matter*, ed. Daniel Miller (London: UCL Press, 1998), 103–20.

44   For the Christian bible see, for example, Matthew Engelke, "Text and Performance in an African Church: The Book, 'Live and Direct'," *American Ethnologist* 31:1 (2004): 76–91.

It might be seen as rather stretching the definition of the geography of the book to include within it the space of the page. Yet the very nature of writing, and of printing, depends on the disposition of things in space, and the interpretation of that patterning depends on the different ways in which it might be done. Take the tiny space before or after a comma or a full stop. Should it be there or not? What can it tell us? Bertrand-Quinquet, in his *Traité de l'Imprimerie*, published in post-Revolutionary Paris in An VII, was certain that these spaces had a national geography. For the comma, 'The English, the Germans and the Swiss always place it immediately after the letter, without space. The Italians and Spanish put it between two equal spaces; the French between two unequal spaces, the one before the comma being less pronounced than the one following it'.[45] The historical bibliographer Donald McKenzie was less sure. Considering the spaces around 13,777 commas in the second edition of John Beaumont's *Psyche* alongside the archival evidence of its production at the Cambridge University Press in 1702, he demonstrated conclusively that there was no pattern, national or otherwise, which connected the particular compositors who composed the text – and who were, as it happens, English, French, Dutch and German – to particular configurations of spacing. More significantly, this exercise in examining tiny spaces – the printed "things" in them and the gaps between them – was, for McKenzie, part of his insistence that the social relations and work processes of the printing house cannot simply be read off from the nature of the text itself.

The printed page can reveal other geographies too. William Sherman's examination of the 'paratexts' of early English printed books – 'the verbal and visual accessories that accompany texts and present them to the public (including covers, title pages, prefaces, and tables of contents)' – reveals a striking array of geographical figures. This 'front matter' becomes a literal threshold leading the reader into the book via 'an architectural space – a gateway, arch, portico, or porch through which the reader entered the text'. As he argues, the new experience of making and reading printed books required, at least at first, ways of presenting them to their audiences which would key into familiar spaces – of church, school house and home – and their entrances.[46]

Giving similar attention to the meaning of the printed page, Robert Mayhew has charted long-term changes in the disposition of the page of geography books which he relates to changes in the sort of knowledge that there was about the world. As he argues, the discovery of the New World produced a crisis in geographical knowledge that was resolved in geography books from Peter Heylyn's *Microcosmus* (1621) onwards via a 'print space' of dense structured paragraphing that allowed the cumulative compilation of new information without disturbing the basic organization of the book. This format for geographical grammars, and their use as

---

45   Quoted in Donald F. McKenzie, "Stretching a Point; Or, the Case of the Spaced-Out Comps," in *Making Meaning*, 94.

46   Willam H. Sherman, "On the Threshold: Architecture, Paratext, and Early Print Culture,' in *Agent of Change*, ed. Baron, Lindquist and Shevlin, 68 and 72.

works of reference, lasted until the late nineteenth century when "the closing of the world", the end of the age of new 'discoveries', required a new geography and with it a new print space. As Mayhew shows, for Halford Mackinder and others this was to involve continuous narrative prose, the integration of text and image, and easy-on-the-eye typography.[47] In his chapter in this collection, Mayhew extends this argument by charting how the editing of the Dutch geographer Bernhard Varenius' texts in England produced his works in a form where the typography, the layout of the printed page and the paratexts all promoted Varenius as a revered classical author. Further, Mayhew locates his interpretation of the editing and production of Varenius' books in the practices of the printing houses – in Cambridge and London – where they were produced and in national typographic traditions which shaped the space of the page.

This emphasis on locality, already highlighted in relation to seventeenth-century London, can be found elsewhere. James Raven has recently used a range of unfamiliar sources for book history – property tax records, fire insurance archives and trade directories – to chart what he calls the topography of the London book trades in the eighteenth century. He maps out the clusters of activity, particularly around St Paul's Churchyard and Paternoster Row, and focuses in to show, street by street, building by building, the space occupied by booksellers and printing houses. Where there are dense patterns of activity he is able to demonstrate how booksellers benefited from flocking together. They shared warehouse spaces and the services of oilmen, tallow chandlers, silversmiths and coffee house proprietors. The variety of premises in such localities, which moved easily between functions – from bookshop to butchers and back again – meant that members of the book trades could stay in the same areas as their fortunes ebbed and flowed. This, Raven argues, allowed Paternoster Row to expand commercial publishing in the eighteenth century, and to lead the rise of periodical and magazine publishing into the nineteenth century.[48]

In other places there were other, similar, localities. Anindita Ghosh shows how the production of cheap popular print in late nineteenth-century Calcutta took its name, Battala ("under the banyan tree"), from the area in the Shobhabazar area of North Calcutta where the first presses producing almanacs, myths and fables, and racy satirical dramas had located. As the number of presses producing these Battala books expanded, the name itself 'soon came to stand for the entire area sprawling

---

47   Robert Mayhew, "Materialist Hermeneutics, Textuality and the History of Geography: Print Spaces in British Geography, *c.* 1500–1900," *Journal of Historical Geography* 33:3 (2007): 466–88.

48   James Raven, *The Business of Books: Booksellers and the English Book Trade* (New Haven: Yale University Press, 2007), particularly Chapter 6: "High and Low: Locating the Trades." See also Robert Mayhew, *Enlightenment Geography: The Political Languages of British Geography, c. 1650–1850* (London: Macmillan, 2000).

across the Garanhata, Ahiritala, and Chitpur localities, and even further'.[49] In her contribution here, Rosa Salzberg turns to the production and sale of popular print in sixteenth-century Venice to demonstrate the neighbourhoods in which printers thrived and the precise streets, squares and bridges where city-dwellers and visitors could expect to find printed broadsheets, ballads and pamphlets on offer. Doing so means identifying new connections between print and performance which animates the distribution of these printed pages with the spoken and sung words of the ballad singers, buffoons and charlatans who hawked them in the streets.

Beyond the city there are other geographies of the circulation of the written word that require attention to the means and meanings of distribution. Harold Love has differentiated the more-or-less centralized scribal production of political manuscripts (something like hand-written pamphlets), by armies of clerks in and around the Inns of Court in London from the chain copying of single versions of politically contentious manuscripts by and for individual readers in what he calls 'scribal communities' stretched across a network of great houses. While the former might rival print production, the latter was non-metropolitan and dispersed, working beyond the view of the Crown. It meant, as Love argues, that 'the scribal text had a politics, and that this politics was also a geography'.[50] More generally, Robert Darnton has suggested that 'The wagon, the canal barge, the merchant vessel, the post office, and the railroad may have influenced the history of literature more than one would suspect'.[51] Yet the nature of that influence needs to be carefully specified. James Raven points out that there was a national market for print before the railway, but that distribution by steam, the railway book edition, W.H. Smith's book stalls and railway circulating libraries certainly made a difference to the size and shape of that market.[52] In her chapter here, Aileen Fyfe, examining the accounts and letters of the Scottish firm W. & R. Chambers, is more cautious. She carefully shows how and when these publishers, selling a popular weekly periodical, saw steam – driving presses and ships as well as trains – as the sensible business option. These new technologies opened up a national market for their Edinburgh-produced wares. But there was no simple railway print revolution. What Fyfe shows above all is the way in which Chambers dealt with local, regional, national and international scales at one and the same time, as they developed a steam press in Edinburgh, fed a lowland Scottish market, developed a national market and sought to broker deals with publishers in Philadelphia

49   Anindita Ghosh, *Power in Print: Popular Publishing and the Politics of Language and Culture in a Colonial Society* (New Delhi: Oxford University Press, 2006), 16 footnote 40. For another example see Mary S. Pedley, *The Commerce of Cartography: Making and Marketing Maps in France and England* (Chicago: University of Chicago Press, 2004).

50   Harold Love, "Oral and Scribal Texts in Early Modern England," in *The Cambridge History of the Book*, vol. 4: *1557–1695*, ed. John Barnard and Donald F. McKenzie (Cambridge: Cambridge University Press, 2002), 109.

51   Darnton, "What is the History of Books?" 77.

52   Raven, *The Business of Books*.

and New York to sell *Chambers's Journal* there. The investigation of local and national geographies always requires an awareness that 'The book was, and is, an international commodity'.[53]

This is an important acknowledgement. It is significant that Febvre and Martin's initial awareness of the many geographies of the book seemed for a time to have given way to a range of studies which took the nation-state as the primary geographical unit of analysis. This has produced important works of synthesis, but the geography of the book can never simply be contained by the national scale.[54] It is evident, for example, that these borders and boundaries were always traversed by books themselves. The book markets, and systems of literary control, of England and France were always open to books brought, legitimately or illegitimately, from other jurisdictions, particularly the Netherlands and Switzerland.[55] With even more extensive movements in view, it is significant that the first volume of *The History of the Book in America* was entitled *The Colonial Book in the Atlantic World*. Its contributors dealt with the works that were available in the North American colonies whether they had been produced there, in manuscript or print, or whether they were part of the extensive export market from Britain.[56]

The transatlantic geography of the book has become an important focus of attention whether that be in terms of the circulation of learned periodicals, the operation of circulating libraries in the American colonies, the printing of reports of crime and criminality in newspapers on both sides of the Atlantic, the manuscript letters and letter books of natural historians and merchants, or letter writing manuals that mixed both script and print.[57] In his chapter, Deryck

---

53   Ibid., 2.

54   For example, Bill Bell, ed., *The Edinburgh History of the Book in Scotland.* Vol. 3: *1800–1880* (Edinburgh: Edinburgh University Press, 2007) and Peter Kornicki, *The Book in Japan: A Cultural History from the Beginnings to the Nineteenth Century* (Leiden: Brill, 1998).

55   Johns, *The Nature of the Book* and Darnton, *The Business of Enlightenment.*

56   Hugh Amory and David D. Hall, eds, *A History of the Book in America,* vol. 1: *The Colonial Book in the Atlantic World* (Cambridge: Cambridge University Press, 2000). See also Giles Barber, "Books From the Old World and For the New: The British International Trade in Books in the Eighteenth Century," *Studies on Voltaire and the Eighteenth Century* 151 (1976): 185–224.

57   Norman Fiering, "The Transatlantic Republic of Letters: A Note on the Circulation of Learned Periodicals to Early Eighteenth-Century America," *William and Mary Quarterly* 33:4 (1976): 642–60; James Raven, *London Booksellers and their American Customers: Transatlantic Literary Community and the Charleston Library Society, 1748–1811* (Columbia: University of South Carolina Press, 2001); Gwenda Morgan and Peter Rushton, "Print Culture, Crime and Transportation in the Criminal Atlantic," *Continuity and Change* 22:1 (2007): 49–71; Susan Scott Parrish, *American Curiosity: Cultures of Natural History in the Colonial British Atlantic World* (Chapel Hill: University of North Carolina Press, 2006); David Hancock, *Citizens of the World: London Merchants and the Integration of the British Atlantic Community, 1735–1785* (Cambridge: Cambridge University Press, 1995);

Holdsworth discusses the counting house libraries of merchants in London and the American colonies to demonstrate the ways in which various *ars mercatoria* – dictionaries, compilations of laws, and maps and charts – brought a world of commerce into view and provided the forms of knowledge through which new mercantile connections and networks could be constructed.

Consideration of such utilitarian texts allows us to begin to see the ways in which written instruments of all sorts were central to the operation of trade and empire across land and sea. Writing was one of the sinews of mercantile and imperial organization as information was exchanged through new networks and across new geographies of metropole and empire. The far-flung geographies of 'knowledge in transit' came to take particular shape as part of the long history of global trade and empire building.[58] As Michael Dorn's contribution shows, however, these knowledge claims were always contested ones. He charts the transatlantic production and dissemination of Constantin-François Chasseboeuf, the Comte de Volney's *Tableau du Climat et du Sols des État-Unis d'Amérique* (1803). Dorn then shows how the challenge posed by the *Tableau*'s presentation of the future possibilities of the American frontier in the Ohio Valley was taken up by the intellectual elite of the new American nation.

As Dorn's chapter demonstrates, by situating Volney's enterprise within the Atlantic intellectual networks of the French liberal émigré communities in the United States, as centred on Moreau de St. Méry's Philadelphia bookshop, working at this global scale increases the need to remain attentive to the particular contexts of book production and circulation. Into each situation that merchants, travellers, natural philosophers, missionaries and empire-builders brought their forms of script and print, and their modes of book making, moving, collecting and dispersing, there also came other forms of communication. Sometimes these were the practices of speech and memory of oral cultures; sometimes they were forms of inscription or recording that – like the knotted *quipu* strings of the Inca or Mexican pictograms – were 'writing without words'; sometimes they were scribal forms of various kinds and degrees of elaboration, or even, as in China and Korea, other modes of printing with wood, clay or metal blocks.[59]

Each of these, just like European practices of script and print, had its own skilled specialists and ongoing historical geographies of change and movement. For Northern India, to take just one example, Christopher Bayly has detailed the elaboration of a varied manuscript culture under Mughal rule which included

---

Eve Tavor Bannet, *Empire of Letters: Letter Manuals and Transatlantic Correspondence* (Cambridge: Cambridge University Press, 2005).

58   See, for example, Ogborn, *Indian Ink*.

59   Tony Ballantyne, "What Difference Does Colonialism Make? Reassessing Print and Social Change in an Age of Global Imperialism," in *Agent of Change*, ed. Baron, Lindquist and Shevlin, 342–52; Elizabeth Hill Boone and Walter D. Mignolo, eds, *Writing Without Words: Alternative Literacies in Mesoamerica and the Andes* (Durham: Duke University Press, 1994).

forms of imperial communication and news-gathering, along with a public sphere without print that hosted debate over religion, politics and poetry.[60] The challenge for histories and geographies of the book is to understand the complex processes whereby these communication practices, these books and manuscripts, and a panoply of speakers, readers and writers, came into contact with each other to undertake trade, fight wars, sign treaties, contest land claims – and to tell stories.[61]

Such contacts were always mixed, involving cooperation and conflict, inclusions and exclusions. For early modern botany, for example, Hal Cook and Kapil Raj have both shown the forms of collaboration between medical and botanical specialists from different traditions, artists, scribes, printers and patrons that needed to be mobilized for books of botany from the Indies to be produced and circulated.[62] Significantly, in Raj's account of the production of a fourteen-volume illustrated botanical manuscript in early eighteenth-century India, the *Jardin de Lorixa*, this meant work with other sorts of books. The work's compiler, Nicholas L'Empereur, a surgeon in the employ of the French Compagnie des Indes, stated that his plan was to buy 'all the books on medicine that the people here have and find out how they use them'. He noted that these palm-leaf manuscripts in a variety of languages were 'very difficult to obtain', and their acquisition clearly required as much care and negotiation as did his personal relationships with ascetic interlocutors and the cloth-painters who depicted the 722 plant species in his collection.[63]

These collaborations were also structured by relationships of power. Cook shows how seventeenth-century Dutch botanists packaged up the knowledge they garnered in the East Indies like so many parcels of spices for transportation to European markets. Raj, in turn, notes that L'Empereur's work did not find favour with the botanical establishment in France, and remained in manuscript form. It never became what he wanted, a work that, 'once printed', would ensure that 'nothing [of Indian medicine] will be left unknown to the European surgeon'.[64]

---

60   Christopher A. Bayly, *Empire and Information: Intelligence Gathering and Social Communication in India, 1780–1870* (Cambridge: Cambridge University Press, 1996).

61   Jill Lepore, *The Name of War: King Philip's War and the Origins of American Identity* (New York: Vintage Books, 1999); Donald F. McKenzie, *Oral Culture, Literacy and Print in Early New Zealand: The Treaty of Waitangi* (Wellington: Victoria University Press with the Alexander Turnbull Library Endowment Trust, 1985); E. Jennifer Monaghan, *Learning to Read and Write in Colonial America* (Amherst and Boston: University of Massachusetts Press, 2005).

62   Harold J. Cook, *Matters of Exchange: Commerce, Medicine, and Science in the Dutch Golden Age* (New Haven: Yale University Press, 2007); Kapil Raj, *Relocating Modern Science: Circulation and the Construction of Scientific Knowledge in South Asia and Europe, Seventeenth to Nineteenth Centuries* (Delhi: Permanent Black, 2006).

63   Quoted in Raj, *Relocating Modern Science*, Chapter 1: "Surgeons, Fakirs, Merchants, and Craftsmen: Making L'Empereur's *Jardin* in Early Modern South Asia," 36 and 41.

64   Ibid., 36.

In his chapter in this volume, Miles Ogborn likewise traces the power-laden relationships between an Indian manuscript culture and the making of a printed book. Using the example of the translation and printing of the *Seir Mutaqharin*, a Persian chronicle of Indian history, he shows the cultural and political relationships that shaped the spaces of print culture in late eighteenth-century British Bengal and limited the ways in which the products of a different manuscript culture could come out in print.[65]

The challenge in reconstructing geographies of the making and moving of books is to work these concerns for context and knowledge in transit together. Doing so means recognizing that oral, scribal and printed forms, and their many specific incarnations and interrelations, were all part of the geographies of the book, both as broad intercontinental histories of communication and within specific locales or as elements in networks stretching across space. It also means working between and across the many scales of the geography of the book to connect them together.

There are different ways of doing this. Fiona Black's chapter argues for an increased attention to the visualization of the many spaces of print culture through the use of Geographical Information Systems. Drawing examples from the history of Canadian print culture, she argues that historical approaches can benefit from a sharpened awareness of geographical data at a range of scales from the individual locations and movements of print shop workers, through national patterns of literacy and transportation, to international migrations of people and texts. Combining these together in a GIS allows new questions to be asked and answered, but must be done through research on the needs of those who research the history (and geography) of the book. By contrast, in his chapter Charles Withers examines the geographies of the works on African exploration by the Scottish explorer Mungo Park. He does so in order to examine the epistemological questions raised by the displacements involved in producing and presenting books about distant places. Detailed examination of the processes of production of Park's texts serve to show how these questions about knowledge on the move were dealt with by those who made and read Park's books. Withers uses the geographies of Park's books to question accepted notions in the history of the book such as 'authorship', 'production', 'reception' and 'audience'. In so doing, he emphasizes the importance of attending to the reading of books as well as their making.

Being Made and Moved by Books: Geographies of Reading

Alongside the idea of the materiality of the text as it appears in the world, the notion of reading as a creative interpretative act has been the other key foundation on

---

65   For the later formal regulation of print by the British in India, see Robert Darnton, "Literary Surveillance and the British Raj: The Contradictions of Liberal Imperialism," *Book History* 4 (2001): 133–76.

which the expansion of the history of the book as a discipline has been built. Ideas from social theory, reader-reception theory and cultural history have combined to stress the ways in which a book is given different meanings in the hands of different readers. This has opened up the possibility of reconstructing a history of reading.[66]

Doing so has moved the study of reading away from the quantitative accounting of changing literacy rates, and away from the identification of grand historical shifts from orality to literacy, or from the 'intensive' reading of one book (especially the Bible) to the 'extensive' reading of many.[67] What these histories of active readers have moved towards is a multiplicity of different histories of reading that stress the variety of this practice over time and space. Reading as a creative interpretative act is necessarily undertaken in different ways in different contexts, and the job of historians of reading is to defamiliarize what is an all too familiar practice in order to reveal how it has fitted into and shaped the processes by which ideas are made and moved, and the ways in which identities are made as people are moved in different ways by their reading.[68]

This grounding of the practices of reading in particular 'interpretative communities', and in the details of specific books read, of modes of making marginal marks, adding to texts, cross-referencing, and discussing reading with others has also been a grounding of the act of reading in space and place: a geography of reading as well as a history.[69] James Secord's extraordinary account of the many readings of the controversial Victorian scientific treatise *Vestiges of the Natural History of Creation* shows in detail that the book was read differently in different places. Drawing on diaries, letters and memoirs, and on the more public pronouncements in periodical reviews and newspaper columns, Secord

---

66   Robert Darnton, "First Steps Towards a History of Reading," *Australian Journal of French Studies* 23:1 (1986): 5–30; Leah Price, "Reading: The State of the Discipline," *Book History* 7 (2004): 303–20.

67   Guglielmo Cavallo and Roger Chartier, eds, *A History of Reading in the West* (Amherst and Boston: University of Massachusetts Press, 1999); David Vincent, *Literacy and Popular Culture: England, 1750–1914* (Cambridge: Cambridge University Press, 1989); François Furet and Jacques Ozouf, *Reading and Writing: Literacy in France from Calvin to Jules Ferry* (Cambridge: Cambridge University Press, 1982); Jack Goody, *The Logic of Writing and the Organization of Society* (Cambridge: Cambridge University Press, 1986); Walter J. Ong, *Orality and Literacy: The Technologizing of the Word* (London: Routledge, 1982); Wolfgang Iser, *The Act of Reading: A Theory of Aesthetic Response* (London: Routledge and Kegan Paul, 1978).

68   Price, "Reading," and, for the inspiration of this section's title, Francis Spufford, *The Child that Books Built: A Life in Reading* (London: Faber & Faber, 2002).

69   Stanley Fish, *Is There a Text in this Class? The Authority of Interpretative Communities* (Cambridge: Harvard University Press, 1980); H.J. Jackson, *Marginalia: Readers Writing in Books* (New Haven: Yale University Press, 2001); David N. Livingstone, "Science, Text and Space: Thoughts on the Geography of Reading," *Transactions of the Institute of British Geographers* 30:4 (2005): 391–401.

shows that the way the text was read in London was not the same as it was read in Liverpool, Cambridge, Oxford or Edinburgh. This is, in many ways, a 'local history' of reading, which accounts for interpretative differences in terms of the fact that 'Victorian town and cities were defined through the character of their literary life, which was in turn shaped by industrial structure, class, population size, and tradition'.[70] London's reading worked through the fashionable conversation of High Society, at dinner parties, clubs and soirees. Liverpool's readings were riven by religious and political faction in a city of rapid growth and massive poverty. Oxford and Cambridge differed from each other as clergymen scientists sought to commend or condemn the challenge to orthodox theology and natural history posed by a work that emphasized progressive development and the transmutation of species. In Edinburgh, *Vestiges* was read yet differently by different readers as part of debates over liberal reform and ecclesiastical politics. What is striking about Secord's work, then, is not a radically new geographical perspective which overturns all that came before. Rather, it is the recognition that if we do not understand that some of the more familiar determinants of reading such as class, gender, religion and politics were geographically differentiated and geographical constituted in particular places – being an Anglican in Liverpool in the 1840s was different from being one in Cambridge – scholars will not be able to explain these, and other, very different readings and interpretations of the same book.

There are yet more geographies of reading at even smaller scales. Secord's descriptions of the reading practices of the many readers that populate his book – from the Halifax apprentice Thomas Archer Hirst to Queen Victoria – show that *how* they read was also a matter of *where* they read – from discussions in taverns and Mechanics Institutes to reading aloud in the drawing room of Buckingham Palace. Just as the study of the production and distribution of books needs to take us into the spaces of the printing house and the bookshop, studies of reading need to understand the ways in which this mode of interpretation is undertaken within particular identifiable spaces with their own distinctive characteristics and histories: the study, the library, the fireside, but also the railway carriage, the counting house, the street and the law court. These questions of geography are not somehow separable from the reconstruction of reading practices – such as John Dee's marginal annotations; Gabriel Harvey's turning of his Livy on a book-wheel; and Anna Larpent's polite sociability – but are, rather, fundamental to their constitution.[71] Dee's readings were inseparable from the construction of his

---

70   James A. Secord, *Victorian Sensation: The Extraordinary Publication, Reception, and Secret Authorship of* Vestiges of the Natural History of Creation (Chicago: University of Chicago Press, 2000), 156–7.

71   William H. Sherman, *John Dee: The Politics of Reading and Writing in the English Renaissance* (Amherst and Boston: University of Massachusetts Press, 1995); Lisa Jardine and Anthony Grafton, "'Studied for action': How Gabriel Harvey Read His Livy," *Past and Present* 129 (1990): 30–78; John Brewer, "Reconstructing the Reader: Anna Larpent in Late Eighteenth-Century London," in *The Practice and Representation of Reading in*

library, the *Bibliotheca Mortlacensis* as a space that was both private and public, 'a place where court, city, and university could meet'.[72] Harvey's political readings of the classics were made within the great houses of grandees for whom these texts might form the basis for action in the world, and Larpent's voracious reading was of a piece with her perambulations of the polite spaces of Georgian London. At the smallest scales, practices of reading are made in and through the spaces of reading.

The appeal of the history of reading is, as Adrian Johns has put it, that it is 'so culturally centrifugal'.[73] These accounts demonstrate difference. Care must be taken, however, not to celebrate the diversity of reading for its own sake. Secord's intention in uncovering not one but many readings of the same book is to show that 'The more completely a case can be situated, the more it reveals wider patterns and structures of response – of competing representations, appropriations, and contests over authority'.[74] He is able, through the reconstruction of what we might see as a map of differences, to address broader questions of the nature of print culture in Victorian Britain, the relationships between religion and science, and between the individual and urban industrial society.

We would suggest that accounts of reading need to look up from the book to consider the broader geographies of reading. This might be a matter of understanding national reviewing cultures or reconstructing what books are being read together (and how they got to be in the same place at the same time).[75] It might be a case of considering the imaginative horizons of the reader, or the broader purposes they have in mind for their reading. Both Gabriel Harvey and John Dee's readings, for example, were intimately linked to England's early modern imperial adventures in Ireland and elsewhere. It might also be a case of considering networks of readers who share practices but not particular spaces. This is easily imaginable in the internet age, but can also be seen in the 'invisible college' of astronomers revealed through Owen Gingerich's meticulous comparison of the marginal annotations in every extant copy of Nicholaus Copernicus's *De Revolutionibus* that he could find.[76] The different expectations of geographically disparate interpretative communities can also be seen in Innes Keighren's chapter in this volume where he carefully disaggregates the reception of Ellen Churchill Semple's *Influences*

Britain, 1500–1900, ed. James Raven, Naomi Tadmore and Helen Small (Cambridge: Cambridge University Press, 1996): 226–45; David N. Livingstone, "Science, Site and Speech: Scientific Knowledge and the Spaces of Rhetoric," *History of the Human Sciences* 20:2 (2007): 71–98.

72   Sherman, *John Dee*, 45.
73   Johns, "How to Acknowledge a Revolution," 109.
74   Secord, *Victorian Sensation*, 519.
75   Livingstone, "Science, Text and Space."
76   For internet book forums visit, for example, www.book-discussion-forum.co.uk, for the search for Copernicus's annotators, see Owen Gingerich, *The Book Nobody Read: In Pursuit of the Revolutions of Nicholaus Copernicus* (London, William Heinemann, 2004).

of Geographic Environment (1911) in order to reveal long-standing networks of critically engaged readers across a range of educational institutions rather than a simple history and geography of the rise and fall of environmental determinism with its making in Chicago and its breaking at Berkeley.

Just as with the production of books, understanding the geography of reading as part of the geography of the book means working between the local contextualization of reading practices and their simultaneous participation in the processes of making and communicating knowledge in transit. The challenge posed is, once again, to combine these approaches and work across and between these scales. The chapter in this collection by Daniel Hopkins seeks, for example, to investigate reading's many geographies through the case of the Danish forts on the West African coast. He uses the traces, in official reports and the inventories of the dead Europeans, of the books that were there to be read. The project of replacing a slaving economy with a plantation economy was, he argues, supported by the many works of natural history, administrative practice and practical guidance. Yet the range of reading – Shakespeare was there, as were Milton and Samuel Johnson – reveals a variety of geographical horizons and contexts for these readers. We cannot know what lessons the botanist Julius van Rohr thought he might draw about Denmark's imperial projects, or those of its European rivals, by transporting to Fort Christianbourg on the Gold Coast a fourteen-volume German translation of Gibbon's *Decline and Fall of the Roman Empire*. But the fact that he did and that it was read and perhaps borrowed or shared by others points to the need to be attentive through the geographies of the book to the consumption of Enlightenment ideas and ideals not in the 'core' of Enlightenment Europe but, as others have shown of Voltaire and his books, or of those of Jedidiah Morse and his geographical works, on the 'margins' of the Enlightenment world.[77] These readings were both locally specific and tied into the far flung networks traced by books and their readers.

***

Although we have divided up this introduction and the contributions that follow in terms of the geographies of the production and circulation of books (understood in terms of their materiality), and the geographies of the consumption of books (their reading), it is evident that this is an overly neat division which is more a matter of convenience than a reflection of the spatial patterns traced out by books themselves. Production, circulation and consumption need to be considered together. For

77 Withers, *Placing the Enlightenment* and Graham Gargett, "Voltaire's Reception in Ireland," in *Ireland and the French Enlightenment, 1700–1800*, ed. Graham Gargett and Geraldine Sheridan (Basingstoke: Macmillan, 1999): 67–89. On this point more generally, see Richard Butterwick, Simon Davies and Gabriel Sánchez Espinosa, *Peripheries of the Enlightenment, Studies on Voltaire and the Eighteenth Century* 1 (Voltaire Foundation, Oxford, 2008) and Neil Safier, *Measuring the New World: Enlightenment Science and South America* (Chicago: University of Chicago Press, 2008).

example, Michael Warner, in his examination of the politics of print in the American Revolution, argues that notions of print technology and questions of reading need to be worked together if we are to understand the conjoined constitution of politics and print from the late seventeenth century onwards. In a complex argument he demonstrates the ways in which the meanings of print technology and the printed object were made together with the meanings of publication and public utterance such that 'mechanical duplication' and the subsequent reading of printed materials was understood as presenting an impersonal public pronouncement and constituting a new sphere of republican politics.[78] This remaking of the meanings of print production, circulation and consumption had a geography that was both locally bounded and far flung. Warner asks 'What did it mean to buy a book, to read a newspaper, or to nail up a broadside in the American colonies?' and answers that such meanings varied 'from context to context'.[79] Indeed, he sketches out a differentiated geography of the book across the different North American colonies. Yet this remaking of print was also undertaken within the networks of Britain's Atlantic world empire whose changing politics meant that these forms of the making and consuming of print had a wider purchase and became understood as 'a natural resistance to imperial sovereignty'.[80]

In thinking through what he calls the 'geographies of textuality', Clive Barnett has also combined the materiality of the book and the constitution of new reading publics. In his interpretation of the making of Heinemann's *African Writers Series* Barnett examines its paratexts – 'everything from titles to typeface and cover design, to reviews and criticism' – to trace out the 'geographical constitution of reading-formations' through the textual and material forms of this book series.[81] The distinctive orange covers, the numbering of each item in the series, the paperback format, author biographies and cover images were all part of the making of a set of geographies which sought to combine readers and books in particular ways. The resultant geographies were multiple ones, involving both the transcultural and transnational geographies of African writing and the embedding of African literature in the specific national contexts of post-colonial educational systems in Africa. Setting out and understanding these geographies brings together concerns with textual production, circulation and consumption.

Fifty years beyond Febvre and Martin's *The Coming of the Book*, there is now hardly a field of textual scholarship that does not take seriously, in one way or another, the questions raised there about the locations of production of print and print culture, about the mobility of print, and about how the printed word has transformed societies across the globe. Adding an attention to the contextually

---

78   Michael Warner, *The Letters of the Republic: Publication and the Public Sphere in Eighteenth-Century America* (Cambridge: Harvard University Press, 1990), 39.

79   Ibid., 11.

80   Ibid., 71.

81   Clive Barnett, "Disseminating Africa: Burdens of Representation and the *African Writers Series*," *New Formations* 57 (2005–6): 81–2.

specific and located reading of texts to this concern for the material production of the printed word has, as we have argued, meant that questions of the geography of the book are central ones within the history of the book. Our concern in what follows is not to offer a rigid definition of what the geography of the book ought to be, depicting it as just one specialist dimension of book history, one part of the jigsaw. Instead, we argue that book history, combining concerns with 'contextualization' and 'dissemination' – the local and the far flung – is necessarily geographical. We therefore allow the various contributors to take their own paths in illustrating how thinking about the geography of the book serves to highlight the importance of place, to signal to the differences that space makes to what a book was or was thought to be, and to show how issues of scale – of local meaning, national frames of reference or questions of transnational reception – matter to what we take books to be and to do. Read together they demonstrate that books cannot be understood outside their geographies.

PART I
GEOGRAPHIES OF
PRODUCTION

Chapter 1

The Amusements of Posterity: Print Against Empire in Late Eighteenth-Century Bengal

Miles Ogborn

In what ways might the geography of a book affect its success? That depends, of course, on what is meant by the geography of a book, and also what is meant by success. If we take the former to be the nature of the places, spaces and networks of production, distribution and consumption we can try to gauge the ways in which they shape the book that is produced, where it goes, and what readers make of it. As for "success," that will have to be assessed differently in different cases. For example, is the successfulness of a work of fiction to be measured in terms of sales, literary prizes or subsequent influence? Matters may be more straightforward in the world of science where notions of objectivity at least mean that subjective judgements cannot be acknowledged to be the final resolution of disputes. What makes one version of the truth of the natural world win out over others, at least temporarily? What part does the geography of knowledge as written and read in books have to play in the process whereby truth is questioned, established and questioned again?[1] Between literature and science we might locate the domain of politics which must be equally committed to both truth and expediency, and which must appeal to both the objective and the subjective. Success here lies in the ability to shape the terms of the debate, and to effect change in ideas, policy or practice. To what extent is that sort of success in politics shaped by the geography of the book?

In each of these situations – the literary, the scientific and the political – it is, of course, the case that the role of books, and therefore the successes that they can hope to meet, differ over time and space. We need to understand what authors' and publishers' expectations were and are before we can judge their successes and failures. It is evident that a literary, scientific or political culture dominated by the publication of printed books is different from one based on the primacy of speech or manuscript. Even where the printed word holds sway we should not

---

1 Adrian Johns, *The Nature of the Book: Print and Knowledge in the Making* (Chicago: University of Chicago Press, 1998); James A. Secord, *Victorian Sensation: The Extraordinary Publication, Reception, and Secret Authorship of* Vestiges of the Natural History of Creation (Chicago: University of Chicago Press, 2000).

expect any simple relationship between politics, print and the public sphere.[2] We can, however, anticipate that the sorts of influence that a book may have would be quite different than in contexts where the spoken or hand-written word is the key to enduring influence.

On a grander scale, it is possible to imagine a complex historical geography of the changing ecology of speech, manuscript and print which would contain these contexts. This variegated landscape would be shaped by the different forms, technologies and uses of print as they developed in Asia and Europe; the increasing dominance of print for some purposes in many places (although always accompanied by speech and other forms of the written word) which has been fashioned, although not determined, by capitalism and Western colonialism and imperialism; and perhaps the subsequent decline, stronger in some places than others, of the power of print to shape literature, science and politics in favour of other media.[3] Each part of this landscape has its distinctive geography of places of speech, writing and print: the marketplace, the coffee house, the print shop, the debating chamber, the scriptorium, the study, the library.[4] These places are linked in complex networks that connect texts, speakers, writers, readers and the materials they need to do their work. Some parts of this landscape are more complex than others, some change faster than others. It is important to know where we are on this map in order to judge the fortunes of any book, and the part played in that by its particular geographies of production, distribution and consumption.

This chapter addresses these issues in relation to a single work produced within the nascent political print culture of late eighteenth-century Bengal. This was an historical treatise in Persian critical of British imperial rule the translation of which was put into print and into circulation in Calcutta in 1789 under the title *Seir Mutaqherin*.[5] It will be shown that the manuscript and then its printed translation should be understood in terms of a particular configuration of speech, writing and print situated at the intersection of two different modes of political communication in imperial north India. Telling the story of the production of the *Seir Mutaqherin* vividly illuminates the changing topography of part of the landscape described above as the extension of the British empire in India meant that spoken, written and printed words shifted their relative positions in the domain of politics depending

---

2   Michael Warner, *The Letters of the Republic: Publication and the Public Sphere in Eighteenth-Century America* (Cambridge: Harvard University Press, 1990).

3   For overviews see Lucien Febvre and Henri-Jean Martin, *The Coming of the Book: The Impact of Printing, 1450–1800* (London: Verso, 1976) and Simon Eliot and Jonathan Rose, eds, *A Companion to the History of the Book* (Oxford: Blackwell, 2007).

4   Peter Burke, *A Social History of Knowledge: From Gutenberg to Diderot* (Cambridge: Polity Press, 2000).

5   Seid-Gholam-Hossein-Khan, *A Translation of the Seir Mutaqherin; or, View of Modern Times*, 3 Volumes (Calcutta: Printed for the Translator, 1789).

on who the audience was and what purpose was intended.[6] Understanding this, and trying to account for the fortunes of this particular text in that context, also means focusing in on the details of the geography of this book. This requires consideration of how the spaces in which it was produced and read, and the distribution networks that it sought to use, worked through the social relations of empire in ways that shaped what the book was and what it could do. The historical geography of this book was, as we shall see, played out as both tragedy and farce. The book's title was, at the time, rendered in English as "A View of Modern Times." Its chequered history will show that its translator's preferred version – "The Amusements of Posterity" – perhaps rather better described its fortunes.

Seir Mutaqherin, Manuscript Culture and the Critique of Empire

The *Seir Mutaqherin*, a chronicle of Indian history, was written in Persian by Ghulam Husain Khan Tabatabai in 1781–82. The author was a high-born Bihari official whose Persian father had served the Mughal emperor and whose mother was related to Alivardi Khan, the *nawab* of Bengal. Ghulam Husain had grown up at Alivardi's court, and held official bureaucratic positions in Delhi and Patna before entering the service of the *nawab* of Awadh. He was part of a class of elite Mughal administrators whose world was profoundly disrupted by the consolidation of territorial rule by the East India Company in Bengal and Bihar after 1757.[7] He was forced to seek the patronage of both the English and their Indian allies and opponents.[8] His history, which may have been written at the instigation of British officials allied to the English Governor General of Bengal, Warren Hastings, offered a stern critique of Company rule in Bengal and, with the exception of Hastings and a few others, its turning away from Mughal traditions of legitimate governance.[9]

Ghulam Husain's criticisms of the 'hat-wearers' highlighted the dangers of according power to the landowning *zamindar*s, the disabling effects of party

---

6   Anindita Ghosh, "An Uncertain 'Coming of the Book': Early Print Cultures in Colonial India," *Book History* 6 (2003): 23–55; Abhijit Gupta and Swapan Chakravorty, eds, *Print Areas: Book History in India* (Delhi: Permanent Black, 2004).

7   C.E. Buckland, *Dictionary of Indian Biography*, 2 Volumes (New Delhi: Cosmo Publications, 1999) s.v. "Ghulum Hussein Khan Tabataba, Syad," vol. I 164; Kumkum Chatterjee, "History as Self-Representation: The Recasting of a Political Tradition in Late Eighteenth-Century India," *Modern Asian Studies* 32:4 (1998): 913–48.

8   Gulfishan Khan, *Indian Muslim Perceptions of the West During the Eighteenth Century* (Karachi: Oxford University Press, 1998).

9   J.S. Grewal, *Muslim Rule in India: The Assessments of British Historians* (Calcutta: Oxford University Press, 1970), 33; Rajat Kanta Ray, "Indian Society and the Establishment of British Supremacy, 1765–1818," in *The Oxford History of the British Empire. Volume II: The Eighteenth Century*, ed. Peter J. Marshall (Oxford: Oxford University Press, 1998), 508–29.

politics, the lack of appropriate justice offered by a British-instituted Bengal
Supreme Court that trespassed on local traditions, and, most importantly, the
distance that had grown between the governors and the governed.[10] Rule by
committee and constant changes in personnel, from the Governor General
downwards, meant that 'this country seems to have had no master at all'.[11] The
hatred British gentlemen showed for appearing in public audiences, and their
'extreme uneasiness, impatience and anger' when they did so, meant that they were
unable to govern either effectively or legitimately since the people 'never see any
thing of that benignity and that munificence which might be expected from people
that now sit on the throne of kings'. He criticized the English for only listening
to the self-interested advice of the 'newmen' whom they had put at 'the summit
of power'.[12] Singled out for particular criticism was Muhammed Reza Khan, the
chief administrator of the *nawab* of Bengal in the 1760s, and crucial intermediary
in the Company's attempts to establish effective territorial rule and revenue
collection. The ups and downs of his political career – including imprisonment
under Warren Hastings, and reinstatement by the Governor General's opponents
– reflected the problems of working between different modes of rule, Mughal and
British, and different conceptions of how they might be brought together.[13] From
the perspective of displaced officials like Ghulam Husain, Reza Khan was part of
the problem. Ghulam Husain called him 'blunt and thoughtless', a man who 'does
not seem to have right notions about truth and falsehood', and one who 'shewed
the utmost disregard to every matter of chastity and decorum; still less did he
know the value of men, of learning, and merit'.[14] In his own attempts to hold the
British to a reworked version of Mughal administrative politics it seems that Reza
Khan did not know the value of men like Ghulam Husain.

This critique of imperial rule needs to be understood as the product of a
particular localized culture of political communication in Bengal. As C.A. Bayly
has argued, northern India possessed a vibrant and well-organized manuscript
culture that shaped literature, religion and politics in the region. The Mughal
emperors had elaborated systems of information gathering that, along with
mercantile and sacred uses of writing, provided the basis for a growing scribal
elite who serviced a population well versed in the uses of the written word. This

---

10   Jon E. Wilson, *The Domination of Strangers: Modern Governance in Eastern
India, 1780–1835* (Basingstoke: Palgrave Macmillan, 2008); Sanjay Subrahmanyam, "On
the Hat-Wearers, Their Toilet Practices and Other Curious Usages," in *Europe Observed:
Multiple Gazes in Early Modern Encounters*, ed. Kumkum Chatterjee and Clement Hawes
(Lewisburg: Bucknell University Press, 2008), 45–81.

11   Seid-Gholam-Hossein-Khan, *Seir Mutaqherin*, II: 509 and 580.

12   Ibid., 408 and 597.

13   Robert Travers, *Ideology and Empire in Eighteenth-Century India: The British
in Bengal* (Cambridge: Cambridge University Press, 2007); Abdul Majed Khan, *The
Transition in Bengal, 1756–1775: A Study of Saiyid Muhammad Reza Khan* (Cambridge:
Cambridge University Press, 1969).

14   Seid-Gholam-Hossein-Khan, *Seir Mutaqherin*, II: 539.

underpinned an indigenous public sphere of rational debate over the interlocking concerns of politics, religion and aesthetics.

This 'Indian ecumene', as Bayly calls it, without the printing press or the formal public meeting, worked through oral and scribal modes of communication. Its mechanisms were news, gossip and opinion passed around and debated at druggists' stalls and sweetshops, around mosques and temples, and in discussions of poetry. In written form it worked through personal letter writing, newsletters read aloud to crowds in the street, and placards posted at significant sites. Within this, and orchestrated by a scribal administrative elite with an ethic of service to the state rather than to any particular ruler, there was a tradition of holding kingship to account by using 'a well-tested arsenal of handwritten media'.[15] Ghulam Husain's historical chronicle was doing just that to the country's new rulers, the British, and to those who worked with them.

Significantly, the forms of political communication at work within this indigenous public sphere also shaped the content of Ghulam Husain's critique, and focused it on communication itself. The British rulers' illegitimate distance from their subjects was, he insisted, effectively produced by their reliance upon certain writing practices:

> [T]he English commenced acquiring a knowledge of the usages and customs of the country: for it was a standing rule with them, that whatever remarkable they heard from any man versed in business, or even from any other individual, was immediately set in writing in a kind of book composed of a few blanc leaves, which most of them carry about, and which they put together afterwards, and bind like a book for their future use.[16]

Englishmen were, he observed, only interested in becoming acquainted with 'noblemen and other persons of distinction' in order to pump them for knowledge of laws or revenue matters, and they 'would immediately set it down in writing, and lay it up in store for the use of another Englishman'. As a result, 'the Books and Memorandums composed by the English upon such interested reports' 'have come to be trusted as so many vouchers: whereas they are only some faint idea of the exterior and bark, but not of the pith or real reason, of those institutions'. These books, with their codifications and reconstructions of the nature of Indian institutions, meant that the Englishmen who used them 'come at last to undervalue the Hindostanies, and to make no account of the natives from the highest to the

---

15   Christopher A. Bayly, *Empire and Information: Intelligence Gathering and Social Communication in India, 1780–1870* (Cambridge: Cambridge University Press, 1996), 182.

16   Seid-Gholam-Hossein-Khan, *Seir Mutaqherin*, II: 402.

lowest'. Indians were shut out of official employment, the British 'esteeming themselves better than all others put together'.[17]

In addition to their recasting of the administration of law and the collection of revenue, Ghulam Husain also recorded how the British had stopped gathering other forms of information. According to him, these new masters had suppressed the offices of the 'Vacaa-nuviss, or Remembrancer, or Gazetteer, and the Sevana-nuviss, or Historiographer, and the Harcara or Spy, [who] were appointed for writing down the events that might happen in the respective provinces, territories, and districts of their residence' for dispatch to the emperor's court.[18] As a result of closing these channels of communication the new imperial rulers could not know what was happening in the territory that they purported to govern.

The sources of Ghulam Husain's critique also suggested a remedy. The failings of the British were to be rectified by more open-ended forms of conversation and dialogue in which manuscripts also had a part to play. In part this was based on a Muslim culture of orality in which it was the spoken word that was to be trusted as authoritative.[19] Under the current government – which had solidified negotiated practices as laws, and had closed the Mughal channels of intelligence – 'the gates of communication and intercourse are shut up betwixt the men of this land and those strangers, who are become their masters; and these constantly express an aversion to the Society of Indians, and a disdain against conversing with them; hence both parties remain ignorant of each other's state and circumstance'.[20] It was only 'conversation' that would allow the British 'to learn what aches these poor natives, and what might given them relief'. Without such social intercourse, and with the problems in understanding each other's languages, 'a Company of Hindians having business with their English rulers, looks very much like a number of pictures set up against the wall'. The English needed to be aware, Ghulam Husain suggested, that without 'communication of ideas', 'no benefit is reaped by either description of men from such an intercourse', and 'no love, and no coalition … can take root between the conquerors and the conquered'.[21] He was well aware of the American Revolution, and warned the Company of the dire consequences of failing to open a conversation with the old Mughal elite about how to govern India properly and legitimately. His treatise, although in manuscript, was to be part of that conversation. It would serve, its author said, 'in the following sheets to shew how managed the antient Sovereigns of this country'.[22]

---

17   Ibid., 404, 406 and 407, see also Jon E. Wilson, "'A Thousand Countries To Go To': Peasants and Rulers in Late Eighteenth-Century Bengal," *Past and Present* 189 (2005): 81–109.

18   Seid-Gholam-Hossein-Khan, *Seir Mutaqherin*, II: 566.

19   Francis Robinson, "Technology and Religious Change: Islam and the Impact of Print," *Modern Asian Studies* 27:1 (1993): 229–51.

20   Seid-Gholam-Hossein-Khan, *Seir Mutaqherin*, II: 545.

21   Ibid., 546, 587–8 and 553–4.

22   Ibid., 546.

The *Seir Mutaqherin* cannot simply be taken, however, as a voice from the 'Indian ecumene'. This was not the only mode of political communication in late eighteenth-century Bengal. If the focus is placed on the British themselves, and on the modes of social and imperial communication between Britain and Bengal as well as within British Bengal itself, then a different culture of communication becomes evident. This included print as well as manuscript and speech; it was based on an imperial circulation of texts of many different sorts between metropole and empire; it included notions of a printed public sphere of politics engaged in by writers and readers; and it placed trust in various forms of the written word.[23] Ironically, in order to open the conversation with the British about their failures to properly manage manuscripts and speech in Bengal, Ghulam Husain's text needed to be part of this form of political communication too. He certainly demonstrated a knowledge (and some appreciation) of European political philosophy and political institutions, and a clear awareness of the power of writing to effect political change in the British empire. This related both to his own ability to compose a 'valuable book, by making use of so inconsiderable an instrument as the slit-tongued reed', and to the uses that might be made of writing and books by others.[24] In the course of his history, he noted how, in the case of the politically controversial hanging of Hastings's opponent the Brahman Nandakumar for forgery in 1775, the crucial testimony was 'written down in the English Language and character; and the whole being bound together in the form of a book, was sent to England, from which such a vast number of copies were drawn out, that this subject became famous, and an object of much curiosity in that nation'.[25] He also noted that Philip Francis, a member of the Bengal Council, had done the same with his accusations against Hastings, drawing 'the whole into a methodical writing which he got bound like a book; and after having obtained it's being registered, he thought himself strong enough with such a piece, and he departed this country'.[26] Although there is no indication of Ghulam Husain's opinion on the matter, the most obvious way of bridging the gap between these modes of political communication, and entering a dialogue with the British, was to go into print. Printing in Bengal was, before 1800, 'entirely the preserve of Europeans', so printing the manuscript could be expected to change the fortunes of this text as well as changing its format.[27]

---

23 Michael J. Franklin, "'The Hastings Circle': Writers and Writing in Calcutta in the Last Quarter of the Eighteenth Century," in *Authorship, Commerce and the Public: Scenes of Writing, 1750–1850*, ed. Emma J. Clery, Caroline Franklin and Peter D. Garside (Basingstoke: Palgrave Macmillan, 2002), 186–202; Miles Ogborn, *Indian Ink: Script and Print in the Making of the English East India Company* (Chicago: University of Chicago Press, 2007).

24 Seid-Gholam-Hossein-Khan, *Seir Mutaqherin*, II: 611.

25 Ibid., 465 and J. Duncan M. Derrett, "Nandakumar's Forgery," *English Historical Review* LXXV (1960): 223–38.

26 Seid-Gholam-Hossein-Khan, *Seir Mutaqherin*, II: 524.

27 Graham Shaw, "South Asia," in Eliot and Rose, *A Companion to the History of the Book*, 131.

Translation, Publication and the Politics of Critique

The guiding hand behind the translation of Ghulam Husain's Persian manuscript into English and its subsequent printing in Calcutta was an enigmatic figure known as both Mustafa and Monsieur Raymond. Having been born in Constantinople and educated in Paris, he had eventually found his way to India in 1751 and worked for the French, before acting in various ways as an intermediary between the British and their new subjects in Bengal.[28] Mustafa operated where the different modes of political communication at work in British Bengal overlapped and became entwined. Although it is difficult to be definitive about his motives for putting this particular manuscript into print, since he himself offered several reasons, there was certainly a desire to critique and change the forms of British rule that chimed with the intentions of Ghulam Husain's work. In particular, Mustafa objected to what he saw as the injustices of a legal system that tried to combine British sovereignty with forms of Muslim and Hindu law. It is also clear that he believed in the power of printed books to effect change in British Bengal.

As Mustafa told it, he had long been involved with works like those of Ghulam Husain. In the late 1760s he had been 'master of an Eastern Library, and of a Cabinet of Eastern Curiousities'.[29] He had, he argued, collected such things at a time when the only other people interested were Warren Hastings and Henry Vansittart, the Governor of Bengal from 1760–64. As Maya Jasanoff has shown, the making of such libraries of Indian manuscripts could form the basis of new cultural identities and new political alliances as the relationships of empire were negotiated between imperial governors, intermediaries like Mustafa, and Indian political elites. These collecting practices were a complex engagement by at least some of the new rulers with India's manuscript culture and its debates over politics, religion and aesthetics.[30] Unfortunately for Mustafa his library was sacked and plundered at Jeddah and Mecca in 1770 and he was forced to return to India 'with a firm resolution never to employ any money in books'.[31] Worse was to come. His new employment, as a seraglio keeper, ended in tragedy when one of the women was killed by the man Mustafa had forced her to marry, and her body left on his doorstep. This, along with the news that Warren Hastings was leaving India, meant that Mustafa teetered on the edge of madness. He was, he said, saved by the discovery of Ghulam Husain's manuscript:

> On my going into one of the Navvab's seats, an old woman, among other articles
> of sale, offered me some broken leaves of a decayed book, in which the author
> talked with encomiums of the English parliament in Europe, and with some

---

28   Buckland, *Dictionary of Indian Biography*, s.v. "Raymond," vol. 2: 353.

29   Seid-Gholam-Hossein-Khan, *Seir Mutaqherin*, I Translator's preface, 17.

30   Maya Jasanoff, *Edge of Empire: Lives, Culture, and Conquest in the East, 1750–1850* (New York: Alfred A. Knopf, 2005).

31   Seid-Gholam-Hossein-Khan, *Seir Mutaqherin*, I Translator's preface, 18.

asperity of the English Government in Bengal. A Persian discourse upon English Politicks! strange indeed! I took the broken leaves, and perused some of them in the Garden; and the style, as well as the matter, having awakened my curiousity, I seized on this opportunity to afford some relief to my wearied mind.[32]

This personal motive for undertaking the translation was soon joined by others that suggested publication. Mustafa said that he wanted to use the profits to fund an English education for his children. More significantly, he also reported that he undertook the task 'as a matter of information, which it is incumbent upon me to impart to my adopted Countrymen (the English); and as a warning which I owe to their prosperity'.[33] He would use his position as a cultural and political intermediary, simultaneously part of an Indo-Muslim manuscript culture and British Bengal's print culture, to hold the mirror provided by a printed translation of Ghulam Husain's manuscript up to the Company's rule. What he presented as information the imperial rulers needed to know could also benefit him and those close to him.

Mustafa certainly understood the role of books in politics, both at the level of theory and in practice. He claimed that 'with a mediocre dictionary and a bad grammar, I learned enough of English in a journey of 19 days from Bombay to Balassor, as I might delight in Bolingbroke's Philosophical works'.[34] Whatever the truth of this language lesson, his reading of Henry St John Bolingbroke would, as Robert Travers has shown, have put him in a position to understand some of the principles that underpinned the political positions taken by Philip Francis and other opponents of Warren Hastings from the mid 1770s. These politicians used Bolingbroke's "country" ideology to articulate a critique of 'the worst features of the British imperial state, including high levels of taxation and debt, standing armies, private profiteering among government officials, bribery, corruption and secret influence'.[35] While, as we shall see, Mustafa shared at least these latter concerns, he was himself a supporter of Hastings, and not an opponent. Moreover, his translation and publication of the *Seir Mutaqherin* in 1789, with its defence of the former Governor General, was initially intended for an audience in Britain at the time of Hastings's impeachment trial (1786–94). The work was dedicated to Hastings and prefaced with a laudatory letter to him expressing 'that reverential awe and that high admiration you have impressed all Hindostan with... and to which you are the only European that ever had any access'.[36] One explicit aim of publication was to support Hastings against his political enemies.

---

32   Ibid., 21–2.

33   Ibid., 22.

34   Letter from Mustapha to William McGuire, Calcutta, 16th March 1761, 3 footnote f, European Manuscripts, Orme OV 6 Item 1, Oriental and India Office Collection (OIOC), British Library.

35   Travers, *Ideology and Empire*, 149.

36   Seid-Gholam-Hossein-Khan, *Seir Mutaqherin*, I Copy of a letter to Warren Hastings, 15th February 1785, 13.

Mustafa, like Ghulam Husain, was well aware of the power of print and the power of the readers of the printed word in British imperial politics. In a letter of 1768 he had defended himself against the charges printed in Luke Scrafton's pamphlet *Observations on Mr. Vansittart's Narrative* (1766) that he was a French spy. He told Scrafton that he intended to clear his name because he wished to travel to England one day. While he recognized the emphemerality of political hackery, noting that 'I have seen the World, and Know what fate soon or Late overtakes pamphlets', he also acknowledged the centrality to British politics in Britain and Bengal – rhetorically at the very least – of a public sphere of critical rational debate in and over print:

> I make no doubt but the English I shall find there [in England], are like the English I see here. Men born thinkers, or made so by the very nature of the Government: Men not to be imposed upon for any course of time; amused for a while with a flying Report, but upon any one Standing against it, ready to listen, and equally able and willing to think for themselves.[37]

It was to this imagined readership that the *Seir Mutaqherin* was addressed. As such, it should be seen alongside Mustafa's other published work. He claimed to have had printed in London a work of futurology entitled *State of Europe in 1800*.[38] Unfortunately, no trace of this Occidentalist text can now be found. What is more certain is that in 1789 he authored a pamphlet printed in Calcutta entitled *Some Idea of the Civil and Criminal Courts of Justice at Moorshoodabad*. This ninety-page tract detailed the lack of justice available though these courts in the old Mughal capital of Murshidabad by setting out the author's own trials and tribulations at their hands since 1782.[39] Mustafa saw this publication as part of a longer tradition of 'complaints about the bad justice to be had in Bengal' that included Ghulam Husain's manuscript and the printed tracts of William Bolts, especially his *Considerations on Indian Affairs* (1772), which had prompted published responses from English East India Company officials.[40]

In his own pamphlet Mustafa complained of the inattentive judges in the civil courts; the ignorant, venal and carelessness *maulvis* (Muslim doctors of law)

---

37   Letter from Mustapha to Luke Scrafton, Calcutta, 2nd January 1768, 38–9, European Manuscripts, Orme OV 6 Item 3, OIOC.

38   Seid-Gholam-Hossein-Khan, *Seir Mutaqherin*, II Appendix: Letter to William Armstrong, 15th May 1790, 14–15.

39   *Some Idea of the Civil and Criminal Courts of Justice at Mororshoodabad, in a Letter to Capt. John Hawkshaw, At Behrampore, of the 30th May 1789* (Calcutta: Printed for the Author, 1789).

40   Seid-Gholam-Hossein-Khan, *Seir Mutaqherin*, II Appendix, 21; William Bolts, *Considerations on Indian Affairs: Particularly Respecting the Present State of Bengal and its Dependencies* (London, 1772); Harry Verelst, *A View of the Rise, Progress and Present State of the English Government in Bengal* (London, 1772).

and *munshis* employed to consult and write legal documents in Persian; and the injustices of being dealt with under Muslim as opposed to British law. When it came to the criminal court he made much the same complaints, but saw them as magnified by the actions and lack of accountability of the presiding judge. Significantly, this was Muhammed Reza Khan, the same figure against whom much of the critique in the *Seir Mutaqherin* was directed. Appointed as *naib nazim* in 1776 when the opposition was in the ascendance over Warren Hastings, Reza Khan had assumed responsibility for administering justice in Bengal's criminal courts (*faujdhar adalats*). He was removed by the Governor General after General John Clavering's death in 1777 and the revival of Hastings's control over the Bengal Council. Finally, Reza Khan was reappointed as *naib nazim* in the early 1780s as part of a compromise between Hastings and Francis.[41] On the basis of his own political views and his experience of the legal system, Mustafa questioned in the strongest terms the nature of the justice that was being dispensed by Reza Khan's criminal courts.

Mustafa reported that he had put these complaints about injustice, and his calls for reform, into print for a wider public because his 'anonimous memorial' of a year before had not been attended to, and his petitions to Governor General Cornwallis had been frustrated by Reza Khan and Mubarak ad-daula, the *nawab* of Bengal.[42] He hoped that making the problems public, and making them a problem for the government as a whole, would lead to their resolution. His suggestion for reform was that the courts run by the British should be extended across Bengal. Again, publicity (although not print) was the key. As he put it, these Englishmen's 'greater regard for their own personal character' would mean that they would hold each other to account for any 'deviation from the rules of equity and abstinence... and [he asked] is it not probable that this very publication and the impending loss of character with all its woeful consequences, would prove a perpetual checkon [sic] every one of them?'[43] Mustafa, in both his suggestions for legal reforms, and his attempt to use a printed pamphlet to achieve them, again, at least at the level of rhetoric, understood the relationship between print, publicity and the workings of the imperial state in British Bengal.

Although Mustafa's critique of imperial rule and the possibilities of reform differed from that of Ghulam Hussain – particularly over the value of extending the powers of the Supreme Court and of British judges and law officers across more legal jurisdictions – he could make political use of the *Seir Mutaqherin*. More specifically, he could offer up Ghulam Husain's manuscript as a warning to the British:

---

41   Seid-Gholam-Hossein-Khan, *Seir Mutaqherin*, II: 493–5; N. Majumdar, *Justice and Police in Bengal, 1765–1793: A Study of the Nizamat in Decline* (Calcutta: Firma K.L. Mukhopadhyay, 1960).

42   *Some Idea of the Civil and Criminal Courts*, 3.

43   Ibid., 88–9.

The general turn of the English individuals in India, seems to be a thorough contempt for the Indians (as a national body). It is taken to be no better than a dead stock, that may be worked upon without much consideration, and at pleasure: But beware! that national body is only motionless, but neither insensible, nor dead. – There runs through our author's narrative, a subterraneous vein of national resentment, which emits vapours now and then, and which his occasional encomiums of the English can neither conceal nor even palliate; and yet he is himself but a voice that has spoken among a million of others that could speak, but are silent.[44]

Most importantly, this 'preaching [of] sedition and anarchy' on behalf of the silent majority was a warning that Mustafa could present as one that he, and not the British, had access to.[45] His own geography, his location in and between the different forms of political communication at work in Bengal, allowed Mustafa to heighten the contrast between them. Doing so emphasized the sense that this critique of imperial rule came from a realm outside British experience, from another place altogether:

Living myself in the center of Moorshoodabad; wearing an Hindostany dress, and making a practice in the evening to walk the streets with only a servant, either to listen to, or to mix with, any company I meet with either there or in the market place; I necessarily get a variety of information, which is often out of the power, and always out of the way, of any other European.[46]

By putting Ghulam Husain's manuscript into print the 'semi-Englishman' Mustafa sought to both develop the critique of British rule that he presented as at large in the Indian ecumene and to argue for reform of the legal system through which he had experienced its inequities.[47] As he put it, the lesson of the *Seir Mutaqherin* was that unless the Governor can secure justice through the courts for his imperial subjects 'then no taxes are due to him: [and] no obedience is due to his commands'. Without justice 'millions of mouthless wretches' will think that 'their whole Government amounts to nothing better than a perpetual scene of sac and plunder'. Holding up this mirror to Company rule would mean challenging them to refute the claim that without attention to the rule of law their motives can be simply stated: 'We are come in India to gather taxes, kill people, and make conquests, ---- and ---- and ---- and ---- care little about all the rest'.[48] Articulating this warning meant Mustafa emphasizing his participation within modes of political communication closed to the British, his access to places that they could never go, and to words and thoughts

---

44   Seid-Gholam-Hossein-Khan, *Seir Mutaqherin*, I Translator's preface, 22–3.
45   Ibid., II Appendix, 21.
46   Ibid., I Translator's preface, 23.
47   Ibid., II Appendix, 17.
48   Ibid., 21, 22 and 25.

they would never hear spoken first hand. He reminded them that when Warren Hastings was thought dead, when there were reports in the streets and in the bazaar that his head and right hand had been seen displayed at Bijigarh, the fortress of Raja Cheyt Singh whose rebellion Hastings had gone to put down, 'then *all, all* Sir, (it is the very word;) *all*, think of rising on the English … [and] the country proves ripe for revolution'.[49] Yet delivering this warning, and his alternative calls for peaceful legal reform, also meant Mustafa leaving the streets and market places. It meant navigating another geography of political communication, a geography of the book which was one made up of the spaces and networks of print and printing. This was, as we shall see, a difficult journey to make successfully.

Printing and Imperial Power

Mustafa's problems began with trying to get the translation of Ghulam Husain's manuscript readied for publication in Britain. He sent it to London under the care of Colonel Allan Macpherson who had been Quartermaster General and private secretary and Persian translator to the Governor General Sir John Macpherson.[50] Hurrying it to England to do service in Hastings's defence meant, however, that 'it unavoidably wanted much correction in the distribution and divisions of the subject, in the punctuation of the sense, and in the style: all which defects required the inspection of a reviewer before it could be sent to press'. Unfortunately, the 'eminent historian' charged with the task was dying and no other person could be found who was qualified to correct the manuscript. These contingencies meant that it would not see a London imprint. Mustafa therefore shifted the geography of this book back across the oceans. As he put it, with 'the original intent of the publication being totally marred and already defeated, there remained no other resource than that of supplicating the British public in Bengal, instead of addressing the British public in London'.[51] There was a reasonable expectation that he might engage the Anglo-Indian reading public with an Indian version of the sort of philosophical history that had 'a vibrant popular market that clearly extended from London and its provinces to Bombay, Calcutta, Madras and their hinterlands'.[52] He would do so under the pseudonym 'Nota-Manus'.

---

49   Ibid., 26.

50   Michael J. Franklin, "General introduction and [Meta]historical Background [Re]presenting 'The Palanquins of State; or, Broken Leaves in a Mughal Garden,'" in *Romantic Representations of British India*, ed. Michael J. Franklin (London: Routledge, 2006), 8 and 34 (footnote 31).

51   Seid-Gholam-Hossein-Khan, *Seir Mutaqherin*, I Proposal for publishing by subscription, 5.

52   Margot C. Finn, "Colonial Gifts: Family Politics and the Exchange of Goods in British India, *c.* 1780–1820," *Modern Asian Studies* 40:1 (2006): 225.

Since Mustafa did not have the capital available to pay for the printing of the manuscript he resolved to publish the *Seir Mutaqherin* in Calcutta by subscription. This was a well-tried publishing strategy by the late eighteenth century, and one which circumvented the risks of the commercialization of print by using existing social connections to establish a market for a book prior to publication, preferably among purchasers that were themselves high profile and well connected.[53] Mustafa estimated that he required a hundred subscribers at sixty four rupees (£6 8s) a piece to fund the publication. To attract them he promised that its three quarto volumes of around three hundred and fifty pages each would be 'neatly printed, upon the best Patna paper'. Unfortunately, the five hundred plates that the volumes should have contained were, he said, still in England. The volumes would have to be produced unillustrated.[54]

As it turned out, the successful gathering of subscriptions required resources that Mustafa simply did not possess, and he had to proceed with less than a hundred buyers having handed over their money. He decried 'the scanty reception which my own work has met with from the public', and compared his failures to the success of Francis Gladwin's *Vocabulary*:

> If then so very small a work has produced so large a sum, it was because the author's reputation, as an author, and a man of letters, is formed, known, and established; whereas no one knows any thing of me. Secondly because he is an Englishman, a man high in station; and of course has many friends: whereas I am next to nobody, and my station is immediately after nothing.[55]

This was, therefore, a social failure. It was a failure of the social networks and connections that Mustafa tried to mobilize in Bengal. He, unlike Gladwin, was not 'an Englishman', not of high station, and had not got enough friends. It was also a political failure. Even more gallingly, another of Francis Gladwin's volumes had only made money because the East India Company's Directors had taken two hundred copies at 120 rupees each. In contrast, without official support Mustafa found himself faced with the prospect of giving refunds to angry subscribers.

The subscribers were up in arms because they judged the book to be defective. Mustafa was unhappy with it too. His attempts to render the pronunciation of 'oriental names' via 'the Italian or rather Scotish Alphabet', using 8s and double dots to represent long 'o' sounds were undone by the printing process: 'the work having been printed by several Printers, this has been disregarded by some, who have adhered to the usual spellings'. Indeed, it was 'teeming with faults', since, as the distraught translator put it, the 'Printers have been guilty of an infinity of

---

53   John Brewer, *The Pleasures of the Imagination: English Culture in the Eighteenth Century* (London: HarperCollins, 1997).

54   Seid-Gholam-Hossein-Khan, *Seir Mutaqherin*, I Proposal, 8.

55   Ibid., II Appendix, 10 and 11; Francis Gladwin, *A Compendious Vocabulary English and Persian* (Malda, 1780).

alterations both through chance and wilfulness'.[56] Farming the text out to multiple printers, a common practice, had multiplied errors, so that, for example, the page numbering of volume two (which contained the criticisms of the British) was hopelessly confused, with numbers missing, misplaced and inverted, and with some pages numbered twice.

Mustafa's readers also questioned his translation, and objected to his orthography and grammar. He was well aware of the epistemological as well as the practical problems that attended translation between languages, the dangers they posed for texts which could not manage the transition, and the social relationships that were being formed between British language learners and Indian teachers and translators in order to bridge the gaps.[57] So Mustafa anonymized himself as 'Nota-Manus' and sought to reassure an audience 'displeased with that wretched performance of mine' that they were hearing the original author's true voice, and that his translation had been assisted by the great Oriental scholar William Jones himself.[58] He also admitted, however, that he had, in search of a larger readership, animated some of the text by elaborating on motivations and emotions in ways which 'shall possibly surprise those Gentlemen employed at the public offices of Calcutta, in translating Persian Letters', but which should appeal to European readers excluded by dry Orientalist scholarship, including 'women; and all those that read for amusement'.[59] Mustafa had also spiced up the text with footnotes alluding to matters such as the 'kinds of intimacies' that existed between 'Indian Eunuchs' and 'sequestrated ladies'.[60]

Doubts over the quality of the translation and the printing clearly raised questions about the reliability of the book and its contents.[61] There was a danger that it would not be taken seriously, with 'no less than twenty pers at table, at a Mr Browne's at Chouringhi' having 'declared that there was no reading two pages of it'.[62] Against this Mustafa argued that any changes to the text which might make it more fluent 'would detract from the genuineness of the translation, and of course from the veracity and integrity of the intended evidence'.[63] In all of this Mustafa suspected that the readers of the book were basing their reading of it on who he was rather than what was in the text itself. An article in the *Bengal Gazette*

---

56 Seid-Gholam-Hossein-Khan, *Seir Mutaqherin*, I Advertisement, 1–2 and II Appendix, 4.

57 Kapil Raj, *Relocating Modern Science: Circulation and the Construction of Scientific Knowledge in South Asia and Europe* (Delhi: Permanent Black, 2006); Michael S. Dodson, *Orientalism, Empire, and National Culture: India, 1770–1880* (Basingstoke: Palgrave Macmillan, 2007).

58 Seid-Gholam-Hossein-Khan, *Seir Mutaqherin*, II Appendix, 30.

59 Ibid., I: 35 and II Appendix, 11.

60 Ibid., I: 536 footnote.

61 Robert Darnton, *The Business of Enlightenment: A Publishing History of the Encyclopédie, 1775–1800* (Cambridge: Belknap Press of Harvard University Press, 1979).

62 Seid-Gholam-Hossein-Khan, *Seir Mutaqherin*, II Appendix, 16.

63 Ibid., 4.

had revealed that he was 'Nota-Manus' and this, he argued, had had disastrous consequences. This was a personal matter. Readers now read the *Seir Mutaqherin* as the work of Mustafa. His British audience in Bengal could interpret, and dismiss, it as a personal gripe about his own mistreatment in the courts rather than as a more general critique of imperial rule. As he put it 'no less than twenty enlightened Englishmen, sitting at table, have unanimously uttered this blasphemy: *what business has a Governor General with Mustepha's quarrel with Mahmed-reza-qhan?* that is in other words, *what business have Supreme Magistrates with a close inspection into the manner of distributing justice to Individuals?*'[64] The latter message, he feared, would be lost in favour of the former. In addition, he thought that a significant and powerful Indian audience was using the knowledge of his authorship to make things difficult for him. He argued that Muhammed Reza Khan and Mubarak ad-daula were angered that he had published a work in which they appeared 'in the most despicable and ridiculous colors' because they now understood the dynamics of reputation and publicity within the politics of British Bengal as well as he did:

> [A]s the great ones of this land, after having been for ages insensible and quite callous to every thing like character, have now learned of the English to pay some regard to it, just as they have learned the use of China-tea, Mahogany-chairs, and Buff-breeches; so, it is natural to think that those men, who have left it as a legacy to their children and sons-in law, to wreck hereafter by every means in their power their vengeance upon the author of that work, have likewise conceived an unconquerable prejudice against the man that has presumed by his translation, to expose them on the Theatre of an English world.[65]

Yet, as Mustafa understood it, these problematic readings were not simply a matter of purely personal politics. This was about broader categories as well. As he said, 'it is only since I have wrote history, that is since, I have thrown away the mask, and given myself for what I was, that I have been taken for a foreigner, and have been thought so greatly defective in language'. This meant that his readers, identifying him as a 'foreigner', were too willing to assume that all the errors in his book were his responsibility alone and not the fault of the printers.[66] They saw the errors and used what they thought they knew about him as a 'foreigner' to come to conclusions about the qualities of the text he had put before them. Just as not being 'an Englishman' of high standing had derailed his search for subscriptions, so being an outsider, a 'foreigner', in British Bengal had shaped the ways in which his text was read.

Mustafa's problem was a simple one, and it was also a geographical one. He did not have sufficient control over the printing offices where his work was being

---

64   Ibid., 24–5.
65   *Some Idea of the Civil and Criminal Courts*, 67–8.
66   Seid-Gholam-Hossein-Khan, *Seir Mutaqherin*, II Appendix, 11–12 and 17–18.

produced. As Adrian Johns has argued for the production of works of natural philosophy in early modern London, authors were often at the mercy of the stationers and printers who undertook the printing of their work. These craftsmen organized the processes necessary for publication in print according to a specific set of rules, customs and practices. They were in control of the spaces of print production. While authors realized that their own direct 'access to the press was vital if their intended meanings were to survive the printing process intact' this was not always easy to negotiate or to establish. It depended upon understanding the organization of the printing process itself, on mobilizing a range of social and cultural resources, and on deploying them successfully. It certainly meant that 'the would-be author was not necessarily dominant when knocking at the door of the printing house'.[67]

In Calcutta, authors and editors like Mustafa were faced with a relatively small number of printing offices. There were usually between three and five presses at work in the city in the 1780s, and between seven and ten in the 1790s. These were organized around the production of a newspaper, but also produced calendars and almanacs, lists of Company servants, works on law, medicine, revenue and horses, and a variety of ephemeral invitations, notices and forms that have rarely survived. They operated in difficult conditions. There was no copyright protection and their production costs, particularly for paper, ink and engraving, were high.[68] As Mustafa found out these cutthroat commercial conditions had direct effects on the quality of the printing:

> No work within my knowledge in Calcutta, has been tolerably printed, but where the author himself was the owner of the printing office or a partner; or where the Printer had purchased the propriety of the work; or at least… made a sharer in the fate of the book being printed, that is, by being promised for his trouble one half of the author's profits.[69]

Commercial printers would, he thought, 'reserve the two or three good hands that may be in a Printing Office' for the most lucrative work. All but one Calcutta printing office was, he noted, 'worked by natives, who print in a Printing-Office, just as they copy in a Counting-House, without understanding the language'. Paying by the sheet or, even worse, paying in advance, only purchased unskilful apprentices and 'supercilious inattention'. Poor composing produced error-strewn proofs. Inaccurate correcting saw 'a dozen of faults corrected, seven or eight preserved, and a novel crop of half a dozen new one's'.[70] All of this produced many errors and delayed the printing by months, angering the subscribers.

---

67   Johns, *The Nature of the Book*, 102.

68   Graham Shaw, *Printing in Calcutta to 1800* (London: Bibliographic Society, 1981).

69   Seid-Gholam-Hossein-Khan, *Seir Mutaqherin*, II Appendix, 4–5.

70   Ibid., 5 and 7.

Just as had been the case with the reading of his book, Mustapha understood that the social relations involved in the management of the printing office were imperial ones. If any author or editor faced problems when knocking at the door of a Calcutta printing office, then some faced more problems than others. As Mustafa understood it his identity mattered, and it damaged his book:

> And what if the Printer despising your quality of a Foreigner, thinks an Englishman *whatever*, has a right to know more of the language than you, and takes the liberty of correcting you, of sneering at your emendations, of substituting his own, and of throwing away without answer some angry notes in which you *inform him that he is paid for printing what is before him, and not for correcting* it?[71]

Mustafa reported that he was constantly told that what he was writing 'may be English, but not English of Europe, only English of Bengal'. To him this was a product of the printers' own ignorance. What they didn't understand, however grammatically correct, became 'Bengalee English' that was despised and expunged.[72] Mustafa's problem was that the printers had the power to make the text speak the way they wanted it to no matter what he intended it to say. Try as he might he could not get them to change it. As he told it, having been 'Shocked at such an enormous liberty, I wrote a note of complaint, backed it by a very angry one, called on the Printer myself, spoke angrily, corrected again.... [H]e made the proof over to his apprentice, who all this while had given evident signs of approbation to his master by his smiles, cast of features, *and* shrugging-up of the shoulders – and the next morning my excellency was saluted with full two hundred ... very ill printed'.[73] In the end the printers even changed the title. Mustafa reported that the printer Joseph Cooper showed it to 'a gentleman skilled in oriental languages' who 'dashed the *Amusements of Posterity* and substituted *Review of Modern times*'.[74] If what had been produced was a translation of Ghulam Husain's manuscript it was hardly the book that Mustafa had wanted at all.

In the end the printers drained Mustafa of all his cash and the subscribers were on his back. Together they reduced him to 'selling trinkets, plates, and books'.[75] Finally, any hopes that the printed text would be able to make political capital back in London were abruptly cut short when the ship carrying the consignment of volumes sank before reaching its destination.[76] Instead of creating a book that would excite the attention of English readers, effect the reforms in their government

---

71 Ibid., 8.
72 Ibid., 8.
73 Ibid., 10.
74 Ibid., 28.
75 Ibid., 30.
76 Reported in *The Siyar-ul-Mutakherin, A History of the Mahomedan Power in India During the Last Century, by Mir Gholam Hussein-Khan*. Revised from the translation of

that he sought, and perhaps make him some money too, Mustafa found himself unable to escape from his text. As he explained to the subscribers who wanted to return unwanted copies to him, the books 'are already lumber upon my hands, and lumber that encroaches so much upon the dimensions of my habitation, and moreover requires so much care and solicitude, that by keeping those Books out of my view, they shall really confer a favour upon me, and render me a service'. The whole affair had, he reported, 'disgusted me with Printers and Books'.[77]

Mustafa was unable to manage the social relations, and the geographies, of print in British Bengal. This mattered because of the role of print in a new form of imperial political communication. Printing had certainly not revolutionized communication throughout India, but it had shifted the ways in which politics was done by changing some of the relationships between speech, script and print for some people in some places. There was a difference between the manuscript culture of the Indian ecumene and the mechanisms of the imperial public sphere which meant that the places and spaces of communication had shifted. How print was managed now mattered to how politics was done. Some of the problems this caused Mustafa may have been contingent. It might have happened to anyone. There were certainly many problems to be had with printers, patronage and readers.[78] Yet, as Mustafa understood it what consistently explained the problems he faced was his identification as 'a foreigner'. As he saw it, the 'adopted countrymen' of this 'Semi-Englishmen' repeatedly positioned him as an outsider. This limited the networks through which he could seek subscribers; it provided the basis for dismissive readings of his book; and by denying him control over the printing process it produced a book that was itself 'teeming with errors', angering his readers, and losing him subscriptions. The spaces and networks which made up the geography of the book in eighteenth-century Bengal were shaped by the social relations of empire, and people like Mustafa could not operate effectively inside them.

Conclusion: The Geography of an Unsuccessful Book

The printed translation of the *Seir Mutaqherin* may not have been the work that Mustafa wanted to produce, and it may not have been able to realize his hopes of political influence and monetary gain, yet it can be read as the product of a specific geography of the book. On the one hand, this book was made at the point where different cultures of political communication in late eighteenth-century Bengal overlapped, mixed and affected each other. This book was not simply a product of

---

Haji Mustefa, and collated with the Persian Original, by John Briggs (London: Printed for the Oriental Translation Fund of Great Britain and Ireland, London, 1832) v.

77   Seid-Gholam-Hossein-Khan, *Seir Mutaqherin*, II Appendix, 31.

78   See, for example, the complaints in John Gilchrist, *A Dictionary, English and Hindoostanee* (Calcutta, 1787–90).

the spaces and forms of writing of the Indian ecumene, although Ghulam Husain's manuscript drew upon long-standing literary conventions, material forms of paper and ink, and modes of political engagement with Indo-Muslim expectations about sovereignty and rule to articulate its critique. Neither was it simply part of the imperial public sphere of print and politics, although through Mustafa's efforts to put it into print it took on a form associated with Enlightenment philosophical history, entered the debate over imperial politics along with other printed works produced in both Bengal and Britain, and was part of geography of reading which took it to London and into spaces in British Bengal such as 'Mr Browne's table at Chouringhi'. As such, the geography of this book was a product of the engagement of different voices using different forms of political communication to enter the debate over empire in British Bengal. The fact that they needed to engage with each other and to bring together different geographies of political communication meant that something new was produced as these worlds intersected. As Mustafa himself had put it, 'A Persian discourse upon English Politicks! strange indeed!'

On the other hand, the historical geography of this particular book is not simply a tale of negotiation and co-construction.[79] The spaces of political communication, and the geographies of speech, writing and print, were not simply open to all. Mustafa played upon his exclusive access to the spoken words of the streets and marketplaces of Murshidabad, but then found himself – as a 'foreigner' in Calcutta – unable to construct patronage networks that would realize subscriptions for his work, unable to manage the readings of his text by those who took him as at fault for its errors, and, at the root of it all, unable to exert control over the printing offices in which his work was being produced in order to make the book that he wanted. These networks and spaces were structured through imperial relationships of race and class as one part of a complex and changing social geography of empire. The many geographies of this book, which must include both connection and exclusion, caution us against either simply condemning the British empire in late eighteenth-century Bengal as built on divisive notions of the imperial self and its indigenous others, or, alternatively, celebrating it as the scene of boundary crossing and hybridization.[80]

If this book can be judged to have failed in that it did not meet its editor's (or even its author's) ambitions for it then perhaps all books are failures. If those that failed are easy to find, it is much more difficult to attribute political success and influence to texts. The law courts of British Bengal were reformed, and in ways that Mustafa might have appreciated, but what influence his ventures into the public sphere of print had on such changes are very hard to trace.[81] What can be

---

79   Raj, *Relocating Modern Science*.

80   For example, Bernard S. Cohn, *Colonialism and Its Forms of Knowledge: The British in India* (Princeton: Princeton University Press, 1996) and William Dalrymple, *White Mughals: Love and Betrayal in Eighteenth-Century India* (London: HarperCollins, 2002).

81   Travers, *Ideology and Empire*.

argued instead is that this book, and the account of its production, circulation and consumption presented here, illuminates the ways in which the configurations of what we see as the geography of the book in any particular period or place favours some books, and those who would produce and circulate them, over others. As the chapter has shown this needs to be investigated in two interconnected ways. Firstly, understanding the context in terms of that long and broad historical geography of the changing relationships between speech, script and print, and the particular conditions that establishes for the forms of communication under discussion. Most specifically, for the role of the printed book in any period, place or network. Here that was a matter of the relationships between the spoken, written and printed word as two forms of political communication intersected in late eighteenth-century Bengal. Second, with a finer focus this 'context' can then be seen to be made up of a complex geography of spaces, places and networks of communication, one part of which is the geography of the book. These spaces are always structured socially, culturally and politically to enable some things to happen and to disable others. In this case these geographies of the book meant that Mustafa's "Amusements of Posterity" could not be born into the world.

Chapter 2
Steam and the Landscape of Knowledge: W. & R. Chambers in the 1830s–1850s

Aileen Fyfe

Introduction

In June 1835, William and Robert Chambers signalled their commitment to the new industrial technologies of printing:

> Nothing, in our opinion, within the compass of British manufacturing industry, presents so stupendous a spectacle of moral power, working through the means of inert mechanism, as that which is exhibited by the action of the steam-press.[1]

The brothers had only recently made the transition from booksellers to magazine editors and publishers, and their additional decision to become printers was more recent still. They argued that the enormous success of their cheap, instructive weekly magazine, *Chambers's Edinburgh Journal*, had forced them to adopt the innovative technologies of steam-printing and stereotyping. Yet their words in 1835 make clear that they regarded the steam-powered printing machine as far more than just a mechanical tool with an economic benefit. Steam-printing was a 'moral power' without which 'the tide of knowledge and human improvement would be forced back, greatly to the injury of society'.[2]

Historians of the book have long recognized that there were significant changes in the production and distribution of printed matter in the nineteenth century. In contrast to the voluminous scholarship on the 'printing revolution' of the fifteenth century, the nature of these changes remains little studied. We know that prices came down and readership went up, and statistical research has highlighted the middle decades of the century as the key point of transition.[3] We assume that the use of steam-powered printing had much to do with this, and we may also suspect that the steam-powered railway system was important to the story. There is an obvious

---

1   "Mechanism of Chambers's Journal," *Chambers's Edinburgh Journal* (hereafter *CEJ*) 6 June 1835, 151.

2   Ibid.

3   Simon Eliot, *Some Patterns and Trends in British Publishing 1800–1919* (London: The Bibliographic Society, 1994); Alexis Weedon, *Victorian Publishing: The Economics of Book Production for a Mass Market, 1836–1916* (Aldershot: Ashgate, 2003).

plausibility to such assumptions. Steam-powered printing made possible faster production speeds, hence larger outputs in a given time, and it reduced the unit-costs on very large print runs.[4] Railways provided a system for the rapid transportation of goods, correspondence and people, and helped to weave Britain's many regional markets into a national one.[5] Historians of reading have also pointed out that railway carriages provided new spaces and times for reading, while the bookstalls on railway platforms offered a new retail outlet.[6] But beyond such broad-brush comments on the impact of new technologies, we actually know very little about the reasons why they were adopted by specific people at specific times.

It is my concern in this chapter to examine one particular publishing firm, and to consider why they adopted, and how they used, these technologies. W. & R. Chambers was pioneering in its early use of the new technologies, and its rich archive allows us to supplement their enthusiastic published statements with an account of what was actually going on behind the scenes.[7] 'Steam' is a convenient shorthand for the new technologies which transformed space and time, and was used as such by Victorian commentators,[8] though it is not a wholly accurate description. The changes in transportation options were indeed steam-powered (railways and steamships), but the innovations in printing and publishing included the process of stereotyping and the creation of edition bindings, as well as the steam-powered machines for printing and paper-making. I will endeavour to focus on the steam technologies, but some of the others will inevitably appear.

My focus on 'knowledge' is, however, no mere shorthand. In the early days of steam-powered printing, its use was distinctly connected to publications which provided some form of useful knowledge. Its first commercial user was *The Times* newspaper, in 1814, and thus steam was harnessed to the provision of political and commercial information. During the 1810s and 1820s, steam-powered printing was too expensive a capital investment to appeal to many printers or publishers, but by the end of the 1820s, it was enthusiastically adopted by a new sector of the book trade. Groups with philanthropic desires to educate and improve the working classes had been campaigning for a better supply of printed matter cheap enough to be accessible to those made newly literate by the Sunday and charity schools. The Bible and tract societies had been set up to meet this demand, and by the late

---

4   Michael Twyman, *Printing 1770–1970: An Illustrated History of Its Development and Uses in England* (London: British Library, 1998); for the USA, Rollo G. Silver, "The Power of the Press: Hand, Horse, Water and Steam," *Printing History* 5:1 (1983): 5–16.

5   Jack Simmons, *The Victorian Railway* (London: Thames & Hudson, 1991).

6   Keven J. Hayes, "Railway Reading," *Proceedings of the American Antiquarian Society* 106 (1996): 301–26; Stephen Colclough, "'Purifying the Sources of Amusement and Information': The Railway Bookstalls of W.H. Smith & Son, 1855–1860," *Publishing History* 56 (2004): 27–51.

7   On Chambers's use of new technology, see Aileen Fyfe, "Information Revolution: William Chambers, the Publishing Pioneer," *Endeavour* 30 (2006): 120–25.

8   See the opening discussion of 'steam, gas and the telegraph' in "Cheap Literature," *British Quarterly Review* 29 (1859): 313–45.

1820s, secular groups were beginning to argue that cheap print need not be wholly religious, but should be more broadly educational.[9] The religious societies were joined by the Society for the Diffusion of Useful Knowledge (SDUK) and by such commercial publishers as Charles Knight and W. & R. Chambers.[10] These publishers all shared a commitment to education and improvement – though they differed over the inclusion of religion – and they all saw steam-printing and stereotyping as a means to produce the large quantities of publications that they deemed to be necessary, while also reducing unit costs. Thus, in Britain, steam-powered printing was initially very closely linked to the provision of knowledge, broadly defined.[11] It was not until the late 1840s that this connection would be over-shadowed by the spectacular effects of the application of steam to fiction publishing, where it created a mass readership for penny story magazines and railway novels.

It should now be clear why 'steam' and 'knowledge' go together, but what about 'landscape'? The firm of W. & R. Chambers was located in Edinburgh, 400 miles north of the centre of the British book trade. New technologies and the expertise to use them were not uniformly available throughout Britain, so the firm's physical location influenced its production options. And Chambers had to overcome some very practical logistical problems about how to deliver its publications to their readers. The issue was not simply distance, but the nature of the intervening landscape and the extent to which it was served by suitable transportation options. Uneven and differentiated transport facilities, combined with different regional cultures of education and literacy, created challenges for Chambers's efforts to sell their wares. I will argue that Chambers had an obvious natural market in central and lowland Scotland, and it was through their careful and imaginative use of new technologies that the brothers were able to transform their firm into a national player, and to make inroads into a transatlantic market.

Further, I would suggest that Chambers, though their use of steam technologies, made a significant impact on what we might think of as the metaphorical landscape of knowledge in Britain and the transatlantic world. Thanks to their efforts, the existing provision of expensive, scholarly books for elite readers was joined by a range of new

---

9   For the religious publishers, see Leslie Howsam, *Cheap Bibles: Nineteenth-Century Publishing and the British and Foreign Bible Society* (Cambridge: Cambridge University Press, 1991); Aileen Fyfe, *Science and Salvation: Evangelicals and Popular Science Publishing in Victorian Britain* (Chicago: University of Chicago Press, 2004).

10   Valerie Gray, "Charles Knight and the Society for the Diffusion of Useful Knowledge: A Special Relationship," *Publishing History* 53 (2003): 23–74; Rebecca Kinraide, "The Society for the Diffusion of Useful Knowledge and the Democratization of Learning in Early Nineteenth-Century Britain" (PhD diss., University of Wisconsin-Madison, 2006); Sondra Miley Cooney, "Publishers for the People: W. & R. Chambers – the Early Years, 1832–50" (PhD diss., Ohio State University, 1970).

11   Aileen Fyfe, "The Information Revolution," in *The History of the Book in Britain, Vol. 6: 1830–1914,* ed. David McKitterick (Cambridge: Cambridge University Press, 2009) 567–94.

publications which were not only cheaper but different in form: things like penny magazines, secular tracts and part-works.[12] The resulting spectrum of knowledge publications varied in their format, level of detail, literary style and content, and when this is coupled with the fact that they all had their own geographical and economic patterns of availability, we can start to see how the provision of knowledge was uneven and differentiated. We might, therefore, think of it as a landscape of knowledge, the charting of which will reveal the patterns of communicative practices which are central to the place of science in modern society.[13]

The Firm of W. & R. Chambers

Most of the population of Britain in the first half of the nineteenth century had only limited access to print and the educational opportunities it offered. The major publishing houses tended to concentrate on the relatively easy profits to be gained from a known market of affluent and educated readers, and paid little attention to the needs of clerks, shop-boys and school teachers, let alone to those of tailors, carpenters and factory workers. In contrast, William Chambers and his younger brother Robert were among the small number of younger men who considered the needs of readers with only a basic education and very limited spare cash. Although the brothers ultimately became major players in the Edinburgh book trade, and substantial employers in their own right, they – especially William – saw themselves as self-made men and this motivated a deep commitment to education, instruction and self-improvement.[14]

They were the sons of a handloom weaver, and grew up in the country town of Peebles, about 25 miles south of Edinburgh. Robert later described his hometown as 'neat and agreeable', but with a 'decidedly dull aspect'.[15] William began an apprenticeship with an Edinburgh bookseller at the age of fourteen. Robert remained in education until he was sixteen, but was forced to abandon his hopes of proceeding to university when his father lost yet another job. He then set up a small bookshop on Leith Walk, the main road from Edinburgh northwards to the port of Leith. A year later, in 1820, having finished his apprenticeship, William opened his own shop on the same street.

---

12   Jonathan R. Topham, "Publishing 'Popular Science' in Early Nineteenth-Century Britain," in *Science in the Marketplace: Nineteenth-Century Sites and Experiences*, ed. Aileen Fyfe and Bernard Lightman (Chicago: University of Chicago Press, 2007), 135–68.

13   James Secord, "Knowledge in Transit," *Isis* 95 (2004): 654–72.

14   The main biographical source is William Chambers, *Memoir of Robert Chambers with Autobiographical Reminiscences of William Chambers* (New York: Scribner, Armstrong and Co, 1872). See also William Chambers, *The Story of a Long and Busy Life* (Edinburgh: W. & R. Chambers, 1882).

15   Robert Chambers, *The Picture of Scotland*, 2 vols. (Edinburgh: Tait, 1827), vol. I, 178.

By the early 1830s, both William and Robert were doing well enough to move to better premises in Edinburgh, and Robert was becoming known in the Edinburgh literary world through his writings on Scottish history. William was watching the progress of the cheap miscellany magazines which had sprung into existence during the political agitation leading up to the 1832 Reform Act. Both in London and Edinburgh, periodicals were being launched at the price of a penny or a penny-and-a-half. William saw the need for such publications – indeed, he and Robert had briefly run one of their own in 1821–22 – but he was critical of their sloppy compilations of extracts and clippings without an overall plan. William Chambers believed that the press had a real power to improve its readers, and he felt that he could produce a better magazine than the others.

The first number of *Chambers's Edinburgh Journal* appeared on Saturday, 4th February 1832. For one and a half pennies, readers would each week receive a large folded sheet with four columns of text on each of its four pages. That first issue made no secret of William's ambitions. The front page announced his conviction that 'a universal appetite for instruction' now existed, and that his new magazine would supply 'a meal of healthful, useful, and agreeable mental instruction' in a format and at a price to suit 'the convenience of every man in the British dominions'. As well as presenting the *Journal* as a medium for the diffusion of knowledge, and ambitiously outlining its audience, William's opening announcement criticized the work of those charitable organizations who also claimed to diffuse knowledge. He mentioned no names, but he must have had the SDUK and the Society for Promoting Christian Knowledge in his sights. According to him, they had failed partly because of their structural inflexibility but especially because of the constraints that their political or religious affiliations placed upon them.[16] Throughout its long life, *Chambers's Journal* would eschew party politics and religious sectarianism in its efforts to appeal to the broadest possible audience.[17]

The Natural Market: Lowland Scotland

William Chambers launched his new *Journal* in the second centre of the British book trade. Being based in Edinburgh, rather than London, had advantages and disadvantages. Edinburgh had a long history of printing, and although it was no longer a seat of government, it continued to be the centre of the Scottish legal system and the home of an eminent university. In the 1830s it was still the centre of the Scottish publishing trade: Edinburgh publishers issued around 4,600 titles in that decade, compared with just 1,500 from Glasgow and 210 from Aberdeen.

---

16   "The Editor's Address to his Readers," *CEJ*, 4 February 1832, 1.

17   Laurel Brake, "The Popular 'Weeklies'," in *Edinburgh History of the Book in Scotland, Volume 3: Ambition and Industry 1800–1880*, ed. Bill Bell (Edinburgh: Edinburgh University Press, 2007), 359–69.

Nevertheless, the Edinburgh publishing trade was small in comparison with London, whose publishers issued over 35,000 titles in the 1830s.[18] Being so far from the London trade would occasionally create problems for Chambers, when supplies, machinery or expertise seemed only to be available in London. On the other hand, being outside London gave the new *Journal* a clearer field in which to seek its market, rather than having to compete directly with the myriad of new publications which seemed to spring up daily in the capital. Britain in the 1830s was not yet an integrated national market and in lowland Scotland, Edinburgh-based publications had the clear advantage of being available more quickly and with lower transport costs than did those from London.

The natural market for *Chambers's Journal* also happened to be an area with a substantial population, characterized by relatively high literacy rates and a thirst for knowledge and self-improvement. The majority of Scotland's 2.4 million inhabitants lived in the lowlands and south of Scotland, particularly in the cities of Edinburgh and Glasgow. Both William and Robert Chambers were convinced of the distinctiveness of Scotland from England, and they had, between them, written three descriptive works on Scotland to this end.[19] As William put it, the institutions and character of Scotland 'differ as much from those of the sister country, as the mountainous and romantic regions of the north differ from the broad luxuriant meadows of the south'.[20] One of the things of which Scots like the Chambers were proud was that they had better access to education than their English neighbours, as demonstrated both by the higher literacy rates of the entire population, and the higher percentage of the population receiving formal education.[21] In reality, as William Chambers knew, the Scottish educational system was not as good as it could be, and the range of educational works which W. & R. Chambers produced can all be seen as helping to improve the education of the people. Nevertheless, it was good enough for Chambers to expect a ready market for his *Journal* in Scotland, and particularly in the lowlands. He wrote in 1830 that:

> To those who are not intimate with the character of the Scottish people in their own country, it would be difficult to convey an adequate idea of that burning desire which almost every parent has, to see his children educated. ... It is not confined to persons in easy circumstances; it descends to the meanest of the

---

18   Figures from searches in *Nineteenth-Century Short Title Catalogue on CD-ROM*, series I and II, 1800–1870 (Newcastle: Avero, 2002).

19   Chambers, *Picture of Scotland*; William Chambers, *The Book of Scotland* (Edinburgh: Robert Buchanan, 1830); Robert Chambers and William Chambers, *The Gazetteer of Scotland* (Edinburgh: Thomas Ireland, jnr, 1832).

20   Chambers, *Book of Scotland*, v.

21   By the mid nineteenth century, barely 10 per cent of Scottish men were illiterate, and only 25 per cent of Scottish women. David Vincent, *The Rise of Mass Literacy: Reading and Writing in Modern Europe* (Oxford: Polity Press, 2000), 9–10.

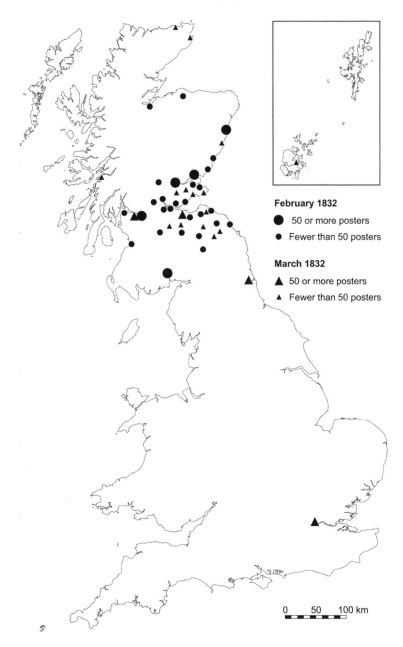

February 1832
● 50 or more posters
● Fewer than 50 posters

March 1832
▲ 50 or more posters
▲ Fewer than 50 posters

0   50   100 km

Figure 2.1 **Distribution of publicity material for *Chambers's Edinburgh Journal* in February and March 1832. Derived from John Johnstone's invoice to Chambers, August 1832 in WRC 312 (LL 1832-40, misfiled under K)**

peasantry, and will be found mingling with the every-day feelings of the poorest family in the land.[22]

In such a context, there was a respect for education and a thirst for learning which would generate a market for the improving and instructional journal Chambers wished to produce. These cultural conditions meant that at its launch, Chambers could expect his *Journal* to have relatively easy geographical access to an appreciative potential audience, and to enter that market with less competition than a London publisher would face.

The extent to which *Chambers's Journal* was originally situated within its lowland Scottish context is graphically illustrated by William Chambers's first efforts at publicity. His existing trade connections helped him make contact with booksellers in an impressive array of Scottish towns. In February 1832, over a thousand posters advertising the *Journal* were dispatched to 28 towns (see Figure 2.1). The largest numbers of posters (a hundred each) went to Glasgow, Perth, Dundee and Aberdeen; then there were forty posters to Galashiels, Falkirk, Hawick, Stirling, Berwick and Greenock; and smaller numbers to other towns. Other than Inverness, Elgin and Aberdeen, all the towns were in central or southern Scotland. It was not until March 1832 that Chambers started to target more distant audiences, sending posters to Wick, Thurso and Kirkwall in the north of Scotland and making an attempt to reach English readers, with two hundred posters for London and fifty for Newcastle. At the end of April, the *Journal* reported that its total print run was now 31,000 copies, but barely 3,000 of those left Scotland.[23]

The wisdom of focusing on local markets can be seen from figures collated by *Tait's Edinburgh Magazine* and reprinted in *Chambers's Journal* in 1834, relating to the sales figures for magazines in 'a certain unnamed country town in Scotland'. Both London and Edinburgh magazines were available, but the Edinburgh ones proved more popular: there were 700 sales of *Chambers's Journal* against 260 of the *Penny Magazine*, and Edinburgh-based monthlies such as *Blackwood's* and *Tait's* were more in demand than such London titles as the *Metropolitan Magazine* or the *Monthly Magazine*. The higher sales of *Chambers's Journal*, despite its extra ha'penny in price, might suggest a distinct preference by readers or illustrate the competitive advantage Edinburgh publishers had in the Scottish market. The figures in *Tait's Magazine* also demonstrated very clearly that low prices were effective at increasing circulation. Although this 'country town' could find over a thousand readers of the various penny periodicals, the most successful monthly (*Blackwood's Magazine*, 2s, 6d.) could find only fourteen readers, and none of the quarterly reviews made it into double figures.[24]

At the same time, *Chambers's Journal* published a geographical breakdown of its own circulation, suggesting that readers could use it as an index to 'the degrees of

22   Chambers, *Book of Scotland*, 370.
23   "Chambers's Edinburgh Journal," *CEJ*, 28 April 1832, 52.
24   "Editorial," *CEJ*, 1 February 1834, 1.

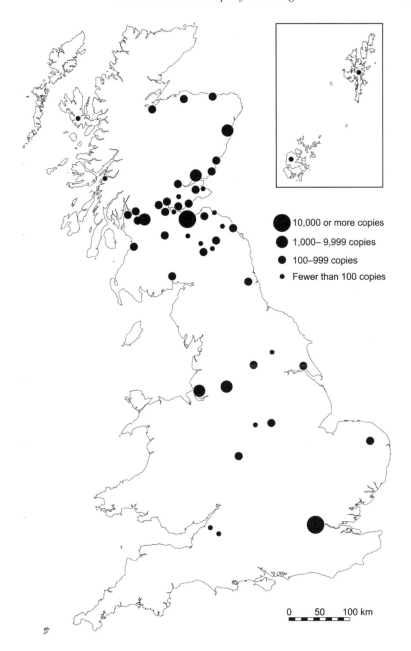

Figure 2.2 **Distribution (each week) of *Chambers's Edinburgh Journal*, as reported in February 1834. Note that about half of the London copies went onwards into the country; and an unknown proportion of the Edinburgh copies went to Dublin for Ireland. Derived from figures given in *CEJ*, 1 February 1834, 1.**

intelligence and appetite for reading which prevail in different parts of the empire'. The figures reiterated Chambers's strong base in lowland Scotland, but also pointed to a growing demand in other parts of the United Kingdom (Figure 2.2). The *Journal* was being distributed to all corners of Scotland, ranging from the large numbers sold in Edinburgh and Glasgow to the few dozen sold on the Isle of Skye or in the Orkneys. The English figures were more difficult to assess, since many copies of the *Journal* were distributed via intermediary agents. London certainly took delivery of the largest number of copies, but about half of those were believed to be forwarded to country booksellers. The available figures suggested that the English readership was rather more patchy than the Scottish readership. Older towns like York and Bristol took no more copies than did Shetland, supporting Scottish prejudices about the lack of enthusiasm for cheap print and self-improvement in most of the English towns. The most marked desire for the *Journal* came from the major industrial cities of northern England, with large numbers going to Liverpool, Manchester, Leeds and Newcastle every week.[25]

Stereotype Plates: A First Attempt at a National System

Both sets of 1834 figures demonstrate that *Chambers's Journal* was performing well in its local, Scottish market, but that England was becoming increasingly important. This raised crucial issues about the practical distribution problems of trying to reach distant locations from an Edinburgh base. The fast communication system of Britain in the early 1830s was the Royal Mail coaches and the privately-run express stagecoaches. On good roads, with regular changes of horses and few stops, these coaches could average more than ten miles an hour. This meant it took two days to reach London from Edinburgh.[26] Neither of these types of coaches, however, had space for cargo: they were used for the mails and for passengers. Newspapers did have the right to be sent through the mails, but *Chambers's Journal* was an unstamped periodical, as this was the only way that it could be sold at so low a price while the stamp duty was four-pence. Although it prevented the *Journal* using the mail coaches, the advantage of not carrying news was that it could be printed three or four weeks in advance of the specified publication date.[27] This gave sufficient time for packages of the journal to travel to London, Dublin, Shetland or Cornwall (by horse carrier or by water) and be released to the public on the same day throughout the entire kingdom. Thus, national distribution was theoretically possible, but there were problems, principally of cost and timing.

---

25   Ibid.

26   Mail coach connections are listed in Chambers and Chambers, *Gazetteer of Scotland*, 400. Journey time was over 42 hours, based on John Langton and R.J. Morris, eds, *Atlas of Industrializing Britain 1780–1914* (London: Methuen, 1986).

27   Three weeks was normal in 1843, see Chambers, *Long and Busy Life*, 64–5.

These problems became apparent even before Chambers began to pay serious attention to the English market. As early as April 1832, they were bemoaning the fact that readers in Inverness and Dingwall were having to pay a surcharge on the cover price, 'owing to the high charges of the Highland coaches', which was seriously hindering the *Journal*'s circulation in the north.[28] Moreover, the advance circulation of the *Journal* required great self-discipline from individual booksellers, who were supposed to keep their packages of the *Journal* closed until the official publication date. In November 1832, Chambers had to insist that all their Glasgow agents signed a formal memorandum promising that none of them would allow the *Journal* to be put on sale before 3pm on the Friday. They would receive their packages on the Thursday, to allow time for orders to be forwarded to more distant booksellers, but Chambers did not dare allow the *Journal* to leave Edinburgh any further ahead of publication, lest the temptation prove too great for the booksellers.[29]

In Scotland, Chambers handled most of the distribution direct from Edinburgh. For England, they decided early on to follow common practice and appoint an agent in London. As the centre of the publishing trade, London was the centre of distribution networks for most of England and Chambers needed someone with access to those networks. They appointed William S. Orr in April 1832.[30] The very first issues of the *Journal* to travel south went by road, but as the demand grew it became obvious that another solution would be needed. One thousand copies of the *Journal* would have weighed about fourteen kilograms and occupied two-thirds of a cubic metre. By March 1832, it was already apparent that perhaps ten or twenty thousand copies would be needed in London each week. Water transport was the only realistic option for bulk of that sort.

There was a well-established shipping route between the port of Leith and the London docks, and Edinburgh publishers routinely transported bales of publications to London by water. Chambers, however, had a different set of requirements from the average Edinburgh publisher. First, their publications were so cheap that any surcharge for transport would be a substantial mark-up. Secondly, to build up a regular readership for a weekly magazine required a reliable method of delivery. With books and with quarterly or monthly magazines, delays of a few days due to adverse wind conditions made little difference, but to a weekly magazine, such delays could be disastrous. The cheapest option for Chambers was sailing ship, but the journey could take anything from four days to over a week. The new coastal steamers offered a journey time of three days no matter what the wind direction, but they cost more. Even worse, the coastal steamers of the 1830s had

---

28   "Chambers's Edinburgh Journal," *CEJ*, 28 April 1832, 52.

29   Memorandum from Glasgow agents, 12 November 1832, W. & R. Chambers Papers, Deposit 341, National Library of Scotland [WRC] 467 (Receipts 1832).

30   Sondra Miley Cooney, "William Somerville Orr, London Publisher and Printer: The Skeleton in W. & R. Chambers's Closet," in *Worlds of Print: Diversity in the Book Trade*, ed. John Hinks and Catherine Armstrong (London: British Library, 2006), 135–47.

side-mounted paddle wheels which could be swamped by high waves. This meant that the Leith to London steamers could not run during the winter months. For two or three months of the year, Chambers would be forced to rely upon the sailing packets with their highly-variable journey times.

Such was the reality of the transport infrastructure of Britain in the early 1830s. At certain times of year, money could buy quicker and more reliable transit times, but a regular year-round service was impossible. Unable to do anything to improve the situation, Chambers instead found a way to work round it. Stereotyping was the key. Chambers decided, at the end of April 1832, to licence William Orr to publish a London edition of the *Journal*. Orr took all the financial risk of the London edition, and kept all the profit beyond a set fee per thousand copies printed. The arrangement recognized that Orr, given his London location and connections, was in a better position to make decisions about the size of the print run than Chambers in Edinburgh and it gave him a strong incentive to make every effort to improve the circulation of the *Journal* in England. In one easy stroke, the entire problem of shipping 20,000 copies of the *Journal* south every week was removed.

Simple as it seems, this was actually a very unusual solution in the book trade of the 1830s because few periodicals faced the same scale of problems. Quarterly and monthly magazines had lower circulations, more relaxed schedules and higher prices which could more easily absorb transportation costs. The only other high-circulation weeklies of the period were London-based and thus, one suspects, took it for granted that Scottish readers would accept occasional delays and extra transport costs as the norm. A Scottish magazine trying to compete with London magazines in their home base could not afford to make the equivalent assumption.

Organizing a London edition of *Chambers's Journal* did not require stereotyping – a single printed copy could be sent by mail coach – but by June 1832, Chambers started to have the *Journal* stereotyped. One set of plates remained in Edinburgh, while the second set was sent to London. The use of plates ensured that the two editions were absolutely identical, and saved 'the expense of setting up the types in London'. The impact was exaggerated by the fact that Orr was working with steam-printers, and so the plates could, within 'a few hours' be made 'to produce twenty thousand or more printed sheets'. By the end of the year, the total circulation had grown to 50,000 copies. Chambers praised stereotyping as 'a wonderful process', which had enabled them to 'extend the circulation of the work on the most liberal principles, and in a very quick manner, all over the country'.[31]

During the 1830s and 1840s, as the range of their publications expanded, Chambers continued to use stereotype plates to extend the geographical reach of their business. Within the United Kingdom, it was only the *Journal* which needed multiple printing centres coordinated by stereotype plates: the other tract and book publications had less tight deadlines, and could be distributed from Edinburgh.[32]

---

31   "Printing and Stereotyping," *CEJ*, 29 September 1832, 278.

32   During 1833, plates of the *Journal* were sent to Dublin, but the Irish edition foundered within the year, because there was insufficient demand for the Irish agent to

Overseas, however, the small size of a set of plates could be advantageous for shipping purposes; and with the United States, plates had the peculiar advantage of being a physical property which could be bought and sold across the Atlantic in the absence of any copyright agreement. Chambers tried to sell a set of plates for the *Journal* to a New York printer in 1833, but could not agree a price. They finally made an arrangement in 1838 with Andrew Jackson, also of New York, who already printed the *Penny Magazine* and the *Saturday Magazine* from imported plates.[33] Chambers also sold plates for the *Cyclopaedia of English Literature* (1844) to Gould, Kendall & Lincoln of Boston and for the *Latin-English Dictionary* (1850) to Lea & Blanchard of Philadelphia.[34]

By the end of the 1840s, however, Chambers were becoming far less keen to use stereotype plates in this way, either within the United Kingdom or beyond. In the British case, the problem was with William Orr. Chambers wanted to take fuller control of their business, and be less dependent on the hapless Orr. With the USA, the problems were more complex. Differences in technological practice between America and Britain meant that selling a set of plates to the USA could involve far more than manufacturing, packaging and shipping the plates. It could require special preparation of the plates to fit the American printing machines, extra illustrations or publicity material to 'assist' the American publisher, or the dispatch of extra equipment and instructions to enable the American printers to produce the same results as their Edinburgh peers.[35] In addition, Chambers had no control over how the exported plates were used: their payment had no regard to subsequent sales, which was galling when sales proved high; and there was a risk that American-printed copies might be sold in Canada, thus interfering with Chambers's own sales there. For all these reasons, W. & R. Chambers by mid century transformed their system from one dependent on stereotype plates to one dependent upon steam power.

A Steam-Powered System

In 1832, William Chambers had proclaimed his ambition to supply 'the universal appetite for instruction' with good food at a price which would 'suit the convenience

---

make a profit from it. Ireland was henceforth supplied from Edinburgh (via Glasgow, to Belfast and Dublin).

33   Entry for 21 June 1838 in WRC 414 (Transaction Notebook). The arrangement with Jackson came to an end in 1840 when he went out of business during the economic depression.

34   Gould to Chambers, 15 October 1845, in WRC 314 (Literary Labour [LL] 1844–47) and succeeding letters, and Lea & Blanchard to Chambers, 27 May 1850, in WRC 121 (Correspondence), fol. 40.

35   Gould to Chambers, 27 February and 31 December 1846, WRC 314 (LL 1844–47).

of *every man in the British dominions*'.[36] This dream was threatened not only by the logistics of distribution, but by the basic question of just how many copies could be produced each week on a hand press. It was a terrible thought that one could potentially create a magazine filled with fascinating material that readers wanted to read, and yet be unable to produce enough copies to satisfy the demand. Stereotyping solved the immediate problem, but the long-term solution lay with steam power. Steam transformed Chambers's production capacity and, once coupled with the railway network, ultimately enabled these Edinburgh publishers to reach a national market without the aid of an intermediary.

The printing for the first issues of *Chambers's Journal* was done in Edinburgh by John Johnstone, 'a genial old man' of a printer whose workshop was at the top of the High Street.[37] He printed 30,500 copies of the first two issues, but his subsequent invoices hint at the strain that such a load was putting on his workshop: he initially printed 20,000 of the third issue, 16,500 of the fourth, and 18,171 of the fifth issue. All of these early issues had to be reprinted to meet demand.[38] The basic problem was the limitation of hand-press production. Johnstone worked two presses, printing the back and front of the *Journal* sheet simultaneously. Assuming each press team could manage 250 sheets an hour, and that they worked without break for the full ten-and-a-half hour day, six days a week, Johnstone's workshop would produce about 15,750 copies of the *Journal*.[39] For many magazines, this would have been adequate, but *Chambers's Journal* already had a circulation of 20,000 a week, growing to 30,000 within a month. The first couple of issues had probably been prepared ahead of time, but once the weekly schedule began, the only way Johnstone could produce substantially more copies was by operating through the night. But night work brought all sorts of extra problems with it. A few years later, Chambers would claim that the workmen could not be persuaded to remain sober throughout the night. It seemed that 'the greater the urgency for the work, and the higher the price paid for its execution, the more extensive were the saturnalia that prevailed'. Time after time, the parcels destined for country booksellers could not be dispatched early enough to ensure their arrival by the Saturday of publication.[40]

---

36   "The Editor's Address to his Readers," 1.

37   Chambers, *Memoir*, 214. In August 1832, Johnstone launched his own imitation of *Chambers's Journal*, later known as *Johnstone's Magazine*. It was absorbed into *Tait's Magazine* (which Johnstone also printed) in 1834.

38   Johnstone invoice, 18 February 1832, WRC 454 (Printing etc 1832). Johnstone charged £36, 11s, 10d. for those two issues. Also, his invoices for 23 March and 4 April 1832, WRC 454.

39   For typical working hours in the Edinburgh print trade, see Bill Bell, ed., *Edinburgh History of the Book in Scotland, Volume 3: Ambition and Industry 1800–1880* (Edinburgh: Edinburgh University Press, 2007).

40   "Mechanism of Chambers's Journal," 150.

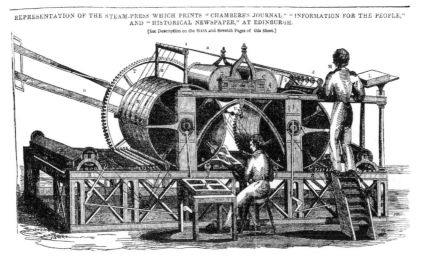

CHAMBERS'S
INFORMATION FOR THE PEOPLE.

CONDUCTED BY WILLIAM AND ROBERT CHAMBERS, EDITORS OF " CHAMBERS'S JOURNAL" AND
' HISTORICAL NEWSPAPER."

No. 35.   PRICE 1½d.

THE ART OF PRINTING.

REPRESENTATION OF THE STEAM-PRESS WHICH PRINTS " CHAMBERS'S JOURNAL." " INFORMATION FOR THE PEOPLE,"
AND " HISTORICAL NEWSPAPER," AT EDINBURGH.
[See Description on the Sixth and Seventh Pages of this Sheet.]

**Figure 2.3   W. & R. Chambers's steam-powered printing machine,
depicted in the tract on 'Printing' in their *Information for the
People* (1842)**

Chambers later claimed that 'the dreadful harassment of mind which ensued'
in those spring months of 1832, 'led us to think of removing the whole of our
printing business to London'.[41] Orr seemed to be having no trouble getting the
London edition printed (variously by William Clowes and by Bradbury & Evans)
at reasonable prices and within the time available. Both of these establishments
were using steam-powered printing machines, which enabled Orr to cope with
the growing English demand for the *Journal* while Chambers and Johnstone were
struggling to meet the Scottish demand.

Chambers did not move to London, but they did become enthusiastic converts
to steam-powered printing. The technology had existed for over 15 years, but
was not yet widely adopted, and there were few steam-powered printing houses
in Edinburgh. By July 1832, Chambers had transferred their printing to James
Ballantyne, whose extensive steam-printing works had been funded by his
connection with Walter Scott. Ballantyne routinely printed about 25,000 copies of

41   Ibid.

the *Journal* each week.[42] Yet Chambers were still not wholly satisfied, and seem to have become determined to take as much control over all aspects of their business as possible. In the spring of 1833, they hired a young engine maker, Robert Gunn, to design a steam-printing machine of their own (Figure 2.3). It came into operation during the following winter, and after teething problems, Chambers were able to print their own publications. The *Journal* hailed the new machine as 'a joyful prospect of future tranquillity'.[43]

The transition to steam-powered printing had two long-term implications for W. & R. Chambers. It encouraged them to expand their activities into a broader range of informative publications, and it also enabled them to replace their multi-centre, stereotype-coordinated system with a centralized system based in Edinburgh. Indeed, the excitement of acquiring their first printing machine can have barely evaporated before Chambers began acquiring additional machines. They bought additional machines in 1837, 1839 and 1841. With each machine costing £300 or £400, it is a clear sign of the success of the business that these improvements could be funded from ready money.[44] At the same time, Chambers were acquiring the equipment and personnel to run their own compositors' gallery, stereotype foundry and bindery. By 1840, the only major department which was still contracted out was the illustrations, both wood-engraving and lithography.

In 1840, Chambers moved their editorial and business offices to 339 High Street, directly opposite St Giles's Cathedral and adjacent to their Roxburgh Close printing house (Figure 2.4). Now that all their departments were on a single site, it became obvious how disorganized and poorly laid out that site was. High Street is a part of the "Royal Mile" which runs from Edinburgh Castle down the slope of the old volcanic plug towards the Palace of Holyrood House. The land drops away sharply on either side of High Street, and Roxburgh Close is one of many narrow alleys which provide a very steep shortcut to the lower ground level. The new Chambers editorial offices were up on High Street, while the various workshops were crammed in behind and below, with entrances on several levels down Roxburgh Close. In 1845, William and Robert Chambers pumped nearly the entire £10,000 profit of that year into improving their business, principally by re-building and re-equipping their production plant.[45]

---

42   The first surviving invoice from Ballantyne is dated November 1832 but relates to July. It is with other invoices for November and December 1832 in WRC 467 (Receipts July–December 1832).

43   "Mechanism of Chambers's Journal," 151.

44   For the two Norton machines, see invoices for 1 February 1838, in WRC 470 (Receipts 1837), and for 1 August 1841, in WRC 474 (Receipts 1841). For Middleton's 1841 machine, see invoice for 9 February 1839, in WRC 459 (Repairs to machinery etc, 1839).

45   Chambers balance sheet for 1845, WRC 304 (Balance-sheet book), fol. 33–6.

Figure 2.4 **The W. & R. Chambers's premises are visible in this rare photograph from the mid 1850s: look for the tall smoky chimney just below and to the left of St Giles Cathedral. Photograph by Thomas Vernon Begbie, *c.* 1854. © Thomas Begbie Cavaye Collection, held by City of Edinburgh Museums and Galleries**

The centre-piece of the new plant was a high-pressure beam engine rated at ten horse-power, five times more than the original engine.[46] It was going to be used to power not only the existing printing machines, but also four brand new cylinder perfecting printing machines (which would print both sides of the paper and get them correctly registered).[47] The brothers wrote in January 1845 that:

> We write at present in a huge building of four stories, flanked by a powerful steam-engine, and with the noise of ten printing machines continually sounding

---

46   The engine was listed against 14 December 1844 in Norton's invoice, June 1845, WRC 477 (Receipts 1844–45).

47   Middleton's invoice, March 1845, WRC 478 (Receipts 1845).

in our ears.... A hundred and twenty persons are required for all the duties which proceed in this large structure, though these have exclusively a regard to works edited by ourselves. Upwards of a quarter of a million of printed sheets leave the house each week.[48]

The result was that Chambers were the proud owners of one of the most high-technology printing establishments in Edinburgh, perhaps even in Britain. The New York *Literary World* mentioned the firm as one of its top five of British publishing houses in an article of 1850, and lavished praise on an establishment which it claimed was 'eleven stories high', employing 'some five hundred persons', and where the presses 'throw off 150,000 whole sheets a day'. Despite the exaggeration, American readers would have been left with no doubt of the size and modernity of the new Chambers's establishment, which was 'considered unrivalled' for its 'extent and completeness'.[49]

It was this printing establishment which gave Chambers the production capacity to take control of their entire British output, both of the *Journal* and of other publications. The London edition of the *Journal* was suspended at the end of 1846, and William Orr became merely a distribution agent (which Chambers hoped would limit their exposure to financial risk).[50] Although the London edition had had the benefit of reducing the strain on production in Edinburgh, it had originally been conceived to ease distribution problems. The fact that Chambers were no longer so worried about distribution reflects the extent to which their transportation options had changed since 1832, thanks largely to the railways.

The first railway line to have a significant impact on W. & R. Chambers was the Edinburgh and Glasgow Railway, which opened in 1842. In addition to the orders for booksellers in Glasgow and the west of Scotland, consignments for Belfast, Dublin and Liverpool (including North America) were dispatched via steamship from Glasgow. Until the opening of the railway, the consignments had usually travelled the 45 miles to Glasgow by road, but Chambers were swift to transfer their business to the railway. The firm's archive contains regular monthly invoices from the railway company for parcels, packages, boxes and bales, most of which were being shipped onward.[51]

From 1846, the North British Railway began building southwards from Edinburgh, to join George Hudson's York, Newcastle & Berwick Railway. Even

---

48   "Address to readers," *CEJ*, 4 January 1845, 1.

49   F.S. [Frederick Saunders], "The Publishing Business," *Literary World* 6, 5 January 1850, 12. For Chambers's own estimate, see Cooney, "Publishers for the People," 214.

50   The new arrangement with Orr is most clearly set out in WC to RC, 28 September 1846 in WRC 314 (LL 1844–47). A letter to Orr (from Chambers's manager, R. Inglis) in February 1847 makes clear Chambers's determination not to allow the London edition to be restarted.

51   Monthly invoices for spring 1845 and spring–summer 1848 survive in files WRC 478 (Receipts 1845) and 484 (Receipts 1848).

before the through-route was completed (with the opening of the railway bridge over the Tweed at Berwick, in August 1850) Edinburgh publishers could take advantage of the railway to send letters and parcels to London, using coastal steamers or stage-coaches to connect with the existing railway lines. It is notable, however, that, although the railway quickly superseded road transport on the Edinburgh-London route, the existence of coastal shipping kept Chambers's options open. The railway was certainly quicker than the coastal steamers (about 12 hours, rather than 3 days), but it was more expensive. On the other hand, its year-round reliability was not affected by winter storms as the steamers still were, despite better designs.

For Chambers, the railways were most important for speeding up communication (via the Royal Mail) with London, which enabled them to keep a closer eye on their agent's activities; and for providing a crucial backup to coastal shipping during the winter months. For most of the year, the steamers were reliable enough and quick enough to transport both the firm's book publications and the time-critical *Journal* and series of weekly and fortnightly part-works. During the winter months, Chambers were willing to pay the slightly higher costs of the railways to guarantee the reliable delivery of their weekly publications.

The extent to which railways had become accepted and reliable parts of the communications and distribution system for British business is clear from a letter from David Chambers in London to Robert Chambers in Edinburgh, which illustrates what happened when things went wrong. Younger brother David had been running the new London branch office for four months, and was in the midst of a big effort to capitalize on the January 1854 re-launch of *Chambers's Journal*. He was supposed to receive his consignments of the weekly publications (the *Journal* and the *Repository of Tracts*) at the start of the week before publication, to give him time to supply the booksellers of the midlands, the south of England and London by the Friday night. Both publications were printed sufficiently far ahead of time that dispatch by coastal steamer was possible, but since it was winter, David specifically requested that his parcels be sent 'by rail until the Steamers are sailing more regular'. He worried that any 'delay is most injurious to the Sale', especially for a new publication.[52]

On the morning of Monday 9th January, David received an invoice from Edinburgh confirming the dispatch of his consignment for the coming Saturday – but no consignment. The next day brought another invoice reporting the dispatch of a further two thousand copies – but still no sign of either consignment. The railway had successfully delivered the invoices overnight, but not the parcels. He was, as he said, 'at a complete stand still', unable to fill any of his orders.[53] The episode illustrates

---

52   David Chambers to Robert Chambers, 11 January 1854, in WRC 109 (Domestic Correspondence).

53   David Chambers to Robert Chambers, 11 January 1854. David attempted to use the new technology of electric telegraphy to sort the situation out, but on this occasion the necessary brevity of telegraphic messages led to further misunderstanding.

that both the Edinburgh and London offices had already become so used to railway reliability that they expected overnight deliveries of both letters and freight to be the norm, and were surprised and flustered when things went wrong.

So far as I can tell from extant evidence, this was an isolated incident. By the 1850s there are far fewer references to distribution problems, and one gets the impression that the combination of steam-powered production in Edinburgh coupled with steamship and railway delivery to London was working well. It enabled Chambers to reach far beyond their natural market in lowland Scotland, to become truly national publishers.

New Information Products

William and Robert Chambers had become steam-printers because of the demands placed on them by the success of *Chambers's Journal*, but the success of the *Journal* and the capacity of the printing machine inspired them to launch a much wider line of instructive publications. It was in this respect that they made their greatest impact on the metaphorical 'landscape of knowledge'. The *Journal* was one of several cheap weekly magazines which aimed to provide information and instruction to the lower middle classes and skilled working classes. The subsequent Chambers publications were even more innovative, and transformed both the sort of knowledge that was available to much of the population, and the format in which it appeared. They were all steam-printed which helps to explain why Chambers proclaimed the steam printing machine 'so stupendous a spectacle of moral power'.[54] For Chambers, the steam printing machine was a tool for education and improvement.

The firm's first steam printing machine worked at about 900 sheets an hour. This meant that the Edinburgh run of the *Journal* took only a third of the working week. To maximize the return on their investment in premises and equipment, Chambers needed to find ways to keep the machine running for more of the week. By 1835, the *Journal* had been joined by two other groups of instructive publications: a series of pamphlets and two series of cheap books. In later years, the firm regularly issued more individual titles, but they would never again launch so many new types of publication in so short a time.

The first of the new serial publications was launched in 1833 while Chambers were still relying upon Ballantyne to do their printing. *Chambers's Historical Newspaper* (1833–35) appeared as a monthly supplement to the *Journal*, and was an attempt to discuss recent events and issues while remaining outside the scope of stamp duty. More significant were the fortnightly tracts or pamphlets issued under the series title *Information for the People* (1833–34). Each pamphlet used the same format as the *Journal*: it was the same size, number of pages, lay-out and price. The key difference was that each pamphlet was devoted to a single topic, such as

---

54   "Mechanism of Chambers's Journal," 151.

'Astronomy', 'Emigration to Canada', 'Moral Philosophy' or 'The American War of Independence'. *Information* sometimes also included illustrations; the *Journal* never did. *Information* was planned as a series of fifty parts, which could be read individually or collected to form a basic encyclopaedia. The aim was similar to some of the publications of the SDUK, but Chambers happily pointed out that each of their issues contained as much text as the *Library of Useful Knowledge* at a quarter of the price.[55] By early 1834, Chambers were pleased to report that *Information* was regularly selling at least 16,000 copies per issue and sometimes twice that. To their relief, the circulation of the *Journal* did not seem to be affected. Chambers had persuaded a substantial number of people to purchase a new piece of reading material without compromising their existing reading habits.[56]

New issues of *Information for the People* ceased in 1834, but the pamphlets were kept in print and continued to sell; and new, expanded and revised editions were issued in 1842, 1848 and 1856. Chambers used the format again for the *Cyclopaedia of English Literature* (1844), the *Miscellany of Useful and Entertaining Tracts* (1844–47), *Papers for the People* (1850–51) and the *Repository of Instructive and Amusing Tracts* (1852–53), although it is notable that some of the later series made the effort to be 'amusing' as well as instructive. These serials were unusual publications, but they were perhaps the most distinctive of the Chambers publications in the 1830s and 1840s. They had impressive sales and made substantial profits for the firm. For instance, by 1848, the 1842 edition of *Information* had sold just over 100,000 complete sets and made a profit of £11,200.[57] As some of the later series titles hint, the closest model was probably the religious tract, hundreds of thousands of which were being distributed annually by city missionaries and scripture readers in Britain's industrial cities. Both religious tracts and Chambers's 'secular tracts' were produced in large numbers at low cost, and aimed to help readers improve themselves. Whereas the religious societies interpreted this to mean spiritual improvement aimed at a future life, Chambers offered instruction for this life.

Tracts could be produced and sold so cheaply because they used very little paper (easily the most expensive production cost), they were written by little-known writers (who were cheap), they were steam-printed in large runs and they were stereotyped to allow for later demand. Chambers would soon apply the same techniques to book publication. Other publishers in the late 1820s and 1830s had attempted to make books available more cheaply, but none had convincingly succeeded. John Murray's *Family Library* and Longman's *Cabinet Cyclopaedia* were only 'cheap' in comparison to the high price of original literature: 5*s*. or 6*s*. a volume was unaffordable for the working classes and many of the lower middle

---

55   "Editorial," *CEJ*, 31 January 1835, 1. In later years, they could also point out that the *Information* had been completed as planned, unlike the *Penny Cyclopaedia* which outgrew its original plan and took ten years to complete.

56   "Editorial," *CEJ*, 1 February 1835, 1–2.

57   Chambers Publication Ledger 1845–67 (WRC 275), fol. 6–7.

classes, and sales were disappointingly low.[58] By the late 1830s, Chambers would demonstrate that it was possible to sell entire books for under two shillings: in their series of *People's Editions* and their *Educational Course*.

The *Educational Course* began in 1835 as a series of graded readers and textbooks which eventually took a student from learning to read and write, through arithmetic, history and geography, to Latin, modern languages and the natural sciences. The series of *People's Editions* (also known collectively as the *Instructive and Entertaining Library*) began in 1837. Both projects involved small expenditure on literary labour, because the *People's Editions* were reprints of out-of-copyright or non-copyright (e.g. foreign) books, while the *Educational Course* books were mostly written by school-teachers for low fees.[59] The books were set in small type to save paper, and the *People's Editions* used such small type that they had to be printed in double columns. They were also steam-printed in large runs. Virtually all steam-printing machines in Britain at that time were used for newspapers or periodicals, because they had the high circulations, and because book publishers were sceptical of the quality of their output. Yet Chambers had Robert Gunn's steam-powered perfecting machine sitting in their premises which – even with the *Journal* and the *Information* – was not fully occupied. In theory, there seemed no reason why they could not print sheets to be bound into a book.

Chambers began steam-printing books in 1835. Usually, they only printed their own publications, but in October 1835 they printed an edition of the *Constitution of Man* (1828), written by Robert Chambers's friend, the leading phrenologist George Combe. *Constitution* had already had three editions, co-published by the Edinburgh firm of John Anderson and the London firm of Longman, and sold at 6*s*. The *People's Edition* sold at 1s, 6d. The initial 2,000 copies were said to have sold out within ten days, and the stereotype plates were immediately put through the machine again to print another 5,000 copies, which again sold out quickly. Chambers printed another 15,000 copies in 1836.[60] Despite the cramped typography and a binding which was nothing more than plain boards, the success of the format was incredible. It was a wonderful proof both that steam-printed books were a possibility, and that large numbers of readers would enthusiastically welcome low-priced books. It inspired Chambers to launch their *Instructive and Entertaining Library*, the volumes of which were sold at prices between 1s, 6d. and 2s, 6d., and routinely sold several thousand copies.[61]

---

58   Murray remaindered the *Family Library* to Thomas Tegg in *c.* 1834. Longman kept the *Cabinet Cyclopaedia*, but sold off its volumes at greatly reduced prices in the late 1840s. See Scott Bennett, "John Murray's Family Library and the Cheapening of Books in Early Nineteenth-Century Britain," *Studies in Bibliography* 29 (1976): 139–66.

59   Cooney, "Publishers for the People," chapter 4.

60   John van Wyhe, *Phrenology and the Origins of Victorian Scientific Naturalism* (Aldershot: Ashgate, 2004), appendix C and 217–20.

61   William Paley, *Natural Theology* (1802, Chambers 1837) was one of the first of these volumes. It sold 10,000 copies in two years, see Publication Ledger 1842–45 [sic]

Together, the instructive tracts and the steam-printed books made knowledge accessible to a large sector of the population which had previously had access to little print other than religious tracts, chapbooks and ballads. Chambers's educational and improving aims had much in common with the goals of philanthropic bodies like the Religious Tract Society or the SDUK. Unlike the first, it was a vision of knowledge which was entirely secular; and unlike the second, Chambers had the commercial know-how to make a success of their ambition.[62]

Steam in the Transatlantic World

At the same time that W. & R. Chambers were transforming their British operations, they were also becoming more aware of the possibilities of the North American market. Chambers had long-established connections with firms in Toronto, Montreal and Halifax, but the real challenge was the United States, where the highly literate population of New England, with its cultural ethos of improvement, seemed an ideal market for Chambers's publications. By the mid 1840s, the United States had come out of a period of economic depression and its publishers had regained interest in doing business with the British book trade.[63] Moreover, transatlantic business was becoming easier to transact thanks to the arrival of steamship services, which carried news, literary gossip and correspondence quickly and reliably.

As discussed, Chambers initially hoped to use stereotype plates as a way of receiving payment for the reprinting of their publications in the United States, but had come to the conclusion that this was not an ideal solution. They sought instead to persuade American publishers to import printed copies from Edinburgh, just as agents in London and Dublin did. With no Anglo-American copyright law to enable them to enforce importation rather than reprinting, Chambers had to make cooperation financially attractive to American publishers.[64] In practice, this meant offering substantial discounts for bulk orders. In 1847, Gould, Kendall & Lincoln were offered a price of £3, 7s, 6d. per thousand if they took a minimum of five

---

(WRC 274), fol. 351–4.

62   As far as I know, Chambers did not comment on the religious publishing societies, but there are comments on the SDUK in Chambers, *Memoir*, 213.

63   On the US book trade in the 1840s, see Meredith L. McGill, *American Literature and the Culture of Reprinting, 1834–1853* (Philadelphia: University of Pennsylvania Press, 2002); Scott E. Casper, Jeffrey D. Groves, Stephen W. Nissenbaum and Michael Winship, eds, *A History of the Book in America, Volume 3: The Industrial Book, 1840–1880* (Chapel Hill: University of North Carolina Press, 2007); on its connections with Britain, see James L.W. West, "Book Publishing 1835–1900: The Anglo-American Connection," *Papers of the Bibliographical Society of America* 84 (1990): 357–75.

64   James J. Barnes, *Authors, Publishers and Politicians: The Quest for an Anglo-American Copyright Agreement, 1815–1854* (London: Routledge, 1974).

thousand copies of the *Journal*.[65] The retail price of a thousand copies was £6, 5s, 0d., and British trade customers routinely received 30 per cent off, paying £4, 6s, 6d. per thousand.[66] By offering Gould, Kendall & Lincoln a further 15 per cent off, Chambers were trying hard to make it possible for an American importer to make a success of the *Journal*. By the early 1850s, Chambers routinely offered their American contacts a trade discount of 40 to 50 per cent for the instructive serials, with a usual minimum order of a thousand copies. Such discounts meant that they would make only a tiny profit from the American sales, but at least they were getting a small sum for each and every copy sold and they were retaining editorial control of the publications which circulated under their name.

As in Britain, extending the transatlantic reach of Edinburgh-printed copies of Chambers's publications required both production and distribution capacity. The steam-powered printing plant provided the first, but for the second, the question was what role transatlantic steamships would play. The first steam crossings of the Atlantic had occurred in 1838 and, from 1840, Samuel Cunard's British & North American Steam Packet Company carried the Royal Mail from Liverpool to Halifax and Boston on a fortnightly service using the Clyde-built paddle-steamer *Britannia* and her three sister-ships.[67] The voyage time was fourteen days to Boston, which was only marginally faster than the best sailing times, with the critical difference that it did not depend upon wind direction. The steamers of this first generation were only 200ft long and carried so much coal that they had little space for anything more than the Royal Mail and around one hundred first-class passengers. Any cargo needed to be small and urgent to justify the expense: newspapers and magazines might fit the bill, but not in large quantities. The faster transit of news and gossip helped bring the British and American book trades closer together, but the real impact of steam on the transatlantic book trade was with the second generation of Cunard ships, from 1848. The *America, Canada, Europa* and *Niagara* introduced a weekly service, began to serve New York (by then, the centre of the American publishing industry), and could carry up to 450 tons of cargo. Using them for cargo was still substantially more expensive than sailing ship, but they gave publishers the option of guaranteed rapid delivery for consignments of books and magazines. The question was when to use them, and for what?

The publication which was most likely to be affected by the decision to use steam-shipping or not was *Chambers's Journal*, with its weekly dated issues. It had a good reputation in the USA: the *Literary World* ranked it alongside Charles Dickens's *Household Words* as an exemplar which American publications would

---

65   Chambers to Gould, 30 September 1847, WRC 314 (LL 1844–47).

66   UK trade price was 1s. 1½d. per dozen, with 13 counting as 12. See, for instance, WRC 600 (1858 Trade Catalogue).

67   Francis E. Hyde, *Cunard and the North Atlantic, 1840–1973: A History of Shipping and Financial Management* (London: Macmillan, 1975).

do well to emulate.[68] Several American publishers had requested copies of its stereotype plates, but Chambers consistently refused. Rather, in 1854 they arranged for Joshua Lippincott of Philadelphia to import it at a discounted price. Seeing the necessity for reliable deliveries, but worried about the cost implications of weekly steam shipping, Lippincott made the decision to market the *Journal* in monthly parts. Given that over 60 per cent of the British sales were now in monthly parts, this decision was unlikely to affect sales significantly.[69] Since Chambers routinely printed the *Journal* two or three weeks ahead of the publication date, steamship delivery would ensure that Lippincott's consignment could arrive in Philadelphia at the same time as the monthly part was due to appear in Britain.

Unfortunately for Lippincott, no sooner had he begun taking the *Journal* than a New York printer by the name of Peter D. Orvis began reprinting it in weekly issues without permission.[70] His source was presumably a weekly issue sent by steam from Liverpool as soon as it was published in Britain. His issues must have been about a fortnight behind the British dates, but two weeks out of four he was beating Lippincott into the marketplace. Even though Lippincott could offer the *Journal* for the same price as Orvis, Orvis's huge advantage was that he was willing and able to pay for steam shipping of the single copy that he needed to reprint his edition. Lippincott, with his substantial order, was not willing to pay for weekly shipments, even though Chambers pointed out that this was the obvious solution.[71] Lippincott asked instead for a larger discount, and eventually managed to persuade Chambers to a barely profitable £3 per thousand, a massive 52 per cent off the British retail price.[72] It was to no avail, however, as Lippincott still claimed to be unable to make money from the *Journal*, and cancelled the order in May 1855. Chambers were disappointed, writing in response that 'We are sorry the Journal cannot be made to pay, but there seems to be no remedy for it, and we shall now discontinue sending any more.'[73]

Apart from the *Journal*, Chambers appear to have been relatively little affected by the fact that unauthorized American reprinters were more willing to pay for

---

68 "Taxes on Knowledge in England," *Literary World*, 1 June 1850, 532.

69 In 1847, ⅝ of the circulation was in monthly parts, see "Booksellers," *CEJ*, 6 February 1847, 88.

70 Orvis was the publisher of the cheap weekly the *New York Journal*. Chambers's refusal (for the second time) to sell him plates is mentioned in Chambers to Lippincott, 2 May 1854, WRC 321 (LL 1854).

71 Chambers to Lippincott, 5 June 1854, WRC 163 (LB 1853–67), fol. 30. There is no record of the actual size of Lippincott's order, but extrapolating from the increased circulation in 1854 (when Lippincott started taking the *Journal*) and its partial reduction in 1855 (when Lippincott cancelled his order partway through the year) it may have been around 16,000 copies.

72 Chambers to Lippincott, 10 March 1854, WRC 163 (LB 1853–67), fol. 45.

73 Chambers to Lippincott, 18 May 1855, WRC 163 (LB 1853–67), fol. 86. Orvis went out of business in mid 1856, but Lippincott did not re-start imports of *Chambers's Journal*.

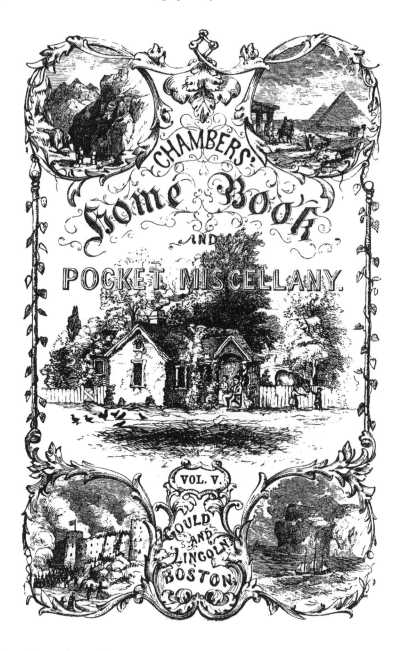

Figure 2.5	Copy of the *Pocket Miscellany* imported into the USA by
Gould & Lincoln issued as *Chambers' Home Book* with a new
title page

steam shipping than were authorized bulk importers. Chambers's publications were highly regarded by American teachers, school inspectors and clergymen – the firm was generally held to have attained 'wide celebrity' for its publication of 'school books, text-books, and some of the best manuals extant for self-education'[74] – but they did not generate the same levels of excitement as the latest Dickens novel. American publishers were not competing frantically to be the first to get a new Chambers work into print, as they were with the latest literary fiction. There certainly were cases where Chambers publications were reprinted without authorization (for instance, A.S. Barnes reprinted some of the science volumes of the *Educational Course* in the late 1840s, and *Information for the People* was reprinted in Philadelphia), but most of the copies of Chambers secular tracts and books which survive in American libraries have Edinburgh imprints or carry the imprints of American publishers who acted as importers (Figure 2.5).

Just as Chambers did not immediately turn to the railways for all their British distribution requirements, but used them for particular routes, publications or times of year, so too in the Atlantic world, they did not immediately use steamships for everything. From the mid 1840s onwards, their business correspondence was going by steam, and the fact that their correspondence with Lippincott in the 1850s was more frequent than with Gould & Co in the 1840s reflects the increased mail service provided by Cunard. But since most of their consignments were of instructive tracts and books, which were both bulky and not time-critical, sailing ship remained the preferred option – especially when we bear in mind that the shipping costs were paid by the American importer and not by Chambers themselves. It was all very well for Chambers to tell Lippincott he should get the *Journal* by steam every week, but they were not the ones who would be paying for it. By about 1856, however, freight prices on the steamships were reduced. From that point, it seems as though steam became the norm for Chambers and their American partners, even for bulkier consignments. The last mention of sail that I have recorded was in October 1856, when Chambers requested Lippincott to return any unsold copies of the *Pictorial History of England* 'by first sailing vessel to Liverpool'.[75]

Conclusions

When William Chambers began to issue *Chambers's Edinburgh Journal*, he had a natural market in lowland Scotland, but faced substantial practical difficulties in making his dream of reaching 'every man in the British dominions' come true. In 1834, Chambers claimed that 'The nature of our publications was, in every respect, so extraordinary, that all the old modes of procedure may be described as having fairly broken down under it.' As a result, they had had to create 'a proper system of

---

74 "Chambers's Papers for the People," *Norton's Literary Advertiser*, December 1851, 92 and "Chambers's Papers for the People," *Literary World*, 12 June 1852, 409.

75 WRC to Lippincott, 24 October 1856, in WRC 163 (LB 1853–67), fol. 185.

printing and publishing'.[76] At first, it seemed as though the cunning use of stereotype plates would relieve the problems of both distribution and production. Ultimately, it was the introduction of steam-powered printing coupled with improving transportation technologies which made the real difference, both in Britain and in the wider world.

The impact of steam-powered printing was the most significant. Railways, coastal steamships and ocean-going steamships each had their problems and their advantages. Chambers had no direct control over the transportation infrastructure and tended to pick and choose particular options for specific purposes. Steam-powered transport was clearly useful, but not straightforwardly so. Steam-powered printing, however, was under Chambers's direct control and, as far as Chambers were concerned, it had no drawbacks.[77] Given the size of print runs they were working with, steam allowed economies of scale which helped keep prices low and created truly innovative instructive products. Steam also gave Chambers the capacity to print for the whole world. No longer limited to the output of a hand-press team, Chambers belonged to the first generation of publishers who genuinely hoped to send their publications in substantial numbers throughout the United Kingdom, the British dominions and beyond. Capacity alone was not enough: the lack of copyright agreement made it far more difficult to do business in the United States than in the dominions; but it was a necessary precursor.

Although the comparison with the United States may seem to have an ambiguous message, I would argue that the fact that most of the copies of Chambers publications which circulated in America were printed in Edinburgh highlights just how effectively Chambers were able to produce and market their cheap informative works. We are well aware that – in a British context – the firm's choice of authors, careful use of paper, and enthusiastic application of the new technologies were innovative and essential to its goal of selling at really low prices. In the United States, low prices had been the norm since the start of the century, and the competition engendered during the American economic recession of the early 1840s had pushed the price of print even lower. It is unsurprising that British publications on science, history and geography should be in demand in the United States, but it is a surprise to find a British publisher managing to be competitive on the grounds of price in the United States. It was an American-based commentator who claimed that 'These enterprising brothers have done more, perhaps, than any other two individuals of the age for the promotion of sound and useful knowledge'.[78] And what made it possible for them to have that impact on the landscape of knowledge in Britain and America was steam.

---

76   "Editorial," *CEJ*, 1 February 1834, 2.

77   Some of their neighbours disagreed, in 1833, and tried to get an injunction to prevent the steam-engine being built. Defenders of traditional hand-press skills might also have objected.

78   F.S., "The Publishing Business," 12. Saunders was the son of a London publisher, but had lived in New York since 1836.

Chapter 3

Construing the Spaces of Print Culture:
Book Historians' Visualization Preferences

Fiona A. Black

Introduction

In spite of Febvre and Martin's early exposition on the "geography of the book",
it is only recently that the discipline of book history has consciously engaged
with theories and methods in support of geographies of the book.[1] This chapter
explores a mode of analysis different from tracing the geography of a particular
book or author. Drawing examples from Georgian Canada and from a long-term
study examining the intersection of print culture, migration and nation-building,
the chapter takes up the question of how scholars might address the challenge of
analyzing large datasets using visualizations and historical geographic information
systems (HGIS). To answer this question the chapter presents the results of the first
usability study to analyse book historians' preferences for interacting with web-
based visualizations of spatially related data in maps, graphs, tables and historical
images.[2]

Two defining traditions of book history scholarship, the French Annales
school and the Anglo-American focus on bibliography (historical, descriptive and
enumerative), have come together in recent years. The Annales approach involves
an holistic view and scholarly awareness of contextual factors affecting the culture
of print in a given time and place. The Anglo-American approach, taking the book
itself as the focus of enquiry, has resulting in detailed scholarly bibliographies
of the production of particular presses, authors, regions, languages, and related
biographical and business information about the rich variety of agents of the press.
The merging of these two scholarly traditions and understandings has produced
national histories for the United States, Canada, Scotland, Wales, Britain (with
a focus on England) amongst others. Some of these, notably the History of the
Book in Canada, have framed their narratives with an implicit acknowledgement
of the relevance of two critical models that illustrate the complexity inherent in

1   Lucien Febvre and Henri-Jean Martin, *L'Apparition du livre* (Paris: Éditions Algin
Michel, 1958), Chapter 6.
2   This study was funded by a grant from the Social Sciences and Humanities Research
Council of Canada. Dalhousie University Graduate Research Assistant, Kathleen Amos,
was of instrumental aid.

book history.[3] Robert Darnton's 'Communications Circuit', based on knowledge of eighteenth-century Europe, took as its premise the claim that printed books have a life cycle generally conforming to 'a communications circuit that runs from the author to the publisher … the printer, the shipper, the bookseller and the reader. The reader completes the circuit because he influences the author both before and after the act of composition'.[4] The model simplifies complexity, focusing primarily on the human agents of print culture, while acknowledging the influences of economic, social, political and cultural factors and relationships. These influences have an implicit geographic or spatial component. Thomas Adams and Nicolas Barker offered an inverted model ten years later, with an emphasis not on the agents involved in the creation, production and distribution of printed materials, but on the books themselves.[5] While offering a different focus, Adams and Barker's model retains references to the crucial influences of economic, political and other factors. As with Darnton's model, Adams and Barker's work is helpful in providing a visual and intellectual reminder of the numerous interrelating factors inherent in book history. In both models, the geographic element is implicit rather than explicit. I argue here for making the geography of the book explicit by the application of technology.

GIS and the Geographies of the Book

Geographic information systems' (GIS) utility for analysis rests on the datasets of variables that may be spatially analysed. In the field of print culture, relevant databases with historical applications include bibliographic files of early printed materials, image databases, guides to newspapers published in various regions and digitized historical censuses. A more recent development is book trade indexes, beginning with the British Book Trade Index.[6] Furthermore, large projects in Canada, the United States and the United Kingdom, have begun to digitize historical censuses in entirety, rather than as aggregate or sampled data. Databases of information relating to demography, businesses and technologies may contain textual and numeric data, such as in censuses, or visual data such as historical photographs, maps, charts, architectural drawings and insurance plans.

---

3   Patricia Lockhart Fleming and Yvan Lamonde, eds, *History of the Book in Canada* 3 volumes (Toronto: University of Toronto Press, 2004–2006).

4   Robert Darnton, "What is the History of Books?" *Daedelus* 111:3 (1982): 67.

5   Thomas R. Adams and Nicolas Barker, "A New Model for the Study of the Book," in *The Potencie of Life, Books in Society: The Clarke Lectures, 1967–1987,* ed. Nicolas Barker (London: The British Library, 1993), 5–43.

6   Now available online at www.bbti.bham.ac.uk, the British Book Trade Index was initially designed by Peter Isaac in the 1980s using dBase software. His prescient use of multiple fields for locational information was instrumental in facilitating the index's use within a pilot GIS developed by Black in 1998.

Sources such as these have long been used by book historians in their enquiries. However, scholars are now beginning to use technologies to help analyse the interrelationships among such sources and variables. GIS technologies might aid such analyses and lead to new knowledge about print culture.

Geographic information systems were first developed in the 1960s. Historical applications of GIS began with anthropology. More recently, GIS has been used by a number of historical researchers in diverse projects, and their work has guided the current project. Ray and Sheehan-Dean used GIS to bring events from American history to life, illuminating the Salem Witch Trials and the Civil War, respectively.[7] Both projects began with the creation of digital archives containing various textual documents, which were subsequently expanded using a GIS to relate this documentary information to location, and used to study the role of geography and to re-investigate previous claims about the events, people and locations involved. Researchers have also utilized GIS to study immigration and other movements of people. Gregory and Southall used census data and historical boundary files to study migration patterns and other demographic trends within England and Wales in the Great Britain Historical GIS.[8] Census information has played a key role in the development of many historical GIS. Fitch and Ruggles, for example, describe the National Historical Geographic Information System, containing census data from 1790 through 2000 and appropriate geographic boundary files, which facilitates historical population research.[9] Beveridge has made use of census data and GIS to investigate the twentieth-century demographic history of New York City, studying the movement and settlement patterns of the population as a whole and of various ethnic groups as they changed over time and exploring their relationship to economic inequality.[10] He further considered the influence of transportation, specifically the subway, on residential patterns. Transportation networks themselves are a focus of Siebert's work. Drawing on census records, historical maps and transportation system records, among other sources, Siebert has developed a GIS to study the development of the rail network in and surrounding Tokyo and the related changes in geography and population.[11]

---

7   Benjamin C. Ray, "Teaching the Salem Witch Trials," and Aaron C. Sheehan-Dean, "Similarity and Difference in the Antebellum North and South," in *Past Time, Past Place: GIS for History,* ed. Anne Kelly Knowles (Redlands: ESRI Press, 2002), 19–33 and 35–49.

8   Ian N. Gregory, "Longitudinal Analysis of Age- and Gender-Specific Migration Patterns in England and Wales: A GIS-based Approach," *Social Science History* 24:3 (2000): 471–503; Ian N. Gregory and Humphrey R. Southall, "Mapping British Population History," in *Past Time, Past Place*, ed. Knowles, 117–30.

9   Catherine A. Fitch and Steven Ruggles, "Building the National Historical Geographic Information System," *Historical Methods* 36:1 (2003): 41–51.

10   Andrew A. Beveridge, "Immigration, Ethnicity and Race in Metropolitan New York, 1900–2000," in *Past Time, Past Place*, ed. Knowles, 65–77.

11   Loren Siebert, "Using GIS to Document, Visualize, and Interpret Tokyo's Spatial History," *Social Science History* 24:3 (2000): 537–74; Idem., "Using GIS to Map Rail Network History," *Journal of Transport History* 25:1 (2004): 84–104.

Urban development is also addressed by Wilson and the Sydney TimeMap project in Australia. Implemented as a public-access kiosk in the Museum of Sydney, this GIS displays a map interface with which individuals can interact to gain access to a variety of textual and visual digital resources.[12]

These studies and projects all relate to contextual information of potential interest to book historians, although the projects themselves do not contain information overtly related to print culture. There has not been, until the current HGIS project for print culture, a GIS that focused on book history variables.[13] Ian Gregory and Paul Ell posit that historical GIS has 'the potential to ... bring many historians who would not previously have regarded themselves as geographers into the fold. This is because GIS ... provides a toolkit that enables the historian to structure, integrate, manipulate, analyse and display data in ways that are either completely new, or are made significantly easier'.[14] Layers in the HGIS of print culture will, when complete, include print production, occupations within the book and allied trades, literacy rates and schooling, political affiliation of agents of the press, religious background and gender, amongst others. Synthesis of these factors will facilitate new understanding that combines the traditions of French and Anglo-American book history and places them in a geographic context. Historical scholarship has tended largely to value words over images and numbers. With an HGIS, it is feasible to look at historical information in its geographical context and to integrate and analyse material from a wide array of sources.

Geography and Visualization

Much geographical book history, including that relating to production, distribution and reception, proceeds successfully without maps. The benefit of developing and using maps, digital and otherwise, relates to their capacity for illustrating patterns that might not be discerned otherwise, and for displaying complexity in a remarkably succinct format. As John Bonnett stated, 'some forms of representation do a better job of communicating patterns of interest than others'.[15] Geography is an inherently visual discipline and Driver suggests that the visual should be

---

12   Andrew Wilson, "Sydney TimeMap: Integrating Historical Resources Using GIS," *History and Computing* 13:1 (2001): 45–69.

13   Fiona A. Black, Bertrum H. MacDonald and J. Malcolm W. Black, "Geographic Information Systems: A New Research Method for Book History," *Book History* 1 (1998): 11–31; Bertrum H. MacDonald and Fiona A. Black, "Using GIS for Spatial and Temporal Analysis in Print Culture Studies: Some Opportunities and Challenges," *Social Science History* 24:3 (2000): 505–36.

14   Ian N. Gregory and Paul S. Ell, *Historical GIS: Technologies, Methodologies and Scholarship* (Cambridge: Cambridge University Press, 2007), 1.

15   John Bonnett, "President's Report," CCHC-CCHI. [undated]. Available at www.cha-shc.ca/cchc-cchi/Francais/ChairReport3.htm

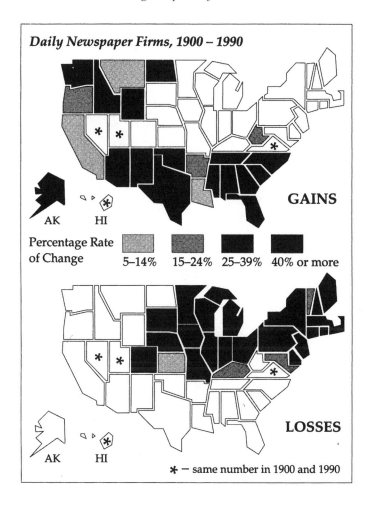

Figure 3.1   Newspaper firms gains and losses in the USA, 1900–1990

Source: Mark Monmonier, *Mapping it Out: Expository Cartography for the Humanities and Social Sciences* (Chicago: University of Chicago Press, 1993). By permission of University of Chicago Press.

a 'subject of enquiry in its own right'.[16] Geographers and cartographers have made contributions to book history scholarship through the related concepts of visualization and spatial relationships.

Journalism and political historians might be said to have taken a lead with visualizing issues of scholarly interest, using a geographic lens. Looking at the development of firms publishing daily newspapers in the United States, these

16   Felix Driver, "On Geography as a Visual Discipline," *Antipode* 35:2 (2003): 227.

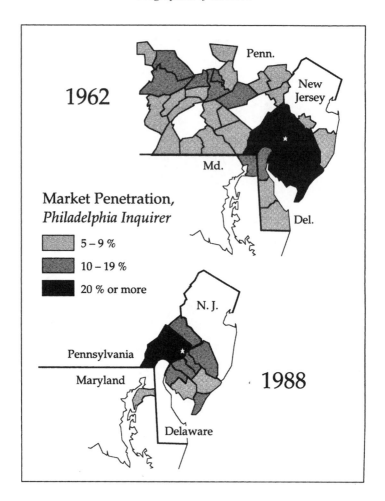

Figure 3.2   Newspaper market penetration comparison across time

Source: Mark Monmonier, *Mapping it Out: Expository Cartography for the Humanities and Social Sciences* (Chicago: University of Chicago Press, 1993). By permission of University of Chicago Press.

juxtaposed maps (Figure 3.1) for gains and losses over a 90-year period make clear the differences between Northeast USA and other parts of the country using a model of clarity and brevity. One map focuses on gains exclusively and the other on losses. Extensive charts, graphs and narrative would have been required to tell the same story. I am not at present commenting on the reliability and validity of the sources and methods used to compile these maps; let us accept them at face value for now. Another issue of interest to newspaper and book historians is market penetration (Figure 3.2). Here again, the simple technique of juxtaposing maps

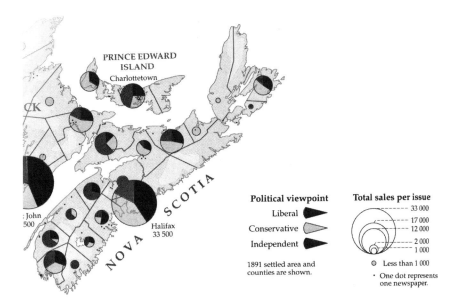

Figure 3.3   Political viewpoint and circulation of Nova Scotia newspapers, 1891

Source: R. Louis Gentilcore, ed., *Historical Atlas of Canada. Volume 2: The Land Transformed, 1800–1891* (Toronto: University of Toronto Press, 1993). By permission of University of Toronto Press.

– here differentiated by year – helps the reader to a rapid understanding of the central point: the newspaper under review suffered a severe reduction in market penetration at some point in the 28-year period covered.[17] There is a caveat in that these maps present bald figures only and we do not know where they were gathered or with what reliability. Furthermore, we have no indication of a larger context or a possible answer to the question 'why?' Nevertheless, this representation of a geographical aspect of print culture has been carefully and clearly presented.

Newspapers, both dailies and weeklies, typically convey particular political viewpoints (Figure 3.3) and reflect a geographic diversity related in part to characteristics that are local or regional. This map from *The Historical Atlas of Canada* displays both production and distribution, in terms of sales per issue, and political viewpoint, for newspapers in Nova Scotia in the late nineteenth century.[18] The use of colour is appropriate for the political interests described and the circles

17   Mark Monmonier, *Mapping it Out: Expository Cartography for the Humanities and Social Sciences* (Chicago: University of Chicago Press, 1993), 197, 195.

18   R. Louis Gentilcore, ed., *Historical Atlas of Canada. Volume 2: The Land Transformed, 1800–1891* (Toronto: University of Toronto Press, 1993), plate 51.

are of relative size to help tell the story of sales. Regarding those sales, one might argue that the concept of 'zones of significant distance' is relevant here.[19] This concept considers how distance from a place or activity shapes the likelihood of engagement with that activity – here involving the purchase of an ephemeral printed product.

Such visualizations of aspects of newspaper history all contain information of a geographic nature of relevance to book historians, information given greater meaning by the forms in which it is presented. The last map (in colour in the original) is more nuanced and sophisticated than the two sets of grey-tone juxtaposed maps, but each has the merit of being clear in its conception and execution, and in analyzing varying complexities underpinning the visualization. There are, nevertheless, potential interpretive distortions and oversimplifications that might occur through the application of 'a highly logical and visual system of analysis to a field of enquiry that is traditionally subjective and textual'.[20]

Information visualization is increasingly studied in its own right. According to Eden, information visualization can be defined as 'the use of computer-supported, interactive, visual representations of abstract data to amplify cognition' and is a method used to handle large datasets of information and increase the ease with which patterns and relationships may be identified. He identifies one subset of information visualization as geographic visualization, 'the graphical representation of geographical and spatial information'.[21] Peuquet and Kraak also focus on geographic visualization, exploring the use of interactive, web-based maps to display spatial information and developing the concept of geobrowsing, 'using geographic visualization to aid knowledge discovery in large spatial databases and to aid interactive analysis and complex problem-solving'.[22] They argue that although maps have long been used to display geographic information, we often use them without really thinking about them, a phenomenon that needs to be altered. Maps should be used in ways that enable interpretation and exploration of geographic databases, thereby facilitating discovery and the development of new ideas; GIS is enabling this transition.

Of special relevance to the current study, information visualization is an important aspect of information retrieval as well. Song considers the use of

---

19   Myron P. Guttmann, preface to *Past Time, Past Place*, ed. Knowles, viii-ix.

20   Anne K. Knowles, "Statement of Significance and Impact of Project" [from a proposal to the National Endowment for the Humanities for the conference 'History and Geography' held at the Newberry Library, Chicago]. Unpublished document, 2003.

21   Brad Eden, "Information Visualization," *Library Technology Reports* 41:1 (2002): 8.

22   Donna J. Peuquet, and Menno-Jan Kraak, "Geobrowsing: Creative Thinking and Knowledge Discovery Using Geographic Visualization," *Information Visualization* 1 (2002): 90.

visualization in information retrieval as it relates to the design of user interfaces.[23] Using a three-level analytic approach comprised of component, technique and interaction level analyses, Song advocates for more research to evaluate the effectiveness of using visualization in information retrieval. Such research should inform the development of quick, easy-to-use and effective visualization tools. Koshman is also interested in the use of information visualization for information retrieval.[24] Focusing on the online environment, she identifies a number of principles of interface design that can facilitate effective visual information retrieval. Collins et al. further address the design of effective interface tools, describing the development of such tools for a specific digital library at the Los Alamos National Laboratory. These authors emphasize the supreme importance of designing interface tools with the user in mind. 'To be most effective, these tools need to leverage the cognitive characteristics of the target users'; the interface and the information retrieval tools should be fit for purpose.[25]

Canadian Book History: Spatial Analyses and Visualizations

The geography of print culture in Canada was not one of steady, continuous spatial expansion and increasing density of networks. As the models of Darnton and Adams and Barker have long reminded us, many external variables affected that culture. Toronto as a centre of print production and distribution was critical by the late nineteenth century, but it was not always so. Earlier in the century, Montreal was the centre of print culture in Canada. It lost its position of dominance due to political, commercial and financial imperatives which affected the location of new rail lines for north-south as well as east-west communication and distribution.[26] Toronto's rise to prominence coincided with a development in mail-order marketing. Mail-order catalogues, notably those of the T. Eaton company, changed the geography of book and periodical acquisition in the more remote parts of Canada. A centre and periphery model emerged in the later nineteenth century, due to the railways and subsequent mail-order marketing. Prior to that, the centre and periphery model did not apply fully to western Canada due to local centres of print production, ethnic presses in numerous western towns, and slow and limited means of transportation.

---

23   Min Song, "Visualization in Information Retrieval: A Three-level Analysis," *Journal of Information Science* 26:1 (2000): 3–19.

24   Sherry Koshman, "Visualization-Based Information Retrieval on the Web," *Library & Information Science Research* 28:2 (2006): 192–207.

25   Linn M. Collins, Jeremy A.T. Hussell, Robert K. Hettinga, James E. Powell, Ketan K. Mane and Mark L.B. Martinez, "Information Visualization and Large-Scale Repositories," *Library Hi Tech* 25:3 (2007): 367.

26   William K. Carroll, "Westward Ho? The Shifting Geography of Corporate Power in Canada," *Journal of Canadian Studies* 36:4 (2001–02): 118–42.

In order to foster a more nuanced understanding of print culture's complexity by making spatial and temporal research possible and convenient for a wide range of scholars, the larger research project, of which this chapter discusses the first phase, is addressing three principal questions: How did regional factors (economic, political and demographic) shape Canada's print culture in the period from the beginning of printing to the end of World War I? What was the geographic spread of agents of the press, such as printers, editors and booksellers and, in particular, how did this change through time in relation to social and economic factors such as demographic and transportation developments? And, how significant is the spatial clustering of book trade activities in relation to technological innovations through time, such as the steam press and the telegraph? While these questions will be addressed by the fully-developed HGIS, this first phase of our work examines two questions: How should a print culture HGIS be designed to facilitate analysis and visualization? And how do historians and literary scholars respond to such "user experience" research?

The development of print culture in Canada is a valuable example for exploring the application of visual methods of analysis because of the country's colonial development and widely diverse physical and human geographies, with growth of settlement from east to west with the coming of the railway in the later nineteenth century. The period 1750–1918 in Canada was particularly rich for the print and allied trades, as they adjusted to numerous significant developments in the broader environment as articulated by Darnton and Adams and Barker: legal frameworks relating to copyright; economic developments such as means of communication and transportation; social connections related to settlement patterns; and the political imperatives of a developing federation of new provinces and territories.[27]

The database underpinning the pilot HGIS of print culture in Canada is the Canadian Book Trade and Library Index (CBTLI), which holds information gleaned from census data, trade directories, newspaper advertisements and other printed and manuscript archival sources. In addition, historical studies were consulted for additional relevant information, including *The Beginnings of the Book Trade in Canada* and *A Dictionary of Toronto Printers, Publishers, Booksellers and the Allied Trades, 1798–1900*.[28] The date range of records in the CBTLI is from 1752, when Bartholomew Green from Boston produced Canada's first known printed item (a government *Gazette*) on his imported printing press in Halifax, Nova Scotia, to the end of World War I in 1918. The geographic range of the 14,000 records currently in the CBTLI is primarily Upper and Lower Canada and the Maritimes, with some coverage of print culture in the Prairies from the 1890s. Records in the database can have up to 45 fields relating to personal (gender, political and religious

---

27   Yvan Lamonde, Patricia L. Fleming and Fiona A. Black, eds, *History of the Book in Canada. Volume 2: 1840–1918* (Toronto: University of Toronto Press, 2005).

28   George L. Parker, *The Beginnings of the Book Trade in Canada* (Toronto: University of Toronto Press, 1985); Elizabeth Hulse, *A Dictionary of Toronto Printers, Publishers, Booksellers and the Allied Trades, 1798–1900* (Toronto: Anson-Cartwright Editions, 1982).

affiliation and other factors), business, geographic and temporal factors, along with detailed notes concerning the sources of the information.[29] The record structure of the CBTLI is designed to facilitate analysis and display of information within a GIS, principally through a group of 15 fields devoted to geographic information, such as province or territory, town or village, street and building where a particular book trade activity took place, such as the 'building at the head of Tobin's Wharf' where David Spence ran his circulating library in Halifax in the 1820s. All of the geographic information is historically sensitive. Scholars wishing information on regions of the prairies before they became provinces can search the database under names such as District of Assiniboia.

Summary data drawn from the CBTLI can be described in several ways, some more fully contextualized than others. For example, tabular form may be used (Table 3.1), presenting nothing more than the current number of records arranged by provincial boundaries. Immediate understanding of the spatial relativity of the numbers is enhanced if viewing the same data in a graph (Figure 3.4). A more nuanced representation of information that clearly has a geographic component is to display it in a map, preferably one where some context is offered beyond location. In Figure 3.5, which was created using a geographic information system, visualization is enhanced by the use of standard cartographic techniques such as denser cross-hatching for greater numbers. Visual representation of geographic information should, to a certain extent, be intuitive to the reader or viewer. This map illustrates more than the raw numbers, it provides the geographic context of the provinces in relation to each other, the major lakes and waterways and the status of the transcontinental railroad by 1882. It offers additional detail for four important centres of print culture: Halifax, Montreal, Toronto and Winnipeg. For these cities, relative percentage figures are supplied for authors, printers, bookbinders and booksellers. Thus, in one map, we have some context for the culture of print that was spreading westward with immigration and settlement. Another geographic framework merely hinted at in this particular map, is the concept of gateway cities – book historians are aware of the critically important role that some towns and cities play in the development of a regional, national or multinational network of print culture.[30] Representing information about those places in map form can help identify the multiple, interconnecting reasons for, and possibly also the effects of, their gateway role. It is not always as simple as that of being a capital city (Montreal is not) or of being on a principal waterway (the Red River does not flow along Winnipeg's main direction for trade).

---

29   Sophie Regaldo, Fiona A. Black and Bertrum H. MacDonald, *History of the Book in Canada Project: Canadian Book Trade and Library Index (CBTLI) Data Entry Manual* 2002–04. Retrieved 19 August 2008, from http://acsweb2.ucis.dal.ca/HBICDB/english/man/CBTI_Manual.pdf

30   A.F. Burghardt, "A Hypothesis About Gateway Cities," *Annals of the Association of American Geographers* 61:2 (1971): 269–85.

Table 3.1   Canadian book trade members west to east, all dates

Province	Number	Percentage
British Columbia	465	3.61
North West Territories	9	0.07
Manitoba	1470	11.40
Ontario	6243	48.41
Quebec	2214	17.17
New Brunswick	540	4.19
Prince Edward Island	153	1.19
Nova Scotia	1803	13.99

Source: Canadian Book Trade and Library Index.

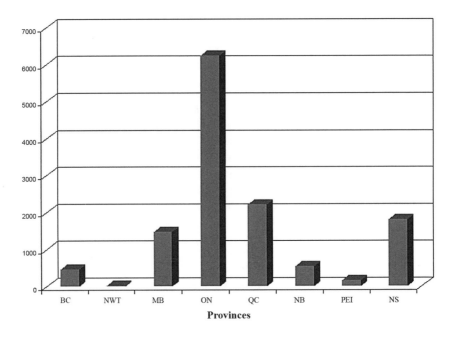

Figure 3.4   Graph of Canadian book trade members, west to east, all dates

Source: Canadian Book Trade and Library Index.

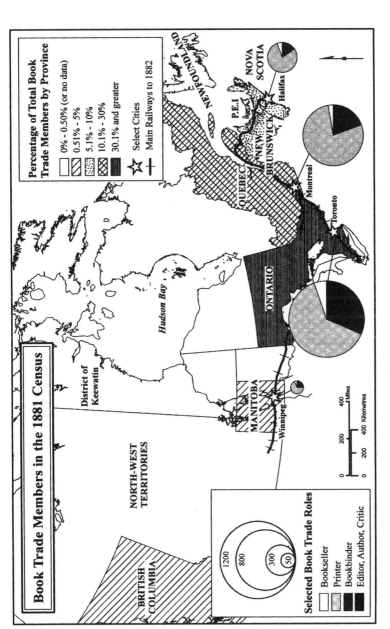

Figure 3.5   Book trade members in the 1881 census of Canada

Source: Yvan Lamonde, Patricia L. Fleming and Fiona A. Black, ed., *History of the Book in Canada. Volume 2: 1840–1918* (Toronto: University of Toronto Press, 2005). By permission of University of Toronto Press.

User Preferences for Visualizing Print Culture in an HGIS

The importance of focusing on the user in the design of interfaces for interactive data analysis necessitates investigation of user requirements and preferences. HGIS is not restricted to desktop applications and can now be used for data retrieval and integration via the web, and user studies provide one means of understanding user requirements. Such studies can be defined as 'any of the wide variety of methods for understanding the usability of a system based on examining actual users or other people who are representative of the target user population'.[31] Such studies can include user or usability testing, surveys, interviews, focus groups and observational research, among other types, and have been conducted by a variety of groups including the National Council on Archives and the National Archives Network User Research Group in the United Kingdom.[32] Julien, Leide and Bouthillier compiled a literature review of 31 controlled usability studies of information visualization tools used to retrieve textual information.[33] They further undertook a meta-analysis using seven of the studies, comparing them to text-only retrieval systems, concluding that there is no evidence that visualization-based systems offer an advantage in performance. They suggest several reasons that may account for this finding, and report that factors other than performance influence users' preference for one system over the other. Building on an earlier usability study, Ahmed, McKnight and Oppenheim developed a five-step user-centred design and evaluation methodology, which includes a competitive analysis, a user task analysis, an heuristic evaluation, a formative evaluation and a summative comparative evaluation, and encourages the iterative development of the information retrieval interface.[34]

Several usability studies have been conducted on specific digital resources with historical or geographical content. Recognizing the value of user testing, the developers of the Alexandria Digital Library have undertaken several iterations of evaluation. This library is a geolibrary containing a wealth of georeferenced information, including maps, text, images and data sets. Hill et al. describe a period of user testing during which three versions of user interface were designed

31   Foraker Design, "Usability Glossary: User Studies," in *Usability First* 2002–05. Retrieved 19 August 2008, from http://www.usabilityfirst.com/glossary.

32   National Council on Archives and National Archives Network User Research Group, *User Evaluation: Report of Findings*. 2002. Retrieved 19 August 2008, from http://www.ncaonline.org.uk/materials/nanurg.pdf

33   Charles-Antoine Julien, John E. Leide and France Bouthillier, "Controlled User Evaluations of Information Visualization Interfaces for Text Retrieval: Literature Review and Meta-Analysis," *Journal of the American Society for Information Science and Technology* 59:6 (2008): 1012–24.

34   S.M.Z. Ahmed, Cliff McKnight, and Charles Oppenheim, "A Study of Users' Performance and Satisfaction with the Web of Science IR Interface," *Journal of Information Science* 30:5 (2004): 459–68; Idem., "A User-Centred Design and Evaluation of IR Interfaces," *Journal of Librarianship and Information Science* 38:3 (2006): 157–72.

and tested using a variety of techniques (online survey, ethnographic observation, target user groups, internal evaluations and analysis of classroom use), with the results of each round of testing informing future development.[35] Jones et al. investigated the usability of the HistoryMap system to provide integrated access to two digital collections, one of historical maps and one of historical newspapers.[36] They found that the majority of users experienced little difficulty searching the collections, whether separately or together, but wanted to be able to interact with the map images as they would on the web, that is by clicking on the map to zoom-in on a location or on a thumbnail to open a full-sized image. Long, Lage and Cronin explored the usability of a prototype digital collection of historical aerial photographs linked to a map interface.[37] During the requirements analysis portion of the user-centred design process, interviews were conducted to assess user preference and needs, while during the user testing portion, information was gathered related to user interaction with the system; this information will guide the redesign of the collection.

Another digital collection that makes use of mapping technology to coordinate resources is found in the VeriaGrid, a prototype virtual map that provides information about the city of Veria, Greece. Developed by the city's Central Public Library, this system is based on an interactive map which is linked to images, videos, photographs and textual information organized thematically. Garoufallou, Siatri and Balatsoukas conducted a usability study of this system, employing both novice and expert users, with overall positive results and have used the results to generate a list of recommendations for improving its usability.[38]

Because print culture permeates all of society, the study of its history has drawn the research attention of scholars from a wide range of disciplines and periods: historians of religion and science, labour, social, cultural and legal historians, literary scholars, sociologists, information scientists and librarians, geographers and bibliographers. Scholars of this diversity bring a wide range of perspectives to the subject and have familiarity with particular sets of archival and other resources. Toms and O'Brien studied scholars from a wide range of disciplines

---

35   L.L. Hill, L. Carver, M. Larsgaard, R. Dolin, T.R. Smith and J. Frew, "Alexandria Digital Library: User Evaluation Studies and System Design," *Journal of the American Society for Information Science* 51:3 (2000): 246–59.

36   Steve Jones, Te Taka Keegan, Matt Jones and Malcolm Barr, "Searching and Browsing in a Digital Library of Historical Maps and Newspapers," *Journal of Digital Information* 6.2 (2005).

37   Holley Long, Kathryn Lage and Christopher Cronin, "The Flight Plan of a Digital Initiatives Project, Part 2: Usability Testing in the Context of User-Centered Design," *OCLC Systems & Services* 21:4 (2005): 324–45.

38   Emmanouel Garoufallou, Ranai Siatri and Panagiotis Balatsoukas, "Virtual Maps – Virtual Worlds: Testing the Usability of a Greek Virtual Cultural Map," *Journal of the American Society for Information Science and Technology* 59:4 (2008): 591–601.

in the humanities in their work on the technology needs of "e-humanists".[39] Their methodology involved an online questionnaire that explored uses of technology in teaching and research. The questions related to research focussed on the use of information and communication technologies (ICTs) in terms of electronic texts, text analysis tools, the research process, collaboration and tools used for analysis. Toms and O'Brien do not indicate any use of GIS by the 169 respondents to their survey; a finding that is unsurprising given the relatively small number of humanities scholars who have embraced GIS to date.

To examine the premise that print culture cannot be fully understood without incorporating the spatial dimension into the picture, established and emerging print culture historians were invited to participate in a usability study. This study engages print culture historians in the design and application of a web-based system that will contribute to their research endeavours. Book historians study history. What do historians do, how do they practice the work of history? The framework embraced by John Unsworth is pertinent for GIS applications in print culture. Unsworth emphasizes the importance of 'scholarly primitives' which are 'basic functions common to scholarly activity' and include: discovering, annotating, comparing, referring, sampling, illustrating and representing.[40] This list may be useful for testing the effectiveness of enquiries within geographies of the book and visualizations accruing from those enquiries.

Gregory and Ell note the challenges of using a positivist methodology such as GIS with historical resources which typically require much greater cross-referencing and authority checking than resources used in quantitative disciplines such as environmental science. They further emphasize that, whilst GIS is certainly software-based, it should also be viewed as a scholarly approach to geographically-related information.[41] As the majority of the participants in the usability study had no prior knowledge of GIS, the technology was essentially transparent to them, thereby facilitating a focus on visualization and search preferences.

Some of the questions posed in this usability study are: How do book historians conceive of the historical geography elements of their searching? How important are maps to historians? Do historians want to be able to search for locations visually on a geographic map or would they rather enter text into a Google-like search box? Should book historians have an option in how they search the HGIS: textually versus visually? Do they want access to images or would they prefer solely textual

---

39   Elaine G. Toms and Heather L. O'Brien, "Understanding the Information and Communication Technology Needs of the E-Humanist," *Journal of Documentation* 64:1 (2008): 102–30.

40   John Unsworth, "Scholarly Primitives: What Methods do Humanities Researchers Have in Common and How Might Our Tools Reflect This?" Paper presented at the Symposium of Humanities Computing: Formal Methods, Experimental Practice, London, UK, May 2000. Retrieved 19 August 2008, from http://www.iath.virginia.edu/~jmu2m/Kings.5-00/primitives.html

41   Gregory and Ell, *Historical GIS*, 1.

facts? Do historians have a preference for photographic over non-photographic images? When searching in an HGIS portal, how important is it for historians to know the source of a visual item at the time of searching? Do historians have a preference for historical or for modern maps?

The methodology involved an interview protocol within a framework that has been tested by usability scholars. Neilsen posits that no more than five subjects are needed at a time for usability studies for web design. He recommends that the design process should be iterative, involving the results from 4–5 subjects, amending the design if needed, asking a further 4–5 subjects, and so on, until nothing new emerges in terms of design comments and suggestions.[42] In order to ensure control over the visuals viewed by the subjects, who were scattered across Canada, selected images were captured from a pilot HGIS and complemented by new visuals developed for the purposes of this study. All of the selected images were imported into a PowerPoint presentation that participants could receive as an email attachment. The interviews were, with one exception, conducted by telephone, with the participants navigating through the presentation in relation to the questions being posed. Each interview lasted approximately one hour with a series of open-ended questions and probes relating to a set of 20 images (Appendix A). The participants were selected based on their knowledge and experience as researchers using a wide variety of sources for print culture research ranging from archival documents to databases of all manner of primary and secondary materials. One participant was a junior scholar who had completed a thesis on historians' and journalists' uses of images in their research; two participants were very experienced book historians with international reputations; one participant was a less-experienced book historian, yet already published and highly regarded; and the final participant in this round was the former project manager for a large print culture project completed in 2006.

A premise underlying the design of the prototype and the captured images is that researchers interrogating an HGIS should be able to move from the general to the specific. The usability study employed a prairie example, using Manitoba information, to illustrate the process of moving through layers of data and through types of information source in an interactive system. In Manitoba and westward into what became Saskatchewan, there were many towns and villages where books and newspapers were either produced or were, at least, made available for borrowing or purchase (Figure 3.6). The principal centre of print production and distribution was Winnipeg. Clicking on that location allows the researcher to move to another layer containing digital photographs, in this case a streetscape of the young city's Main Street, complete with a hanging sign indicating a bookselling and stationery business (Figure 3.7). The researcher may move further through the layers of data to a search of the CBTLI database which produces the record

---

42   Jakob Nielsen, "Why You Only Need to Test With 5 Users," in *Jakob Nielsen's Alertbox* 9 March 2000. Retrieved 19 August 2008, from http://www.useit.com/alertbox/20000319.html

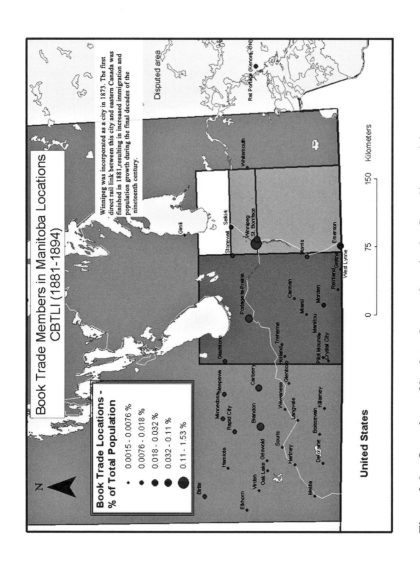

Figure 3.6　Locations of book trade members in the Canadian prairies

Source: Canadian Book History GIS Pilot, Dalhousie University, 2002.

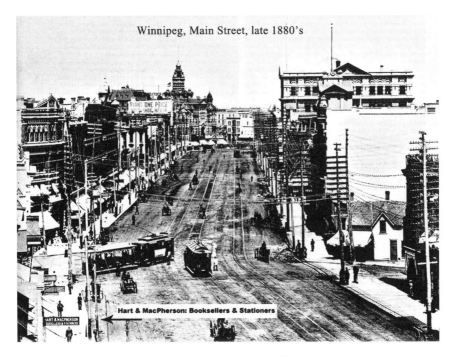

Figure 3.7   Winnipeg streetscape in the late 1880s

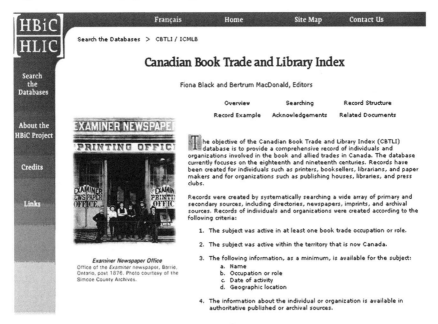

Figure 3.8   Canadian Book Trade and Library Index home page

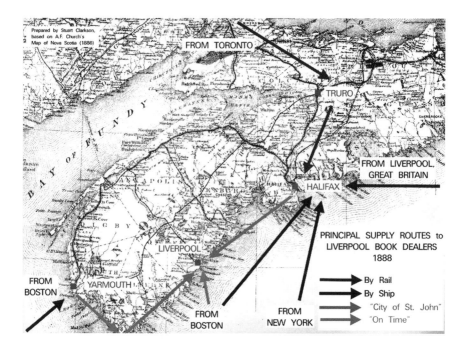

Figure 3.9   Sources of book supply to Liverpool, Nova Scotia in the late nineteenth century

Note: Map developed by Stuart Clarkson.

relating to that business (Figure 3.8). As with all CBTLI records, the information includes the name of the business, location, known dates of operation and the source or sources of the information. The scholarly reliability of the information used in compiling the large scale map can thus be checked in the same way as historians typically check footnotes and endnotes.

Historical geographers often stress the importance of using contemporary maps, and details pertinent for a geography of the book can be layered onto an historical map. An examination of import data for Nova Scotia in the 1870s and 1880s, along with archival research by Stuart Clarkson, brought to light insights regarding imports and routes of transportation. Publishers in Toronto would supply books and periodicals to wholesalers and retailers in Halifax, who in turn would supply other towns. Truro, small compared to Halifax, acted in part as a hub for the book trade due to its role as a hub for the railways. Although a railway line would eventually be built from Halifax down the coast as far as Yarmouth, for several decades thereafter books were still distributed by coastal shipping. The ships were, perhaps astonishingly, faster than the railway. Figure 3.9 illustrates the greater number of book shipments by sea versus land. This map was created without a GIS, and it displays the results of research concerning just one bookseller over a period of a few years in one small coastal town in Nova Scotia. The Liverpool supply route map offers direct and suggestive information about a) oceanic versus coastal shipping, b) water transport versus railways, c) details concerning direction as well as relative distances and, d) the complexities of national, colonial and regional loyalties regarding communication and trade. Coastlines and ports of entry are particularly fruitful spaces of enquiry for geographies of the book. In addition, this map is a very good example of the differences of scale that can be adopted within any geography of the book.

Observations from Print Culture Historians

The five hour-long interviews provided results that will be used in honing the interface design for a Canadian HGIS of print culture prior to the next phase of usability testing. One general, and consistent, finding was that participants like to have control of the options for visualizing print culture information. When presented with tables, graphs and maps, it was challenging for participants to make clear which they might prefer. The potential context of their search was paramount in their thinking, and the ability to have all three formats was important to them. The fact that many interactive, web-based tools provide such options, was not known to some of the participants. This indicates some of the interrelated barriers to book historians thinking visually, in addition to the relative complexity of the technology. Book historians' own scholarly education has neither equipped them with appropriate technological competence nor provided them with frameworks for thinking about technological applications.

Participants tended to view the images with remarkably fresh perspectives, drawing on their prior focus on fixed, printed materials, rather than interactive digital materials. In spite of this difference in prior experience, several findings were consistent across all participants.

The image relating to newspaper publication (Figure 3.3) was developed by Byron Moldofsky drawing on data within a GIS at the University of Toronto. Participants in our study were asked to describe an incident from their own research when this type of visualization was of aid to them. Print culture historians are deeply familiar with the type of information represented here. It was feasible, then, to ask them to comment on the visualization of such information – we were not asking them to speculate, but to describe, with evidence. This can be a critical feature of successful usability studies. We further asked if participants required data or information in a visual way in order to achieve their research goals. Spatial representation had value, based on their responses, and the use of colour was mentioned in relation to ease of comparison, for example with the political affiliations indicated here.

The GIS query framework slide (Table 3.2) was enthusiastically received as we posed the question: which, if any, of these questions is reflective of the types of questions to which you seek answers in your research? The first four questions were of interest and, as with all good questionnaire designs, we asked

Table 3.2   GIS query framework for a geography of the book

Name	Query	Definition	Example
Location	What is at?	Determines the attributes of a location	How many wholesale stationers were within two miles of Port Glasgow in the 1780s?
Condition	Where is?	Seeks locations fulfilling certain conditions	Which Canadian towns had bookshops with stocks of more than 1,000 titles by 1880?
Trend	What has changed since?	Determines changes in attributes over time	What was the increase in the literacy rate in Nova Scotia between 1881 and 1911?
Pattern	What is the spatial distribution of?	Investigates the spatial distribution of selected attribute	Were there more specialist bookshops in 19th century Canadian towns with higher proportions of British immigrants?
Projection	What if?	Explores potential patterns based on past data	Based on book export figures from Scottish ports to North America in 1750–70, what would the figures have been in the 1790s if the same trend had continued? What were the actual figures in the 1790s?

Note: First published in Fiona A. Black, Bertrum H. MacDonald and J. Malcolm W. Black, "Geographic Information Systems: A New Research Method for Book History," *Book History* 1 (1998): 11–31.

a variety of follow-up questions to test the validity of our instrument. Whilst all claimed that questions about location were important to them, when asked to describe an example from their research when they needed information about a particular place one respondent declined. The other responses offered details about how they searched for information about a place. They tended to search for references to a particular place in personal, proprietary or open access databases such as those relating to archival resources. Their use of maps or atlases, digital or otherwise, was limited. Concerning the fifth query, all respondents expressed a lack of interest in projections and models. This is an anticipated response from those whose interests are primarily historical, although John Bonnett and others have demonstrated the intellectual value of projections and "what if" scenarios for more nuanced understandings of historical events and trends.

The majority of participants were familiar with the Canadian Book Trade and Library Index, and a screenshot of the home page was used to remind them of the resource. While it is feasible for someone to interact with an HGIS without any knowledge of databases, familiarity with such research tools can inform scholars' confidence with IT and may potentially increase the likelihood of their use of an HGIS. The protocol included a series of questions around the participants' prior use of electronic resources, databases and websites, and their uses of databases of information in any format, including images, and their assessment, if any, of the reliability and authority of the sources of information contained in the databases they access. All of these factors influence the planned design of our print culture HGIS.

All participants were, not surprisingly, familiar with the common methods provided for searching such databases, for example by key word or authority lists, the use of field options. For an HGIS that will engage a wide range of users, we are not surprisingly interested in how scholars might wish to interrogate or otherwise explore a geospatial interface. The screen capture from an earlier GIS project at Dalhousie (Figure 3.6), illustrates the administrative boundaries and settlements in part of the Canadian prairies in the late nineteenth century. Questions to the participants focused on how they might want to query such a map-based interface. Did they want mouse-over interactivity, such as in Figure 3.6, and/or a search box, Google style, for entering a keyword or query? This image and the questions related to it led to rich discussions around search possibilities, such as when features like drop-down menus would have value. Many issues for design consideration accrued from these discussions, such as the potential importance of links to historical gazetteers for place name changes as with an online portal.

As already mentioned, earlier studies with GIS have included databases of historical images and a Winnipeg streetscape (Figure 3.7) was presented to study participants with queries related to their use of images for information, context and illustration purposes, and how they would prefer to search, retrieve and use such images. The question about interrogating the image itself for further details met with a strongly positive response and a series of screen captures showed them how this might work in practice, leading the user through just a couple of clicks directly to the source of information, in the CBTLI, about a particular book trade business at the particular location and time represented in the photograph.

The usability study also included questions around print culture historians' preferences regarding the display and visualization of numerical data related to an HGIS. The participants, for example, appreciated tabular data arranged west to east to have a geographic logic, or east to west to have an historical logic (as Canada was settled principally from the east westward). Their interests in simple charts as visualization tools were mixed as they all required more detail (such as raw numbers). Judging from the extensive published literature in print culture, few scholars in this field generate either tables or charts in their own work, yet they refer to them with a degree of comfort when they are specifically related to their research interests.

One of the images used was a relatively dense greyscale map from the *History of the Book in Canada, Volume 2*.[43] The objective series of usability questions resulted in more positive responses than anticipated. It might be easy to dismiss enthusiasm for such a map as due to a lack of knowledge by the participants of

43   Lamonde, Fleming and Black, *History of the Book in Canada, Volume 2*, 199.

cartographic principles. Good practices in information studies focus on less rather than more density, and good practices in cartography suggest that colour helps the eye distinguish more classes and ranges of information. The responses were, however, instructive about the potential tolerance by these humanities scholars for dense information presentation.

Another factor related to visualization concerns qualitative aspects of the maps themselves.The visual for the participants was that referred to above developed by Stuart Clarkson (Figure 3.9). Comments from participants related to authenticity and historical context, although they were puzzled by aspects of the overlay. Having seen simpler visuals from the GIS, they offered a variety of comments about the pros and cons of clarity versus historical context. These are highly pertinent comments for an HGIS design. How such participants would engage with Unsworth's scholarly primitives is still to be explored in detail, though some tentative conclusions are possible. As Gregory and Ell make clear, maps often 'suggest patterns or relationships but are rarely capable of identifying the processes causing them'.[44] Participants in the study, without probing, regularly offered comments on potential causes drawing on their own specialist knowledge. It was thus clear that they had interest in using a geographically-based interactive system for analytical work in addition to 'discovering, annotating, comparing, referring, sampling, illustrating and representing'.[45] In this way, the HGIS was perceived to have the potential to add understanding to historical questions about print culture. This is an interesting finding in itself as an interactive site such as an HGIS is counterintuitive to the normal scholarly practices of book historians. These practices emphasize the use of publications that present the results of analysis of data, rather than the use of systems where the onus for data analysis is on the user.[46]

Toward Collaborative Geographies of the Book

Book historians are book geographers to the extent that their research questions deal with interrelationships among location, time period and print culture. Whether interacting with an HGIS or not, it is essential for book historians to use diverse sources and methods appropriate for analysing them. Historical applications of GIS have been implemented for over a decade. Such systems have enormous potential to benefit scholarly enquiry and analysis. One aspect that has been missing from the literature of the various projects is user studies or needs assessment exercises to determine what would be meaningful as an aid to scholars. Using Nielson's iterative model of usability testing, the HGIS of Canadian print culture will be developed in line with the results of that testing.

---

44   Gregory and Ell, *Historical GIS*, 19.
45   Unsworth, "Scholarly Primitives."
46   Gregory and Ell, *Historical GIS*, 116.

An HGIS should not be presented as a panacea. It cannot provide definitive answers for a field in general. It can help provide details and context and to identify pertinent additional questions.[47] With the increase in availability of census and other large spatiotemporal datasets at the individual level, the GIS community is challenged to develop analytical tools that 'can efficiently help researchers explore the datasets in order to uncover useful information'.[48] Until humanities computing becomes a core subject for doctoral study in book history, however, expectations for complex quantitative applications of GIS should be tempered by careful analysis of book historians' research and visualization preferences. Driver praises Gillian Rose's 'suggestive attention to the interaction between particular technologies, spaces and audiences'.[49] An HGIS is a web space rather than the physical spaces referred to by Rose, though of no less importance for studying such interactions. New, potentially more intuitive tools, such as Google Earth, may reduce some of the technological barriers to book historians' uptake of GIS.

This chapter has discussed selected aspects of the contribution of an HGIS to an investigation of book history by examining practicing scholars' visualization and search preferences. Following Gregory and Ell's urging, book historians using GIS should be asking what are the geographical aspects of their research questions, rather than asking what they can do with their datasets using GIS software.[50] The utility of an HGIS to enable enquiries relating to the movement of agents of the press, such as printers, booksellers and newspaper editors, will be enhanced through scholarly collaboration. The fact that books, and their agents, paid little heed to physical or political borders suggests strongly that the next stage of large scale book history must be at the international level rather than the national. Collaboration can take many forms, from scholars in Scotland, Canada and New Zealand sharing information about migrating printers, to research-based verifications and enhancements to web-based research tools.[51] The advent of social media, and their cautious adoption in selected libraries and scholarly communities, has already produced results relating to verification and corrections to web-published information relating to events in place and time. If, as Patrick Geddes strongly urged, geography is to have an effective role as social practice, additional uses of social media such as shared image sites (e.g., Flickr) and blogging will help foster collaborative scholarship.[52] Furthermore, there

---

47   Deryck W. Holdsworth, "Historical Geography: New Ways of Imaging and Seeing the Past," *Progress in Human Geography* 27:4 (2003): 486–93.

48   Shih-Lung Shaw, Hongbo Yu and Leonard S. Bombom, "A Space-Time Approach to Exploring Large Individual-Based Spatiotemporal Datasets," *Transactions in GIS* 12:4 (2008): 425.

49   Driver, "On Geography as a Visual Discipline," 229.

50   Gregory and Ell, *Historical GIS*, 1.

51   A new collaborative project, relating to print trade members' migration throughout the British Empire, is planned by David Finkelstein of Queen Margaret University, Edinburgh, Sydney Shep of the Victoria University of Wellington, NZ and Fiona Black.

52   David Matless, "Regional Surveys and Local Knowledges: The Geographical Imagination in Britain, 1919–1939," *Transactions of the Institute of British Geographers* 42

are formats that complement the visual, such as audio, that can help examine the geography of the book and audio files can also be layered within an HGIS.[53] While Geddes urged geography's role in a social practice fostering citizenship, collaborative web-based approaches to the geography of the book – using an HGIS – might help book historians fulfil his motto that every town or village was not merely a 'place in space, but a drama in time'.[54] There are dramas relating to the spatial contexts of print culture waiting to be explored.

Appendix A: Book History HGIS Usability Study Telephone Interview Protocol

1.   Preamble spoken over Slide (Data layers in an HGIS of print culture)
Thank you for agreeing to participate in our study. This project is investigating the potential uses, by print culture scholars, of an integrated online collection of historical resources. Our research focuses on historical geography and its influence on print culture. We call this online collection an "historical geographic information system" or an HGIS. The elements of such an information system involve different categories or layers of data as shown here. The HGIS allows scholars to pose queries about any or all of the categories available within the system. The element that connects categories is location. The layers may include information in any available format, including images. This interview focuses on your own research interests in images and the visualization of information relevant to the historical study of print culture.

2.   Slide (1891 newspapers)
Can you describe an incident from your research when this type of visualization was of aid to you?
PROBES:
What were you researching?
Did you require data or information in a visual way in order to achieve your goals?

(1992): 464–80; "Remarkable Outreach: Flickr Project Draws 100 Million Views," *Library of Congress Information Bulletin* 68:1–2 (2009): 22–4. Library of Congress photographs, made available via Flickr at "The Commons" have been enhanced with additional and corrected information provided by Flickr commentators. Library staff has amended metadata based on this new information.

   53   Mei-Po Kwan and Guoxiang Ding, "Geo-Narrative: Extending Geographic Information Systems for Narrative Analysis in Qualitative and Mixed-Method Research," *The Professional Geographer* 60:4 (2008): 443–65. An example for a Scottish HGIS of print culture might include the audio files related to a recent oral history of the book trade, see Alistair McCleery, David Finkelstein and Jennie Renton, eds, *An Honest Trade: Booksellers and Bookselling in Scotland* (Edinburgh: John Donald, 2009).

   54   Matless, "Regional Surveys and Local Knowledges," 473.

How did the visual representation of the information aid you in your research enquiry?

3. Slide (Query framework for geographies of the book)
Which, if any, of these questions is reflective of the types of questions to which you seek answers in your research?
PROBES:
Can you describe an example from your research when you needed information about a particular place?
What type of information were you seeking and how did you look for it?
Did you use any maps, historical or modern, to help in your research?

4. Slide (CBTLI opening page)
Introduce CBTLI: The Canadian Book Trade and Library Index is one of several under development around the world. It is the first such index of book trade workers and other "agents of the press" to facilitate detailed research around the locations of activities and hence how place might affect the spread of print culture. The database was designed and developed as a part of the History of the Book in Canada project, funded by the Social Science and Humanities Research Council of Canada. A key element of the CBTLI is the authority of the sources used to compile the data. All sources are contemporary and all have been seen and checked by a member of the CBTLI team. Sources include archival business records, business directories, census returns, etc.
Do you use electronic resources, databases, or websites in your research?
What kind of information do they contain?
What methods do you use to ensure the reliability and authority of the content of the databases?

5. Slide (Manitoba book trade members)
This is an example of the output from an HGIS that shows book trade members in Manitoba as a percentage of the total population. In such a system there are several ways to search for information.
How might you want to use this map to retrieve information from the HGIS?
PROBES:
Do you want a text box for entering your own words? Why? How might that aid your research?
Do you want a drop down box of categories to guide your search? Why? How might that aid your research?
Do you want to be able to click on any portion of the map for pop up information about the book trade at that location? Why? How might that aid your research?

6. Slide (Manitoba map – Winnipeg pop-up)
This is an example of a mouse-over that provides information about a geographic area. Would you find this type of feature helpful when doing research?

Why would it be useful in terms of your research? How else might you wish to view summary information about locations?

7.   Slide (Arrow on Winnipeg)
A further example of interrogating the system through the map would be clicking on Winnipeg to move to a menu of further available information relating to that place. Further sources might include historical photographs, for example.

8.   Slide (Winnipeg streetscape)
Do you use historical images in your research and, if so, for what purposes?
PROBES:
Would you characterize your use of images as primarily to obtain information or to illustrate text?
Would it help support your research if you could query the photo for additional details by clicking on relevant components?

9.   Slide (Winnipeg streetscape – pop-up)
This is an example of a mouse-over that provides information about a book trade member. Would you find this type of feature useful when doing research?
PROBES:
Does this provide enough information for your research needs?
Would you need to know the source of such information?

10.   Slide (CBTLI entry for Hart and MacPherson)
This is the entry from CBTLI that the mouse-over information in the previous slide was taken from. We are interested in how you might use this information. Would the mouse-over shown in the previous slide be sufficient for your research needs or would you want the system to take you to the full record? With respect to the record, would you want to print it? Would you want to download it? If so, would your research preference be to have the photo (from the previous slide) printed or downloaded automatically with the CBTLI record? If there were several images related to a particular CBTLI entry (for example, if there were photos of the inside of their store, or photos of the two men themselves), would you wish all images to be displayed together as thumbnails?

11.   Slide (CBTLI table west to east)
Tables are a traditional way for scholars to display data. As a method of visualizing search results, is a table such as this meaningful to you?
PROBES:
Is the arrangement west to east of any relevance to the types of enquiries you might pose?
Tell me about your preferences around raw numbers versus percentages.

12.   Slide (CBTLI bar chart)

Would you compare your impression of this chart to the earlier table in terms of its potential usefulness to you as a means of representing information?
PROBES:
Do you generate charts in your own research?
Do you seek charts in scholarly publications?

13.   Slide (1881 census map)
This map illustrates the same data, while adding new information. This map was published in Volume 2 of the History of the Book in Canada.
What is your response to this visualization of information?
PROBES:
Greyscale maps can be challenging to understand. Does this map make clear sense to you?
What features of this map are meaningful for your own research?
What additional features would you require for your own research?

14.   Slide (Ethnicity table for NS) – raw numbers and percentages
Part of our research includes a study of national influences on Canadian print culture. Drawing on the 1881 census, we have compiled this information for Nova Scotia.
What are your views on this way of presenting the information?

15.   Slide (Ethnicity bar chart for NS)
This bar chart displays the percentage information. Is this a more helpful visualization for you? In what way?
PROBES:
Would you want access, through an HGIS, to both tabular and charted information?
If you had to choose one format over the other, which would you choose? Why?

16.   Slide (Ethnicity map for NS)
What information, if any, in this map is meaningful for your research?
What additional information do you require for your research?
PROBES:
Would you prefer numbers and percentages added to the pie charts?

17.   Slide (NS supply routes historical map)
This map was created by Stuart Clarkson when he was a graduate student at Dalhousie University. What meaning does this map have for you?
PROBES:
Do you make use of this type of map in your research?
Is the level of detail in this map useful to you? Why?

18.   Slide (Knowles Book Bindery)

This is a photograph of a street corner in late nineteenth-century Halifax. The business illustrated here is the Knowles Book Bindery.

What would you want to know about the book trade member or the business in terms of geography?

PROBES:

Do you want to know where he is from originally?

Do you want to know the exact address of the business or is knowing that it is in Halifax enough for your research?

Do you want to know the geographic details of his trading partners or business contacts?

19. Slide (CBTLI Advanced search with field names showing)

This drop-down menu shows the searchable fields in CBTLI. The geographic information available in CBTLI records (when known) includes those items presented here:

- Geographic Origin
- Province/Territory/District
- Town
- Street Address
- Trade Contact Locations

Which, if any, geographical fields would be useful to you in your research?

How would you use that information?

20. Slide (Steam Printing House, Montreal – Burland Lithographic business)

This illustration is a contemporary lithograph from earlier in the nineteenth century. For many book trade businesses, there are no photographs. We are interested in your assessment of lithographs, engravings, etc., compared with photographs as sources of historical evidence.

What is your reaction to this image?

PROBES:

Do you have a preference for photographic over non-photographic images as sources of historical information?

PART II
GEOGRAPHIES OF CIRCULATION

Chapter 4

'Per le Piaze & Sopra il Ponte': Reconstructing the Geography of Popular Print in Sixteenth-Century Venice

Rosa Salzberg

To historian Martin Lowry, the rapid growth of Venetian printing in the late fifteenth century was a 'sudden explosion in [the city's] vitals'.[1] When knowledge of the new technique of printing with moveable type spread across the map of Europe, it fell on particularly fertile ground in Venice where the first printed editions appeared in 1469. The city quickly became the largest producer of printed material in Europe, a position it would hold for nearly a century.[2] The superiority of the Venetian Republic's established trade network meant that a sophisticated distribution system was easily set up. Books from Venice soon reached across Italy and across Europe. By the early sixteenth century, however, another significant development was gathering pace: the rapid expansion of vernacular printing, much of it in smaller and, thus, cheaper formats. Ephemeral printed matter such as posters and decrees, and cheap, 'popular' pamphlets destined more exclusively for the Venetian market, made up an important element of the output of many printers.[3]

---

1 Martin Lowry, *Nicholas Jenson and the Rise of Venetian Publishing in Renaissance Europe* (Oxford: Blackwell, 1991), 177.

2 Ugo Rozzo, *Linee per una storia dell'editoria religiosa in Italia (1465–1600)* (Udine: Arti Grafiche Friulane, 1993), 21–2, estimates that Venice produced as many as 50,000 to 60,000 editions in the sixteenth century, out of an estimated 400,000 for all of Europe. Horatio F. Brown, *The Venetian Printing Press* (London: Nimmo, 1891) remains an essential reference point on the general history of the industry.

3 I use the term 'popular' as outlined in Paul F. Grendler, "Form and Function in Italian Renaissance Popular Books," *Renaissance Quarterly* 46:3 (1993): 453. That is, works mostly in the vernacular and in small formats (chiefly octavo or quarto), appealing and accessible to a wide range of readers. Roger Chartier, among others, has drawn attention to the dangers of equating 'popular' print exclusively with a specific (lower-class) readership in this period (see his "Culture as Appropriation: Popular Cultural Uses in Early Modern France," in *Understanding Popular Culture: Europe from the Middle Ages to the Nineteenth Century*, ed. Steven L. Kaplan (Berlin: Mouton, 1984), 229–53). However, the term is still of use in defining this sort of material, if not its readership.

Much work has explored the expansion and impact of popular print, especially in northern Europe, but Venice remains an essential point of focus, as it was among the first places to feel the full measure of this major development.[4] There is still a good deal to be understood about how the flood of print was experienced within the unique urban environment of Venice, and not just by the intellectual elite but by the wider population, both literate and non-literate.[5] This chapter seeks to apply a broadly geographical lens to a consideration of print in the city in the sixteenth century, considering both the movement of popular print, and of the people that made, sold, performed and consumed it in urban spaces. This approach is an extremely useful way to bring to light the dissemination of popular print in particular, an aspect of print culture that tends to be hidden by more traditional approaches to book history.[6] Street selling of print, which has not been studied systematically for this period in Italy, is an especial focus. Although popular print was also sold in bookshops, street selling played a significant role in the dissemination of this kind of material.[7] Study of how available, visible – and audible – print actually was in the urban environment can help to illuminate how the new technology entered the lives of all Venetians, and influenced the culture of the city at this crucial moment in its history.

Encounters with Print

In 1509, during Venice's engagement in the disastrous war of the League of Cambrai, the patrician diarist Girolamo Priuli noted that '*frottole*, verses and

---

4   Pioneering works on the dissemination of popular print include Robert W. Scribner, *For the Sake of the Simple Folk: Popular Propaganda for the German Reformation* (Oxford: Clarendon Press, 1994; first published 1981); Margaret Spufford, *Small Books and Pleasant Histories: Popular Fiction and Its Readership in Seventeenth-Century England* (Cambridge: Cambridge University Press, 1985; first published 1981); Roger Chartier, *The Cultural Uses of Print in Early Modern France*, trans. Lydia G. Cochrane (Princeton: Princeton University Press, 1987); Tessa Watt, *Cheap Print and Popular Piety 1550–1640* (Cambridge: Cambridge University Press, 1991). No major work has surveyed the expansion of popular print in Venice in the sixteenth century.

5   Paul F. Grendler, *Schooling in Renaissance Italy: Literacy and Learning, 1300– 1600* (Baltimore: Johns Hopkins University Press, 1989), 46–7, estimates levels of full literacy for later sixteenth-century Venice based on schooling rates (circa thirty-three per cent of boys and twelve to thirteen per cent of girls), although functional literacy probably extended into a relatively broad segment of the male population at least.

6   Filippo De Vivo's *Information and Communication in Venice: Rethinking Early Modern Politics* (Oxford: Oxford University Press, 2007) demonstrates the value of considering acts of communication – via oral, hand-written and printed media – with close reference to their specific urban contexts.

7   Angela Nuovo, *Il commercio librario nell'Italia del rinascimento* (Milan: Franco Angeli, 1998), 105–10.

songs' about recent developments in the war 'were being sold in Venice in the *piazze* and on the Rialto Bridge'. There was nothing particularly unusual about this – it was, indeed, 'just as usual'. But Priuli was perturbed that the Venetian authorities should permit the circulation of such pamphlets, damning their enemies and celebrating temporary successes, since constant reversals of fortune meant morale in the city was declining.[8]

In fact, one month earlier, the city's powerful Council of Ten had acted to remove from sale a song against Venice's enemies, printed within days of the declaration of war, worried it would cause offence. Yet others soon took their place on the market.[9] The upheavals of these years prompted the famous scholarly printer, Aldo Manuzio, temporarily to close up shop and leave Venice.[10] They evidently did not impede some enterprising others – popular printers, hack poets, ballad singers, ambulant sellers – from launching their wares onto the market. These wares were cheap: those confiscated by the Council of Ten had been selling for a *bezzo*, or half a *soldo*, a price theoretically within reach of a large number of Venetians.[11]

Because so few of the kinds of small pamphlets and printed sheets mentioned by Priuli survive from this period, of the large quantities that are known to have been produced, their part in the history of early printing and growing presence in the day-to-day life of Venice rarely is given adequate recognition.[12] The perceived

---

8   Girolamo Priuli, *Diarii*, Venice, Biblioteca del Museo Correr (henceforth BMCV), MS. Prov. Div. 252–c, vol. 5, cc. 55r–v (end of December 1509). *Frottola* is a type of popular song from the period. I would like to thank Krystina Stermole for alerting me to this reference. The Venetian calendar began on 1 March. Unless otherwise stated, I have adapted *more veneto* dates to the modern style.

9   Marino Sanudo, *I diarii (1496–1533)*, ed. Rinaldo Fulin et al. (Venice: Visentini, 1879–1903), vol. 9, col. 335 (22 November 1509): 'Era stampado una canzon si chiama: *La Gata di Padoa*, con una altra in vilanescho di Tonin: *E l'è partì quei lanziman*, qual, per non offender il re di romani, cussi chome si vendevano un bezo l'una, fo mandato a tuorle per li capi di X, adeo piu non si vendeteno. Tamen, vene fuora altre canzon fate contra Ferara numero tre, et sono lassate vender.'

10   Martin Lowry, *The World of Aldus Manutius: Business and Scholarship in Renaissance Venice* (Ithaca: Cornell University Press, 1979), 159.

11   A useful source for prices is Ferdinand Columbus's pamphlet collection which is frequently annotated with price and date of purchase. Like Sanudo's testimony (quoted in note 9), this suggests that the going rate for a four-page quarto pamphlet in the 1510s and 1520s was about one *quattrino*, or one-third of a *soldo*. See Klaus Wagner and Manuel Carrera, *Catalogo dei libri a stampa in lingua italiana della Biblioteca Colombina di Siviglia* (Ferrara: Panini, 1991). The lowest paid shipyard worker in Venice in the sixteenth century was paid about twenty ducats per annum, with 124 *soldi* to the ducat, while a loaf of bread cost two *soldi* in 1534; see Paola Pavanini, "Abitazioni popolari e borghesi nella Venezia cinquecentesca," *Studi veneziani* 5 (1981): 68 and 71.

12   Neil Harris, "Marin Sanudo, Forerunner of Melzi. Parte I," *La bibliofilía* 95 (1993): 1–37; Ugo Rozzo, "Fogli volanti," in *Il libro religioso*, ed. Ugo Rozzo and Rudj Gorian (Milan: Bonnard, 2002), 137–46.

lack of literary merit in such material likewise has made it, until recently, a less common focus of study. The city's restricted literary salons and the houses and *botteghe* (workshops) of the most celebrated representatives of the new printing trade have received the most investigation by scholars, not least because the most abundant evidence survives for them.[13]

Yet if we are to understand the wider impact of the printing expansion, there is a need to consider the full spectrum of places in which people produced and came into contact with the new output of the press, as well as a need to consider different types of printed matter. The operations of less eminent but still markedly prolific producers and distributors of popular printed material merit further investigation. There is no doubt that some less-studied figures produced extraordinary quantities of work, especially if one takes into account the low rates of survival for ephemeral material.[14] Crucially, too, one must look not just at printing shops but outdoors: into the streets and squares and on the bridges of the city. The cheapest printed products were particularly suited to street sale because they were light and portable. Printers designed them with appealing, eye-catching title page illustrations and bold titles that echoed the advertising patter of the street seller. Street selling of printed material was a natural extension of a wider urban street trade which was a pervasive feature of city life in Renaissance Italy. While cities like Venice hosted ranks of skilled artisans with established workshops and retail shops, a whole other array of commerce was transacted daily from temporary stalls and baskets.[15]

As a large printing and bookselling industry developed into one of the most important commercial facets of the city, Venetians and visitors to the city were exposed to an ever-widening panorama of print. In the early 1490s, the humanist Sabellico famously described a friend becoming distracted and entranced by the many bookstalls as he walked down the great commercial artery, the Merceria, from the Rialto Bridge to Piazza San Marco.[16] Shops exhibited their goods out of

---

13   Important works include Amedeo Quondam, "'Mercanzia d'onore', 'Mercanzia d'utile.' Produzione libraria e lavoro intellectuale a Venezia nel Cinquecento," in *Libri, editori e pubblico nell'Europa moderna. Guida storica e critica*, ed. Armando Petrucci (Rome: Laterza, 1977), 51–104; *idem*, "Nel giardino dei Marcolini: Un editore veneziano tra Aretino e Doni," *Giornale storico della letteratura italiana* 157 (1980): 75–116; Claudia Di Filippo Bareggi, *Il mestiere di scrivere. Lavoro intelletuale e mercato librario a Venezia nel Cinquecento* (Rome: Bulzoni, 1988), especially 123–45; Angela Nuovo and Christian Coppens, *I Giolito e la stampa nell'Italia del XVI secolo* (Geneva: Droz, 2005). See also Nuovo's discussion of bookshops as important places of intellectual exchange in *Il commercio librario nell'Italia*, 266–72.

14   Lorenzo Baldacchini, "Chi ha paura di Nicolò Zoppino? Ovvero: La bibliologia e una 'coraggiosa disciplina'?" *Bibliotheca. Rivista di studi bibliografici* 1 (2002): 187–91.

15   Evelyn Welch, *Shopping in the Renaissance: Consumer Cultures in Italy 1400–1600* (New Haven: Yale University Press, 2005), especially 32–60.

16   M.A. Sabellico, *De latinae linguae reparatione* (1493), discussed in Lowry, *World of Aldus Manutius*, 37.

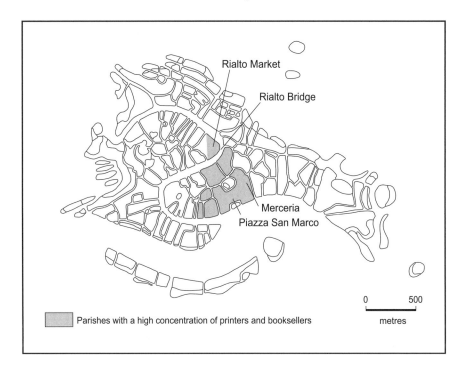

Rialto Market

Rialto Bridge

Merceria
Piazza San Marco

0   500
metres

Parishes with a high concentration of printers and booksellers

Figure 4.1   Principal places for the production and sale of print in sixteenth-century Venice

doors, or stuck up stock lists outside to lure in customers.[17] As the sixteenth century progressed, print penetrated the city ever more. By 1533, there were at least sixteen dedicated booksellers in the city, and at least thirteen printers who may also have sold books, catering to a population of over 120,000, plus visitors.[18]

In 1549 the newly-established Holy Office questioned Angelo Lion, a young merchant, regarding his possession of prohibited books. Lion testified that, 'wishing to read some nice things' (*qualche bella cosa*), he had both borrowed and bought some prohibited works. He had purchased them in one of the city's key civic spaces: Piazza San Marco, the nexus of government and civic and religious ritual,

---

17   Nuovo, *Il commercio librario nell'Italia*, 114.

18   A list of booksellers compiled in 1533 records sixteen *librarii* with shops, alongside thirteen *impressores* (printers) many of whom probably also had some kind of bookselling operation. Archivio di stato, Venice (henceforth ASV), Avogaria di Comun, Notatorio, registro 2054, c. 40v (29 March 1533). However this does not seem to be a comprehensive list as it omits several figures thought to have been active in this period. It also includes no stall-holders or street-sellers, who appear in some quantity on another list of 1567 discussed in note 36. The population of Venice at 1540 was around 129,000 people.

specifically under the clocktower and under the porticoes of the grand buildings that surrounded the square, 'from those booksellers who sell on feast days'. Others of his books, he noted defiantly, he had bought 'publicly in the Merceria' (see Figure 4.1).[19]

The testimony of Priuli, Sabellico and Lion highlights the range of possible places and methods to obtain printed material in Venice during the first century of the industry's development. Below the top level of merchant booksellers, with sophisticated national and even international distribution systems, a rank of lesser known figures toiled to meet the needs of a more exclusively local market. At the bottom end of this spectrum existed the booksellers and publishers without their own shops who operated in the streets from stalls, benches or baskets. For example, there were those who sold on feast days under the porticoes of Piazza San Marco, who may have been working for themselves or sent out by established shopkeepers. At certain times, public spaces such as the piazza were co-opted for a variety of ephemeral sales operations – temporary but nevertheless important sites for the dissemination of printed texts to the city's reading public. Just as Lion bought works 'publicly in the Merceria', buying, reading and listening to printed texts could be very public activities in sixteenth-century Venice.

The Established Trade

Walking through Venice before the commencement of printing there in 1469 the presence of books would have been far less apparent than later on. The city had no large-scale book trade before the arrival of the press, a situation which is thought to have made it easier for printing to flourish there as rapidly as it did.[20] Books were expensive and prized, and moved within networks of scholars and the social elite by means of lending and borrowing.[21] Certainly, in the fifteenth century one might have witnessed auctions of books from deceased estates taking place in Piazza San Marco and at Rialto. But these were held sporadically and the manuscripts sold were costly. There were also some bookshops and *cartolai* (stationers) who sold books located around Rialto and Santa Maria Formosa, the parish to the north of San Marco.[22] At least partially, the dissemination of printed books seems to

---

19   ASV, Sant'uffizio, busta 7, fasc. 18, c. 1v.

20   Mary A. Rouse and Richard H. Rouse, *Cartolai, Illuminators, and Printers in Fifteenth-Century Italy* (Los Angeles: Department of Special Collections, University Research Library, University of California, 1988), 32. In Florence a very strong manuscript trade seems to have stifled the beginnings of the printing industry.

21   For a nuanced reading of Venetian manuscript culture in the era preceding the introduction of print, see Lowry, *Nicholas Jenson*.

22   Susan Connell, "Books and their Owners in Venice, 1345–1480," *Journal of the Warburg and Courtauld Institutes* 35 (1972): 163–86; Anna Melograni, "The Illuminated Manuscript as a Commodity: Production, Consumption and the *Cartolaio*'s Role in

have grown from and fed upon an existing circuit of dissemination of manuscripts, as happened elsewhere. The two kinds of book distribution could co-exist, even collaborate, because there continued to be a market both for manuscripts and print. When a section of the Rialto Bridge collapsed in 1524, Sanudo recorded the great damage done to the shops on the bridge that had thereby lost much of their stock in the water, which included two *cartolai* and two sellers of 'libri a stampa'.[23] Yet as the production of printed books escalated rapidly towards 1500 the press necessitated great changes in book distribution, especially for the growing numbers of small, cheap, unbound pamphlets and *fogli volanti* (literally, 'flying sheets') that had fewer precedents in the manuscript era.

Romano's study of Venetian society in the fourteenth and early fifteenth centuries found that practitioners of the same trade could be spread widely across the city, rather than residing in 'closed occupational neighbourhoods'. Nevertheless, he and other scholars noted clusters of certain trades in certain areas, particularly by the sixteenth century.[24] The location of printers and booksellers would seem to support this latter proposition. After the advent of printing, the shops of printers and booksellers rapidly came to be clustered around and between Rialto and San Marco (Figure 4.1). Printers and print-sellers who did not actually live above their *botteghe* or shops, tended to reside in the same parishes in which these sites were found, and thus in the central Venetian *sestieri* of San Marco and San Polo (administratively, Venice is still divided into *sestieri*, or sixths). Marino Zorzi noted twelve printers active in Venice in 1473, nearly all German and mostly living in the parishes of San Paternian and San Zulian, thus around the Merceria. These parishes had an established German community and were proximate to the German trading house, the Fondaco dei Tedeschi.[25]

Rialto, San Marco, and the zone in between them were the chief loci of trade and commerce in early modern Venice. In these areas merchants and noblemen gathered to negotiate and gossip, along with a great variety of other Venetians and foreigners come to trade, buy and transport goods around the city. The fact that the city's two chief food markets were located at Rialto and San Marco also made them essential stops. Venice made use of the university in the mainland subject

---

Fifteenth-Century Italy," in *The Material Renaissance*, ed. Michelle O'Malley and Evelyn Welch (Manchester: Manchester University Press, 2007), 71–84.

23   Sanudo, *Diarii*, vol. 36, col. 526 (14 August 1524).

24   Dennis Romano, *Patricians and Popolani: The Social Foundations of the Venetian Renaissance State* (Baltimore: Johns Hopkins University Press, 1987), 79–81. Élisabeth Crouzet-Pavan highlighted the localization of certain trades in *Venise triomphante. Les horizons d'un mythe* (Paris: Albin Michel, 1999), especially 228–48, as did Richard Mackenney in *Tradesmen and Traders. The World of the Guilds in Venice and Europe, c.1250–c.1650* (London: Croom Helm, 1987).

25   Marino Zorzi, "Stampatori tedeschi a Venezia," in *Venezia e la Germania* (Milan: Electa, 1986), 122. See also Martin Lowry, "The Social World of Nicholas Jenson and John of Cologne," *La bibliofilía* 83 (1981): 193–218 and Cristina Dondi, "Printers and Guilds in Fifteenth-Century Venice," *La bibliofilía* 106 (2004): 229–65.

town of Padua at this time, but the city's official schools were located at Rialto and San Marco. It is still disputed as to what degree Venetian women used these central zones of the city, given their closer confinement to household and neighbourhood. Yet, it has been argued that lower class *popolane* women at least ranged more widely across the city than often has been supposed and many expeditions would have taken them through these central places.[26]

The activity of producers and distributors of books and pamphlets should be considered in the intimate parochial or neighbourhood context in which it took place. For example, in the 1520s and 1530s, a person walking through the small parish of San Moisè, a minute or two to the west of Piazza San Marco, could have encountered the shops and workshops of some of the most productive printers and publishers of popular material active in the first half of the sixteenth century. These included the fruitful partnership of Francesco Bindoni and Maffeo Pasini at the sign of the Archangel Raphael, Giorgio Rusconi and Guglielmo Fontaneto. Francesco Bindoni's uncle Bernardino was in the Frezzaria, the main thoroughfare through San Moisè. This area continued to be a focal point for the production of popular print throughout the century. From the 1540s, Matteo Pagan, one-time partner of Bernardino's brother Agostino, worked as a printer and an engraver at the sign of Faith in the Frezzaria in San Moisè, while from the 1550s the printers and booksellers Pietro and Domenico de' Franceschi ran a *bottega* in the same street, at the sign of the Queen.[27]

The densely-packed lattice of streets that made up San Moisè, so narrow at times that little light penetrates, was in the early sixteenth century an area devoted to the production and sale of fine artisanal goods, particularly arrows (*frezze* – hence, Frezzaria). As in much of Venice at this time, the parish housed some grand patrician palaces in proximity to dilapidated buildings and courtyards reputed for their poverty and crime. This parish was presumably attractive to producers and sellers of print because of its closeness to the commercial strips of

---

26   Monica Chojnacka, *Working Women of Early Modern Venice* (Baltimore: Johns Hopkins University Press, 2001); Dennis Romano, "Gender and the Urban Geography of Renaissance Venice," *Journal of Social History* 23 (1989): 339–53; Robert C. Davis, "The Geography of Gender in the Renaissance," in *Gender and Society in Renaissance Italy*, ed. Judith C. Brown and Robert C. Davis (New York: Longman, 1998), 19–38.

27   Bindoni and Pasini were active from around 1525 until the 1550s; see Marco Menato, Ennio Sandal and Giuseppina Zappelli, eds, *Dizionario dei tipografi e degli editori italiani. Il Cinquecento* (Milan: Editrice Bibliografia, 1997), i, *s.v.* (only one volume (A–F) has thus far been completed, henceforth referred to as *DTEI*). Rusconi is documented in the parish from 1514 and was active until the 1520s; see Ester Pastorello, *Tipografi, editori, librai a Venezia nel secolo XVI* (Florence: Olschki, 1924), 76. Fontaneto was active from around 1514 until the 1540s, and recorded at San Moisè in the 1533 list cited above (note 18). Bernardino Bindoni was active *c.* 1532–62; see *DTEI*, s.v. Pagan was active from *c.* 1543–60; see Fernanda Ascarelli and Marco Menato, *La tipografia del '500 in Italia* (Florence: Olschki, 1989), 383. On the Franceschi brothers, active from *c.* 1557–76, see *DTEI*, s.v.

the Frezzaria and the Merceria, renowned for the glittering array of goods for sale there. In the seventeenth century, San Moisè became the focal point for writers of manuscript newssheets, because of its proximity to the hub of political power at San Marco.[28]

In the early sixteenth century, a number of printers also established their homes and businesses in the parishes directly bordering San Moisè, such as San Fantin, San Salvador and San Luca. Among these were Giovan Antonio Nicolini da Sabbio and his brothers, Nicolò Aristotile de' Rossi nicknamed 'il Zoppino', Benedetto Bindoni and his brother Agostino and, no more than a few minutes walk away, the engraver and printer Giovan Andrea Valvassore. In the Merceria, probably in the parish of San Zulian bordering Piazza San Marco directly to the north, was the *bottega* of the successful publisher and bookseller Melchiore Sessa, who farmed out printing jobs to many of these smaller printers.[29]

Another cluster of printers, publishers and booksellers was located around the end of the Merceria, at the south-eastern side of the Rialto Bridge, and on the north-western side and on the bridge itself, crowded with shops. These included Giulio Danza, indicted with Bernardino Bindoni and others in 1544 by the blasphemy magistrates, a 'seller of books and paper next to the church of San Giacomo di Rialto', and his brother Paolo, a printer active from around 1511 to 1543. The publisher and *libraio* Giovan Antonio Pederzano, who financed editions printed by Agostino Bindoni, the brothers Nicolini and others, informed readers in his colophons that he could be found at the sign of the Tower at the foot of the Rialto. Later in the century, the publishers Stefano Alessi and Giovanni Bariletto were among those who worked off Campo San Bartolomeo, on the San Marco side of the Rialto, in Calle della Bissa and Calle dei Stagneri respectively.[30]

---

28   Mario Infelise, *Prima dei giornali. Alla origine della publica informazione (secoli XVI e XVII)* (Rome: Laterza, 2002), 25–6.

29   The Nicolini worked in San Fantin from *c.* 1512 until mid century; see Ascarelli and Menato, *Tipografia del '500*, 354. Zoppino was publishing in Venice (located at least some of that time 'sul campo della Madonna di San Fantino') from 1507 until the mid 1540s; see Neil Harris, *Bibliografia dell'*Orlando innamorato (Ferrara and Modena: Istituto di studi rinascimentali/Panini, 1988), 2:87. Benedetto Bindoni was recorded in San Fantin in the 1533 list (note 18) although later relocated to San Geminiano next to Piazza San Marco. He was active *c.* 1520–41. Agostino Bindoni had his shop in San Paternian, immediately north-east of San Fantin, according to the 1533 list. He was active *c.* 1523–58. For both, see *DTEI*, s.v. For the location of Valvassore's shop near the Ponte dei Fuseri in San Luca, see J.D. Passavant, *Le peintre-graveur* (Leipzig: Weigel, 1864), 5:88–9. Valvassore was active *c.* 1530–72; see Ascarelli and Menato, *Tipografia del '500*, 363. Sessa was active from 1505 to *c.* 1562, when his heirs carried on in the same shop; see Ascarelli and Menato, *Tipografia del '500*, 327.

30   For the Danza, see *DTEI*, s.v. For Pederzano, see Ascarelli and Menato, *Tipografia del '500*, 360. Alessi, active from *c.* 1551–60, had previously had a shop at the sign of San Moisè in the Merceria. Bariletto was active from at least 1559–74. For both see *DTEI*, s.v.

Aside from a few printers located in more peripheral parts of the city, such as Aurelio Pincio at San Giovanni in Bragora in the eastern *sestiere* of Castello and Tomaso Ballarin at San Giacomo dell'Orio, west of Rialto, the majority were thus located in Venice's most central, frequented areas.[31] Printers who specialized in images and maps, from large and expensive ones to cheaper varieties, were also, significantly, clustered around the Merceria and the Frezzaria.[32] It is important to stress this clustering of purveyors of print in these central neighbourhoods for several reasons. From the customer's point of view, it suggests that the presence of print in Venice in the sixteenth century was difficult to ignore, encountered whenever one passed through the central arteries of the city. Printed matter from the most elegantly illustrated and richly-bound tome to the flimsiest pamphlet could be found readily. For the printers and print-sellers, the rather intimate, parochial context in which many of them lived and worked mirrored the close-knit nature of the industry from its early days. Their lives were complexly interwoven, held together by ties of kinship and marriage, neighbourhood and shared provenance, friendship and partnership.

Printers pursuing exclusive privileges for their work from the government complained of acute competition and rampant piracy rife in this rapidly growing and most 'capitalistic' of industries.[33] Yet close examination of the editions, of colophons and shared use of woodcut images for example, unearths evidence of an ever-changing constellation of collaborations and informal partnerships, which sometimes must have been concluded verbally.[34] The survey provided above also suggests that printers would have been aware of the work of their 'rivals' via personal contact and observation in the neighbourhood. Moreover, the shifting mass of press-workers moving from shop to shop created immediate links and promoted the flow of information between publishing operations.[35]

It is also vital to note that there was no apparent spatial dislocation between the shops of the more prestigious publishers and printers and those who produced

---

31　Both recorded in the 1533 list (note 18).

32　Gert Jan van der Sman, "Print Publication in Venice in the Second Half of the Sixteenth Century," *Print Quarterly* 17 (2000): 235, suggests this proximity reflected the symbiotic relationships between print and book publishers; see also David Woodward, *Maps as Prints in the Italian Renaissance: Makers, Distributors and Consumers* (London: British Library, 1996).

33　See the petitions collected in Rinaldo Fulin, "Documenti per servire alla storia della tipografia veneziana," *Archivio veneto* 12 (1882): 84–212.

34　Dennis Rhodes, *Silent Printers: Anonymous Printing at Venice in the Sixteenth Century* (London: British Library, 1995), viii. The practice of verbal agreements is signalled by Neil Harris, "Nicolò Garanta, editore a Venezia 1525–1530," *La bibliofilia* 97 (1995): 105–106.

35　In the 1534 blasphemy trial of the press operator Iseppo of Carpenedolo he claimed to have worked 'in all the printing shops in this land' (c. 462r) and painted a vivid picture of the strong interconnections between members of the industry (Archivio storico patriarcale, Venice, Archivio segreto, Criminali Inquisizioni, busta 1, cc. 455r–500r).

significant quantities of popular material. The bookshop of the esteemed publisher Gabriele Giolito was near Rialto, in the parish of Sant'Apponal, while the powerful family of publishers and book merchants, the Giunta, was based on the Merceria, in San Zulian.[36] Geographical proximity may have brought a stronger sense of identity and community to a trade that encompassed men spread over a broad spectrum of specializations and socio-economic levels, and that until the later sixteenth century had no official guild to bring them together. The concentration of members of the print trade in these neighbourhoods also provided strong bases for new arrivals to access, particularly networks of fellow migrants from the same regions of the mainland who were connected further by ties of marriage and business.[37]

Print in the Streets

As suggested, printers and booksellers with their own shops were not the only source of print, particularly of popular print, in Venice. They were supplemented by street-sellers, and we might now consider where they operated in the city. First though, it is important not to draw too great a distinction between bookshops and the street trade in print. In the words of Evelyn Welch, this distinction between shops and the street in Italian cities in this period was a 'defined yet permeable' boundary, not absolute.[38] As we have already seen, bookshops could be open to the street, and advertise their wares outside. Furthermore some stalls were semi-permanent fixtures of the urban landscape, occupying the same position for a number of years, such as those of Giacomo da Trino under the portico at Rialto and Battista Furlan in Piazza San Marco in the 1560s and 70s.[39]

Commercially speaking, it has been estimated that street or ambulant selling occupied only the 'marginal spaces' in the diffusion of print. Angela Nuovo, author of one of the few investigations of Italian bookselling in this period, considered that this mode of commerce was confined to an 'inferior circuit' of dissemination, as it was incapable of distributing books of the most commercial importance, such

---

36   On Giolito, active *c.* 1538–78, see Ascarelli and Menato, *Tipografia del '500*, 374. The Giunta, active throughout the century, were located in the Merceria according to a list of booksellers in ASV, Sant'uffizio, busta 156, unnumbered sheet dated 13 September 1567.

37   Many in the industry came from the Riviera del Salò region on Lake Garda; see Ennio Sandal, ed., *Il mestier de le stamperie e de i libri. Le vicende e i percorsi dei tipografi di Sabbio Chiese tra Cinque e Seicento e l'opera dei Nicolini* (Brescia: Grafo, 2002).

38   Welch, *Shopping in the Renaissance*, 97.

39   For Giacomo (recorded on the 1567 list (note 36) and still working in the same location in 1571), see ASV, Sant'uffizio, busta 156, c. 27r. Another seller, Tomaso Zanier, was recorded in the 1567 list and in 1575 was picked up by the Sant'uffizio for selling unlicensed *fogli volanti* orations; see ASV, Sant'uffizio, busta 39, fasc. 7.

as legal and liturgical editions, generally larger and heavier.[40] Nevertheless, this often very public mode of sale, whereby sellers displayed their works and used their voices to attract customers, hold their attention, and convince them of the merits of a text, is central to understanding the presence of popular print in city life. Susan Noakes, for example, has proposed that street sellers, especially those who combined the sale of pamphlets with some kind of performance, may have been more successful at bridging the gap between printers and customers newer to book buying than more established booksellers and *cartolai*.[41] Noakes based her argument particularly on sources such as the logbook of the Ripoli press in Florence, that shows charlatans and ballad-singers commissioning runs of up to 1000 copies of pamphlet or broadsheet devotional orations and chivalric poems, which they collected in batches presumably to hawk on the road, and which they appear to have sold more quickly than those distributed through shops.[42] The rapid distribution of topical print was made possible precisely because street sellers were not fixed and immobile, but rather free to range through the city's public spaces.

There is considerable evidence that the 'Ripoli model' also applied to some extent in Venice.[43] One need only look at the law passed by Venice's Council of Ten on 12 February 1543. Reflecting a delayed recognition of the potential threat of printing in light of its role in the spread of Lutheranism in the north and reformist ideas in Italy, the law stated that no one could print or sell a book without first obtaining a licence from the Ten. This was because some were printing and also selling books 'many of which are against the honour of God and of the Christian faith, and very dishonest, setting a bad example and [causing] universal scandal'. The councillors especially singled out those 'who sell such books and works, prognostications, stories and songs, letters and other similar things on the bridge of the Rialto, and in other places of this city'.[44] It is also noteworthy that while regular printers and booksellers who contravened the law were to be fined twenty-five ducats, the street vendors were rather to be whipped publicly along the prominent route between the Rialto and San Marco, and then imprisoned for six months. This may have been because they were expected to be too poor to pay the

---

40   Nuovo, *Il commercio librario nell'Italia*, 108–109. However, Ugo Rozzo suggests that *colporteurs* of heretical material in the mid sixteenth century were capable of shifting quite large quantities of sometimes heavy works, leaving deposits of books in certain places, see "Pietro Perna colportore, libraio, tipografo ed editore tra Basilea e l'Italia," *Bibliotheca. Rivista di studi bibliografici* 1 (2004): 51.

41   Susan Noakes, "The Development of the Book Market in Late Quattrocento Italy: Printers' Failures and the Role of the Middleman," *Journal of Medieval and Renaissance Studies* 11 (1981): 46–7.

42   Melissa Conway, *The* Diario *of the Printing Press of San Jacopo di Ripoli, 1476–1484: Commentary and Transcription* (Florence: Olschki, 1999).

43   Martin Lowry, "La produzione del libro," in *Produzione e commercio della carta e del libro, secc. XIII–XVIII*, ed. Simonetta Cavaciocchi (Prato: Le Monnier, 1992), 371, 385.

44   ASV, Consiglio dei Dieci, Parte comuni, filza 32, fasc. 234.

fine, as well as to make a very public example of them in precisely the spaces in which they tended to operate.

The Ten singled out the Rialto as a particular site of print-selling. As was suggested by the earlier quote from Priuli, this was perhaps the city's key location for such activity from very early on in the history of the printing trade. The bridge was an 'obligatory passage' of the city, then the only way possible to cross on foot the Grand Canal that cut the city in half.[45] It was an obvious place for selling and begging, while the nearby Rialto market area was also a focal point for ambulant selling of all sorts of goods. Pamphlet sellers on the Rialto can be found in contemporary poems that depict the bustle of life in this most frenetic part of Venice. In *The Zaffetta's Thirty-One* (*Il trent'uno della Zaffetta*), the courtesan-victim lamented that the news of her malicious gang-rape described in the poem was being spread rapidly around the city, for already she could 'hear the loud cry of the boys/ on the Rialto Bridge [shouting]:/ "Who wants [to buy] the story of the Zaffetta?"'[46] Later in the sixteenth century, the poet Maffio Venier described sellers on the Rialto hawking devotional booklets, prognostications and cheap pamphlets of poetry, while another work described among the many items for sale on the Rialto Bridge – 'so crowded / with people and shops and stalls' – the printed 'song of San Martino / twenty for a *quattrino*'.[47]

Information recorded by the Holy Office or Sant'uffizio, established in Venice in the 1540s, as well as other censoring bodies, is helpful in respect of the prevalence of street vendors and in noting where they were located in the city. In 1551, for example, the printers Giovan Antonio and his father Bernardino Bindoni were punished for printing an anti-clerical story, which the magistrates noted had been sold 'in this city on the Rialto Bridge, and in the Piazza [San Marco]'.[48] A list of bookmen to whom a censorship law was communicated in 1567 included many well-known printers and booksellers, identified by the signs and locations of their shops, followed by a significant number of less familiar characters, such as 'Jacomo di Simon of Venice, bookseller who sells under the portico at Rialto'; 'Julio Bressanin di Bortolomeo the Brescian, has a bench in Piazza San Marco'; 'Nicolò di Bortolomeo Pierio Toschan from Bergamo, sells books on feast days in the Merceria'; and 'Zuane de Anzolo *erbariol* [*erbarolo* = seller of greens] ... sells books at San Salvador, without a shop'.[49]

---

45   Donatella Calabi and Paolo Morachiello, *Rialto: Le fabbriche e il ponte (1514–1591)* (Turin: Einaudi, 1987), 176.

46   Lorenzo Venier, *La Zaffetta* (Catania: Libreria Tirelli di F. Guaitolini, 1929), 47.

47   'Sul ponte de Rialto chi ghe cria: / ... A chi dàghio sti bei officietti? / Un pronostico nuovo ... / che ve mostra i pianetti /... La barceletta de Missier Sbruffaldo!' Maffio Venier, *Canzoni e sonetti* (Venice: Corbo e Fiore, 1993), 177–8. *Viaggio de Zan Padella, Cosa ridiculosa e bela, dond es descrif tug le cose ches vende sul punt de Rialt in Venesia...* (Modena, c. 1580), c. 2r.

48   ASV, Esecutori contra la bestemmia, busta 61, c. 33r.

49   See note 36.

Print was frequently sold alongside many other goods, such as soap, gloves and perfumes.[50] The ballad-singer Ippolito Ferrarese, who commissioned works and probably sold them in Venice and other cities, was apparently as famous for selling soap as for his songs.[51] A sometime publisher and seller of Pietro Aretino's works, Jacopo Coppa, was a charlatan who also sold medicinal drugs.[52] This selling of print alongside other merchandise is difficult to trace in any detail, but it should be kept in mind when assessing the cultural significance of a printed pamphlet in its possible original contexts. Popular print could be a somewhat quotidian commodity, which is easy to forget when one encounters an example of it now in a richly-bound miscellanea in the rare books collection of a library.

The activity of street sellers of print could be greater not just in certain spaces but at certain times, as in the case of Nicolò da Bergamo who sold books on feast days. Street sellers were drawn to the times and places when buying and selling were most common, particularly market days and fairs. Sanudo records that great markets were held weekly at the city's two largest open spaces, Campo San Polo (Wednesdays) and San Marco (Saturdays). Every May, the Ascension festival, the Sensa, was accompanied by a famous fair, during which Piazza San Marco was filled with temporary stalls, and it also attracted many street vendors.[53] In 1565, the magistracy of the Giustizia Vecchia, which controlled commerce, ruled that on feast days under the portico of the Rialto known as the Drapparia, it was permitted only to sell 'images of Saints, books of the epistles, the evangelists, and legends of the saints, Offices, bibles, and similar devout works' and not 'dirty books, plays, and [works] of any other sort that be profane'. Down the Merceria and under the portico at San Marco there might be sold only images of saints and other 'honest and devout' subjects, and not of 'dishonest and shameful things'.[54] Later in the century, when the guild of printers and booksellers had been established, it would attempt to prohibit the activity of vendors of books on holy days, as they cut in on the business of guildsmen who were not supposed to operate during *feste*. The guild complained that these people 'that have not matriculated in this guild' could thereby 'enjoy the fruits belonging to it, without sharing in its burdens and expenses, besides that they are permitted against human and divine laws to

---

50   See the records of ambulant sellers working in sixteenth-century Florence in Gustavo Bertoli, "Librai, cartolai e ambulanti immatricolati nell'Arte dei medici e speziali di Firenze dal 1490 al 1600," pts 1 and 2, *La bibliofilia* 94:2 and 3 (1992): 125–64, 227–62.

51   Vittorio Rossi, "Di un cantastorie ferrarese del secolo XVI," *Rassegna emiliana* 2 (1890): 441.

52   On Coppa, who also published the first edition of Ariosto's *Rime*, see David Gentilcore, *Medical Charlatanism in Early Modern Italy* (Oxford: Oxford University Press, 2006), 302–6.

53   Welch, *Shopping in the Renaissance*, 166.

54   Law copied into the *Matricola dell'Arte dei stampatori e librari di Venezia*, BMCV, MS. Cicogna 3044/Mariegola n. 119, c. 42r.

sell prohibited books, and others that are against God, and to the shame of this guild'.[55]

These last examples demonstrate how printing and bookselling were coming under increasing surveillance and control in the later sixteenth century in the climate of the Counter Reformation.[56] Street selling was especially subject to restriction, also as a result of a general drive on the part of the Venetian authorities to clean up the streets of the city and eradicate (or hide) less respectable forms of activity.[57] Concern about street selling may have been amplified because sellers who combined the sale of cheap pamphlets with some kind of public declamation of their contents were disseminating the works orally as well as in print, making them accessible to anyone who passed in hearing distance.

After the 1543 law made special mention of street sellers, the earliest printing trials brought before the enforcing magistrates (the Esecutori contro la bestemmia) implicate them in a way not hitherto recognized. For instance, the publishers Giovan Antonio and Bernardino Bindoni were punished by the Esecutori along with Paris Mantovano, known as 'il Fortunato', who later registered in Florence as an ambulant seller, commissioned publications in various locations around Italy, and probably was also a performer.[58] While the Bindoni had had the work in question printed, Mantovano was condemned 'for having sold the said letters on the *piazze*, and also for having had printed other works of bad quality'.[59] In 1545, in only the second press-related trial of the Esecutori, a Francesco Faentino 'canta in banco' (literally, 'bench singer') was fined along with two well-known printers (Giovanni Padovano and Guglielmo de Monferrato) for selling a 'very dishonest' work about the God Priapus from his *banco*, and thus in the street.[60]

---

55   Ibid., c. 40v (16 July 1598). The guild was ordered to be established in 1549, although its gatherings are only recorded from 1571.

56   Paul F. Grendler, *The Roman Inquisition and the Venetian Press, 1540–1605* (Princeton: Princeton University Press, 1977).

57   Calabi and Morachiello, *Rialto: Le fabbriche e il ponte*, 25–35, 183 and Welch, *Shopping in the Renaissance*, 34–41. See also Renzo Derosas, "Moralità e giustizia a Venezia nel 1500–1600: Gli Esecutori contro la bestemmia," in *Stato, società e giustizia nella Reppublica veneta (sec. XV–XVIII)*, ed. Gaetano Cozzi (Rome: Jouvence, 1980), 431–528.

58   'Paris di Matteo alias Fortunato of Mantua who sells stories and goods in and around the city of Florence' matriculated on 24 September 1554 in the Florentine guild of doctors and spice-dealers; see Bertoli, "Librai, cartolai e ambulanti", pt. 2, 232. Some of the works Paris commissioned are listed on the database Edit 16 (Censimento nazionale delle edizioni italiane del XVI; http://edit16.iccu.sbn.it/web_iccu/ihome.htm), entry under *editori*: 'Mantovano, Paride.'

59   ASV, Esecutori contro la bestemmia, busta 61, c. 33v (26 September 1551).

60   ASV, Esecutori contro la bestemmia, Notatorio, busta 56, vol. 1, c. 49r (12 August 1545). Faentino was also a publisher of many pamphlets around Italy. Some are listed on Edit 16, entry under *editori*: 'Faentino, Francesco'.

Performing Print

The link between performance and the selling of popular print, manifest in the case of Francesco Faentino, has been alluded to several times. Performance, effectively anything that extended from the hawker's cry to staged songs and recitals, would have drawn more attention to the activity of street-selling. It was a natural development when the advertised item was a printed pamphlet, often containing stories or songs that might hook the attention of listeners. Selling printed pamphlets quickly became a common feature of the appearances of charlatans and other street performers. Evoking the appeal of the performer's expert pitch, the writer Pietro Aretino asked who, 'at the first stroke of their lyre, at the first sound of their voice, at the first advertisement of their merchandise', would not be seduced into buying immediately the remedies and stories that they sold.[61]

Indeed the content of printed pamphlets often was taken directly from the established repertoires of street performers such as ballad singers, buffoons and charlatans. These performers in turn mined the printed corpus for new material, often already of proven popularity with audiences.[62] As in the case of Faentino, it was these same people who sold the works to a gathered audience, possibly after a performance sung or recited sometimes of the same material published in the pamphlets. Such complex and creative borrowings, thefts and collaborations between the producers and disseminators of printed and oral culture, signalled by historians such as Peter Burke, have never been systemically studied in the case of Venice.[63] Considering the oral performance of print in the most frequented urban spaces of Venice reminds us how texts did not just move physically around the city, from printers and sellers to customers. They also moved between oral and literate forms of communication, and from restricted social circles out to wider publics.

Contemporary with the expansion of printing, Venice was experiencing an exceptionally vibrant period of development in public spectacle and performance in the early sixteenth century. Frequent, lavish festivals and processions involved large numbers of the populace as participants or observers, and increasingly worked to frame and direct attention towards Piazza San Marco as the chief site of

---

61   Pietro Aretino, *Lettere* (Rome: Salerno, 1999), 3:326 in a letter dated 1545 to the charlatan-publisher Jacopo Coppa, who had been selling Aretino's works on the piazza in Ferrara.

62   Antonio M. Adorisio, "Cultura in lingua volgare a Roma fra Quattro e Cinquecento," in *Studi di biblioteconomia e storia del libro in onore di Francesco Barberi*, ed. Giorgio de Gregori and Maria Valenti (Rome: Associazione italiana biblioteche, 1976), 24–5.

63   Peter Burke, "Oral Culture and Print Culture in Renaissance Italy," *ARV: Scandinavian Yearbook of Folklore* (1998): 18. See also Gentilcore's brief discussion of the 'symbiotic relationship' between charlatans and printers in *Medical Charlatanism in Early Modern Italy*, 338–9; and my "The Lyre, the Pen and the Press: Performers and Cheap Print in Cinquecento Venice," in *The Books of Venice*, special issue of *Miscellanea Marciana*, ed. Craig Kallendorf and Lisa Pon (New Castle: Oak Knoll Press, 2008), 251–76.

secular and spiritual power.[64] Even theatrical performances held in private palaces or official spectacles within the Ducal Palace often spilled out into the streets, squares and the canals, opening themselves up to a wider audience. There were also abundant designated street performances, particularly during the period of Carnival and the Ascension fair.

The coincidence of performance spaces with some of the city's most important sites for politics and religious ritual was a matter of periodic concern to the Venetian powers. Throughout the sixteenth century, there seems to have been tacit permission for various kinds of street performers to operate in Piazza San Marco, although, as with the street pedlars, there were attempts to restrict their activity to appropriate times and places. A law promulgated in early 1543, within weeks of the Council of Ten's printing regulation, circumscribed the places in which entertainers could operate in the Piazza, instructing them to mount their benches towards the clocktower at the northern edge rather than in the Piazzetta that gave onto the waterfront at the south, closer to the important sites of the Ducal Palace and the Basilica of San Marco.[65] Tommaso Garzoni's *Piazza universale*, first published in the 1580s, famously described charlatans and performers competing for attention on platforms in Piazza San Marco. Visitors to Venice in the early modern period often reported on this phenomenon. According to Garzoni, buffoons and charlatans could be seen in the piazza 'every evening from the twenty-second to the twenty-fourth hours of the day, acting out tales, inventing stories, making up dialogues ... singing impromptu, getting angry at each other, making peace, dying from laughter, becoming angry again, brawling on the stage, making a fuss together, and finally passing out the collection boxes, and getting around to the coins, which they want to swindle out of you with this most polite and courteous chatter of theirs'.[66] In the early seventeenth century, the Englishman Thomas Coryate described five or six stages erected in the Piazza for such performances, plus more humble performers working on the ground.[67]

There were many points of contact between this performance culture and the printing industry. Many of the performers Garzoni mentioned can be identified as authors or publishers of pamphlets in the printing annals of the few decades prior

------

64   Edward Muir, *Civic Ritual in Renaissance Venice* (Princeton: Princeton University Press, 1981), 154, 210–11.

65   'Alcuno che canta in bancho per l'avenir non debbi più montar in bancho per cantar, o altro, dalla piera del bando verso le colone in loco alcuno, ma debano star da li verso il relogio.' Law of 4 January 1543, noted in ASV, Provveditori alla Sanità, Notatorio, busta 729, c. 21r.

66   Tommaso Garzoni, *La piazza universale di tutte le professioni del mondo* (Florence: Olschki, 1996), 2:910.

67   Thomas Coryate and George Coryate, *Coryate's Crudities: Hastily Gobled up in Five Months Travels in France, Savoy, Italy* (Glasgow: MacLehose, 1905; reprint of 1611 edition), 410.

to his writing.[68] Presumably they sold or even gave away such works in the context of performances in these most public spaces of the city. Coryate reported that the mountebanks he witnessed peddled 'oyles, soveraigne waters, amorous songs printed, apothecary drugs, and a commonweale of other trifles', while another restriction on their performances specifically listed their activities including singing, giving out stories and pulling teeth.[69] The records of the Venetian health magistrates throughout the century confirm that charlatans gave away or sold printed copies of their remedies and 'secrets' as a key promotional strategy.[70]

In 1518, during the festival of the Sensa, the Florentine ballad-singer l'Altissimo ('the greatest' or 'highest') visited Venice, where he performed improvised verses with his lyre before a great crowd near the Ducal Palace.[71] A year later, l'Altissimo took advantage of his success in Venice and of the city's burgeoning printing industry when he requested and received a privilege from the Venetian Senate to have some of his works printed.[72] Another telling intersection of print and performance occurred in 1533. At this time, the buffoon Zuan Polo, probably the most famous performer in Venice from the first years of the sixteenth century, set up a platform next to the clocktower in Piazza San Marco, 'dressed up as a poet with a laurel wreath on his head, ... [and] made a speech to all'. At the end of his performance, Polo distributed a printed work he had composed, a chivalric parody called the *Book of Rado Stizoso*, for which the buffoon secured a printing privilege.[73] As we have seen, Piazza San Marco, even specifically under the clocktower, was a noted location for bookselling in Venice at this time. As the example of Polo demonstrates, it was also a site for performance and the more occasional sale of pamphlets.

Performers and sellers of print knew what they were doing when they erected their stages and *banchi* in Piazza San Marco or launched into a performance on the Rialto Bridge or around the market. Visitors to Venice were dazzled by the teeming, cosmopolitan crowds they witnessed in these places. They were the primary locations to seek news and information in this period.[74] Popular print rapidly insinuated itself into the existing matrix of oral and written communication

---

68   Robert Henke, *Performance and Literature in the Commedia dell'Arte* (Cambridge: Cambridge University Press, 2002), 117–18.

69   Law of 2 May 1543, noted in ASV, Provveditori alla Sanità, Notatorio, busta 729, c. 26r.

70   The charlatan Latino de' Grassi was reprimanded by the magistrates in 1551 for selling and giving away a remedy and 'two printed recipes' that it was feared could harm unwary purchasers, in ASV, Provveditori alla Sanità, Notatorio, busta 729, cc. 216r–217r.

71   Sanudo, *Diarii*, vol. 25, col. 391 (10 May 1518).

72   Fulin, "Documenti per servire," 193–4. It is unclear whether l'Altissimo followed through and had these works printed, as no edition survives that can be clearly linked to this privilege. However, several of his works were printed in Venice in the succeeding years.

73   Sanudo, *Diarii*, vol. 58, col. 542 (10 August 1533).

74   See De Vivo, *Information and Communication in Venice*, 89–119, for an exploration of Venice's 'information centres' in the sixteenth and seventeenth centuries.

that centred on these strategic city spaces. Laws, just beginning to be printed in the early sixteenth century, were routinely posted at Rialto and San Marco and also read out loud, along with court summons and banishments. It was presumed that a good number of Venetians would frequent these places in person, or that the news would rapidly make its way from there to the more outlying quarters of the city. The printer Paolo Danza, who wrote and printed songs about recent events of the kind that were sold on the Rialto, was also employed in the 1520s and 1530s to print some government laws, several of which survive preserved in Sanudo's diary.[75] Somewhere between news, commerce and entertainment, public auctions (with auctioneers standing on the columns where laws were also posted and announced) took place at Rialto and San Marco, as did lotteries.[76] In 1522, Sanudo described an upcoming lottery being advertised one morning in Rialto, with 'trumpets and pipes'; printed notices were also used.[77]

Printed items were not only frequently sold or posted on the streets, but produced in close relation to the ebb and flow of daily events. In Florence, the Ripoli logbook shows that charlatans and ballad singers were commissioning pamphlets with a close eye to what was topical or current; for example the *Oration of San Rocco* in time for the saint's feast day, or a news-poem about a recent battle.[78] There are many indications that print was used in the same way in Venice. The testimony of Priuli and Sanudo showed how quickly pamphlet writers and sellers were responding to events in the war of the League of Cambrai. Likewise, the speed of the pamphleteers to spread news of local interest was a recurring literary trope, as shown in the example of the poem about the courtesan La Zaffetta. Pamphlets tapped into popular enthusiasm for events such as the Sensa festival and the day of San Martino.[79] As the century progressed, there were huge outpourings of cheap printed pamphlets, maps and images informing the public about, and commemorating, major events such as the Venetian victory over the Turks at Lepanto or the French King Henry III's visit to Venice.[80] So too, print

---

75 Sanudo, *Diarii*, vol. 50, cols 140–41, 306–7; vol. 58, cols 107–14.

76 Welch, *Shopping in the Renaissance*, 191–208.

77 Sanudo, *Diarii*, vol. 33, col. 408 (8 August 1522).

78 Conway, Diario *of the Printing Press of San Jacopo di Ripoli*, 193, 199. Many such news-poems, often printed with amazing speed in relation to the events they recounted, were printed in Venice. See the collection in *Guerre in ottava rima*, 4 vols. (Ferrara and Modena: Istituto di studi rinascimentali/Pannini, 1989).

79 See, for example, the pamphlet, thought to have been produced in Venice in the early decades of the sixteenth century, entitled *El triompho e festa che fanno le Garzone alegrandosi de la Sensa*, listed in Emilio Picot, "La raccolta di poemetti italiani della biblioteca di Chantilly," *Rassegna bibliografica della letteratura italiana* 2 (1894): 122. Many pamphlets were produced containing songs for the festival of San Martino, a popular celebration in Venice, for example *Canzone e barcellette de san Martino. Con la vita del Pizinino* (Venice: Agostino Bindoni, [before 1558]; listed on the database Edit 16).

80 Iain Fenlon, *The Ceremonial City. History, Memory and Myth in Renaissance Venice* (New Haven: Yale University Press, 2007); Bronwen Wilson, *The World in Venice:*

preserved the texts of successful public performances, even if we do not always possess the testimony of a diarist like Sanudo that allows us to link pamphlet and performance together, print with orality.

Print thus flowed into and intermingled with existing currents of oral and written communication, while a growing wave of printed ephemera, for the most part now vanished, became part of the temporary architecture of the city. The pamphlets and broadsheets that now survive, often only in single copies and scattered through libraries and archives, have been unpicked from the fabric of city life in which they were once closely woven.[81]

Conclusion

Scholars such as Roger Chartier have pointed to the importance of printed matter being displayed in the streets and read out loud in various group settings in the process of 'acculturating' the illiterate or partially literate to the written word in the early years of printing. Reconstructing the presence of these activities in the urban environment encourages us to consider how, at street level, the expansion of printing impinged upon the lives of all Venetians. The ubiquity of ephemeral printed matter and its connections to public performance confirm the need to move beyond calculations of literacy rates and book buying practices if we are to understand this development.[82] If the same material being sold in printed form was also performed independently or recited as part of the process of sale, it would suggest that even if a majority of the Venetian population did not possess full functional literacy, or did not wish to part with the few coins necessary to buy a pamphlet, they were experiencing something of a shared culture via public performance.

This chapter has suggested, too, how the mechanisms of producing and disseminating popular print insinuated themselves into the economic, political and ritual heart of Venice by the early sixteenth century. The city's canals, bridges, public squares and alleyways provided and described an urban geography for the production, circulation and consumption of printed texts. More 'ephemeral' means of diffusion were far from marginal in Venetian life. While the shops of prolific printers and publishers of popular material were clustered tightly around the main commercial thoroughfares of the city, the sellers of print that were probably more visible and familiar to the majority of city-dwellers were those now most invisible to modern scholars. Stall holders and street sellers who sold a variety of goods,

---

Print, the City and Early Modern Identity (Toronto: University of Toronto Press, 2005).

    81   Rose Marie San Juan, *Rome: A City Out of Print* (Minneapolis: University of Minnesota Press, 2001), 26.

    82   See Roger Chartier, "Publishing Strategies and What the People Read," in *idem.*, *Cultural Uses of Print*, 145–82, which disputes overly pessimistic conclusions about the dissemination of printed material based on estimates of literacy rates and the evidence of archival records.

including print, operated in strategic, central locations, and so put themselves in the path of Venetians going about their everyday business. The performance of sacred and secular celebrations additionally drew the city's inhabitants towards the city centre and, as we have noted, ambulant sellers and performers did not fail to take advantage of the crowds on these occasions. The ever-wider dissemination of the printed text cannot be explained in abstract from this urban social setting, and Venice in the sixteenth century cannot be rendered fully without reference to its print culture and to its intricate geographies.

Acknowledgements

I would like to acknowledge the support of the Bennett Travelling Scholarship, the Gladys Krieble Delmas Foundation and the University of London Central Research Fund in funding the research necessary for this chapter, and Claire Judde de la Rivière and Innes Keighren for their advice and suggestions.

Chapter 5

The Counting-House Library: Creating Mercantile Knowledge in the Age of Sail

Deryck W. Holdsworth

Introduction

In the age of sail, trading in words shaped the world of trade. Correct forms of writing and correspondence were central to proper conduct in the world of eighteenth-century commerce, as the many books on writing and commercial conduct testify.[1] In an economy and culture in which accountancy, registration and book-keeping had enormous significance, print culture became a key means of regulation, a basis to the moral conduct of merchants and to the capacity of states as political bodies.[2] Merchants' counting-houses were important repositories for the private and collective knowledge such practices produced. Confidential intelligence from merchants' supercargoes and their agents in ports, notes of conversations with other merchants in coffee houses and at the exchange, and volumes of letter books that recorded copies of instructions and contracts: all these were accumulated in the private realm. More publicly-available knowledge took the form of charts, maps, atlases and merchant law books in the counting-house library. Knowledge about places, products and trade regulation had to be kept up-to-date. This was difficult in the age of sail when communications were unreliable and when, for commercial knowledge as for scientific enquiry, credibility was rooted in gentlemanly status,

---

1   For discussions of this literature, see Miles Ogborn, "*Geographia*'s Pen: Writing, Geography and the Arts of Commerce, 1660–1760," *Journal of Historical Geography* 30 (2004): 294–315; David Hancock, *Citizens of the World: London Merchants and the Integration of the British Atlantic Community, 1735–1785* (Cambridge: Cambridge University Press, 1995); Miles Ogborn and Charles W.J. Withers, "Knowing Other Places: Travel, Trade and Empire, 1660–1800," in *A Concise Companion to Restoration and the Eighteenth Century*, ed. Cynthia Wall (Oxford: Blackwell, 2004), 14–36; Larry Stewart, "Other Centres of Calculation, Or, Where the Royal Society Didn't Count: Commerce, Coffee Houses, and Natural Philosophy in Early Modern London," *British Journal for the History of Science* 32 (1999): 133–53; Miles Ogborn, *Indian Ink: Script and Print in the Making of the English East India Company* (Chicago: University of Chicago Press, 2007).

2   Perry Gauci, *The Politics of Trade: The Overseas Merchant in State and Society, 1660–1720* (Oxford, Oxford University Press: 2001).

in trust in the spoken and written word of others and in the moral not just financial credit worthiness of distant sources and informants.[3]

Crucial among the many printed documents were large dictionaries designed and produced "for the use in counting houses". Produced, consulted and reworked by a constellation of authors, publishers, subscribers and merchant users, this genre of commercial books represented the cultural and economic contours of the mercantile world in their frontispieces and foldout maps. The geographies of knowledge embedded in these dictionaries reveal uneven and changing topographies of information about places and regions, as well as their uses within mercantile communities and networks at work within and between particular cities and across the oceans. Others have shown how different genres of English geography books in the eighteenth century – geographical grammars, gazetteers and encyclopedia or compendiums – had particular textual conventions, and that their incorporation of new information affected their commercially viability: new knowledge, or claims to it, could advance reputations as well as sales.[4] My concern here is with the making and content of commercial dictionaries and with the uses of knowledge that are revealed in this genre of book. Two cities are studied in detail: London in the "long" eighteenth century and Philadelphia in the early nineteenth century.[5] It is my aim overall to make connections between the printed and epistemic spaces of commercial dictionaries as guides to trade and commerce, and the social and calculative space of the counting-house through which these books and their readers circulated. I argue that they are connected via the making of epistemic communities in both cities: groups of merchants and those associated with them which were defined by what they knew and how they knew it.

---

3   Steven Shapin, *A Social History of Truth: Civility and Science in Seventeenth-Century England* (Chicago: University of Chicago Press, 1994); Simon Schaffer, "Defoe's Natural Philosophy and the Worlds of Credit," in *Nature Transfigured: Science and Literature, 1700–1900*, ed. John Christie and Sally Shuttleworth (Manchester: Manchester University Press, 1989), 13–44.

4   Robert J. Mayhew, "Geography Books and the Character of Georgian Politics," in *Georgian Geographies: Essays on Space, Place and Landscape in the Eighteenth Century*, ed. Miles Ogborn and Charles W.J. Withers (Manchester: Manchester University Press, 2004),192–211; Idem., "Materialist Hermeneutics, Textuality and the History of Geography: Print Spaces in British Geography, 1500–1900," *Journal of Historical Geography* 33 (2007): 466–88. The ways in which British geographical textual culture was modified in later eighteenth-century American books is central to Martin Brückner, *The Geographic Revolution in Early America: Maps, Literacy and National Identity* (Chapel Hill: University of North Carolina Press, 2006).

5   In this respect, this study parallels the attention to networks of commercial and scientific information flooding into early modern Amsterdam in Harold J. Cook, *Matters of Exchange: Commerce, Medicine, and Science in the Dutch Golden Age* (New Haven: Yale University Press, 2007).

Ars Mercatoria

The private correspondence of individual merchants is perhaps the closest record of their thinking, tactics and emerging geographical sense of trade possibilities, but few complete series survive. Francesca Trivellato has examined the persistence of letter writing, drawing on over 13,000 letters written by the Sephardic merchant firm Ergas & Silvera, between 1704 and 1746 from their base in Livorno, Italy. Their correspondence network connected them to merchants across Europe and the Mediterranean and as far east as Goa, linking Mediterranean coral and Indian diamonds and re-exporting many other goods.[6] Trivellato argues that the transition from medieval private practice to the more public and open ways associated with early modern capitalism came later than is conventionally argued by Jurgen Habermas who marks it as occurring in the early seventeenth century. Since private merchant records are rare, we need to rely more on printed works, *ars mercatoria*. This broad spectrum of literature pertaining to and written by merchants included treatises on law, dictionaries of commodities, manuals on record-keeping practices, books on penmanship and conversions of prices, weights and measures from one region to another.[7] Of concern in this section are those written in the seventeenth and eighteenth centuries by various British authors that accompanied the emergence of Britain as a global trading nation.

The material symbol of London's ascendancy was its new bourse, mimicking those in Antwerp and Amsterdam. Built in 1585 at the corner of Cornhill and Threadneedle Street by long-time Antwerp merchant Thomas Gresham, London's bourse was given the appellation 'Royal Exchange' by Queen Elizabeth I. In the streets around the Exchange, dozens of coffee houses provided venues for merchants to share information, make deals and seek insurance for their wares and voyages. Newspaper accounts of ships arriving and departing and reports of market conditions in various ports and regions were available. Some coffee houses provided settings for specific regional merchant interests such as the Senegal Coffee House for African slave traders, the Virginia Coffee House for tobacco merchants and Garroway's Coffee House for fur traders.[8] The proximity of these

---

6 Francesca Trivellato, "Merchant Letters Across Geographical and Cultural Boundaries," in *Correspondence and Cultural Exchange in Europe, 1400–1700*, ed. Francesco Bethencourt and Florike Egmund (Cambridge: Cambridge University Press, 2007), 80–103.

7 Pierre Jeannin, *Merchants of the Sixteenth Century* (New York: Harper and Row, 1972); John J. McCusker and Cora Gravetseijn, *The Beginnings of Commercial and Financial Journalism: The Commodity Price Currents, Exchange Rate Currents, and Money Currents of Early Modern Europe* (Amsterdam: NEHA, 1991).

8 Charles Wright and C. Ernest Fayle, *A History of Lloyd's From the Founding of Lloyd's Coffee House to the Present Day* (London: Macmillan, 1928); Bryant Lillywhite, *London Coffee Houses* (London: Allen and Unwin, 1963); Anne Saunders, ed., *The Royal Exchange* (London: London Topographical Society Publication No 152, 1997); Stewart, "Other Centres of Calculation".

information nodes was invaluable to stock jobbers, as noted by Daniel Defoe in his account of Exchange Alley, but visits to these coffee houses were also part of the routine business practices of many merchants.[9] In 1766, for example, the partners of the House of Herries, importers of Spanish wheat and brandy, would walk each Tuesday and Thursday from their Jeffrey Square counting-house off St Mary Ax to visit five specific coffee houses (Tom's, Jerusalem, Garroway's, Lloyd's and John's) at set times, before and after spending two hours at the Exchange. The Herries's directions on how to run a counting-house also included instructions on what clerks then did with that exchange-and-coffee-house information, and what letters were then to be written to which correspondents.[10]

Some merchants had accumulated knowledge of specific trades after a lifetime of work, which they wrote down to pass on to their children, and then decided to publish. One of the earliest examples of this is Gerard Malynes in 1622, writing as a 70-year-old. His intended audiences are signaled in the title: *Lex Mercatoria: Or, the Ancient Law-Merchant. In Three Parts, according to the Essentials of Traffick. Necessary for Statesmen, Judges, Magistrates, Temporal and Civil Lawyers, Mint-men, Merchants, Mariners, and all Others Negotiating in any Parts of the World.* Over 380-pages long, the guide was revised several times throughout the 1680s. It became even larger when the works of Richard Dafforne were incorporated. Dafforne, a bookkeeper who had spent over two decades in Amsterdam with 'real experience among Merchants', shared this knowledge as *The Merchant's Mirrour* (first published in 1636).[11]

Malynes provided pages of information on weights, measures and coinage in the form of tables detailing how products were dimensioned in different places. English merchants would already know that a "Newcastle ton" was 53 cwt, for coal, and that a Winchester wine gallon defined a standard, but what of measures abroad? Take the example of a yard of cloth:

> wheras all Commodities, Wares or Stuffes made of wooll, linen, silke, or haire,
> are measured by the Elle, or Yard, which was taken upon the measure or length
> of the arme, accounting the halfe Elle for Cubits, divided into foure quarters,
> and every quarter into foure inches: wee shall also follow the elle at Antwerp,

---

9   Daniel Defoe, *The Anatomy of Exchange-Alley: Or a System of Stock-Jobbing* (London: Printed by E. Smith, 1719).

10   Jacob M. Price, "Directions for the Conduct of a Merchant's Counting House, 1766," *Business History* 28 (1996): 134–50.

11   Michael F. Bywater and Basil S. Yamey, *Historic Accounting Literature: A Companion Guide* (London: Scolar Press, 1982), 96–8; Richard Dafforne, *The Merchants Mirrour. Or Directions for the perfect Ordering and Keeping of his Accounts. Framed by way of Debitor and Creditor, after the (so termed) Italian Manner: Containing 250 Rare Questions, with their Answers, in forme of a Dialogue* (London: Nicholas Bourn, 1660); Idem., *The Apprentices Time-Entertainer Accomptantly: or A Methodical means to obtain the Exquisite Art of Accomptantship: Digested into Three Parts...Now Reviewed, and in several places Rectified.* Third Edition. (London: W. Godbid, for Robert Horne, 1669).

generally known and observed in all places, in the correspondence and buying of forraine commodities by it, reducing the same afterwards to our elle or yeard, the hundredth ells of Antwerp make in England London 60 Ells for linen cloth, 75 for woolen but 72 in Scotland.

This is followed by a list of 156 mainly European places that between them used 74 different measures of that Elle or Yard. Thirty-six of those places were in the Low Countries, and they used fourteen different measures. There were also eleven different places with six variants in France, 67 places and 35 variants in 'Italie' (a region that stretched to Tripoli, Alexandria and Aleppo), 21 'Eastland' (that is, Baltic) ports with nine variants of measure, nineteen ports and cities in Germany with thirteen variants, ten places in Spain with six measures, and two in Portugal (one of these for its North African outports).[12] Here is one reason why commercial dictionaries were vital books of trade: they provided information that helped manage the problem of geographical variation in the units of commerce.

The same dazzling list of variants came on adjacent pages for weights, volumes and liquid measures. These calculation tables were a vital part of the *lex mercatoria,* but so too were all the rules and regulations that governed trade in certain regions. These led to additional text, editorials on what should be and ought to be, and digests of other authors' opinions. Typical is a chapter devoted to 'the Fishing Trade'. At first glance an account of the expanding Newfoundland trade, Malynes' account includes a summary of the ideas of 'Tobias Gentleman, Fisher-man', who advocated a more aggressive presence for Britain's fishing interests, building ships and manning them, rather than reliance upon foreign supplies of fish. The treatise of Tobias Gentleman, *England's Way to Win Wealth, and to Imploy Ships and Mariners* (1614) included an exhortation for the 'strengthening of his majesties dominions with two principall pillars, which is, with plentie of coine brought in for fish and Herrings from forraine nations, and also for the increasing of mariners against all forreine invasion, and for the bettering of trade, and setting of thousands of poor and idle people on worke'.[13]

A second influential tract was written by Thomas Mun who was involved in the English East India Company. In 1664 he offered advice to his son on the twelve different qualities required in 'a perfect merchant of Forraign trade':

1. A good Penman, a good Arithmetician, and a good Accomptant, expert in the order and form of Charter-parties, Bills of Lading, Invoices, Contracts, Bills of Exchange, and Policies of Insurance.

2. Know the Measures, Weights, and Monies of all foreign Countries, and the Monies not only by their denominations, but also by their intrinsique values in weight and fineness.

---

12   Gerard Malynes, *The Merchant Almanac* (London, 1622), 28–30.

13   Ibid., 174.

3. Know the Customs, Tolls, Taxes, Impositions, Conducts and other charges upon all manner of Merchandize exported or imported to and from the said Foreign Countries.

4. Know in what several commodities each Country abounds, and what be the wares which they want, and how and from whence they are furnished with the same.

5. Understand rates of Exchanges by Bills, from one State to another, and remit over and receive home his Monies to the most advantage possible.

6. Know what goods are prohibited to be exported or imported in the said foreign Countries, lest otherwise he should incur great danger and loss in the ordering of his affairs.

7. Know upon what rates and conditions to fraight his Ships, and ensure his adventures, and be well acquainted with the laws, orders and customs of the Insurance office.

8. Have knowledge of building and repairing of Ships, and for Masts, Tackling, Cordage, Ordnance, Victuals, Munition, and Provisions; together with wages of Commanders, Officers, and Mariners.

9. He ought to have indifferent if not perfect knowledge in all manner of Merchandize or wares, which is to be as it were a man of all occupations and trades.

10. By his voyaging on the Seas to become skilful in the Art of Navigation.

11. Speak diverse Languages, and be a diligent observer of the ordinary Revenues and expenses of foreign Princes, together with their strength both by Sea and Land, their laws, customs, policies, manners, religions, arts; be able to give account thereof for the good of his Country.

12. Lastly, although there be no necessity that such a Merchant should be a great Scholar; yet required, that in his youth he learn the Latin tongue.[14]

Such qualities only came with time, of course, and any commercial dictionary could only register the salient elements. Other insights after a lifetime of trade overseas came from the Levant merchant Dudley North who had learned his trade in Constantinople and in Smyrna. In *Discourses on Trade* (1691) he offered opinion

---

14   Thomas Mun, *England's Treasure By Forraign Trade* (London: Printed by J.G. for Thomas Clark, 1664), 2–5. On Mun and writing upon trade, see Ogborn, *Indian Ink*.

on monetary policy and exchange issues, and drafted what is widely regarded as one of the first statements advocating free trade. This was not the idle ramble of a retired merchant, but someone who was Sheriff of London, and who had served as commissioner of the customs and of the treasury under Charles II and James II.[15]

During the late seventeenth century, the state apparatus on trade and on foreign territories was being codified. In 1660, two important groups were formed, the Council of Trade and the Council of Foreign Plantations. In 1675 they were consolidated as the Committee of the Privy Council of Trade and Plantations; committee members being known as the Lords of Trade and Plantations. William Blathwayt was appointed as their clerk. He compiled a comprehensive atlas of regions covered by these plantations, including one of the first examples of an explicit economic geography in the 'Map of Barbadoes' (1678) which included every sugar plantation, all the windmills and the summary volumes of crops. In 1696, this Committee became the Council for Trade and Plantations and stayed in existence until 1782. Writing on the power of the Lords of Trade and Plantations, the American historian Windham Root noted that 'the close connection of the merchants with the statesmen and officials in the dual task of building and governing the empire is a striking factor in British expansion'.[16]

Overseas trade and domestic commodity production were closely linked. The connections between the English manufacture of woolens and foreign trade is evident in the thorough listing of commodity chains provided by Defoe in his *Compleat English Tradesman* (1727). This identified not only the British towns, counties and regions where specific segments of the English woollen (and metal) industries were based, but also identified the export destinations of these products for traders so 'that he may know what particular Merchants to apply himself to for the sale of those particular Goods which he has occasion to sell'. That knowledge would have taken the trader to particular "walks" on the Exchange where merchants congregated, clustered either by region or commodity. Defoe also identified the regions served by the Muscovy, Turkey and East India Companies or by private traders. He noted, for example, that after arriving in Smyrna or Scanderoon, English woolens were 'carried from thence to Aleppo, and sold there to the Armenian and Georgean Mercants for raw-silks, gauls and drugs and by them carried to Isfahan in Persia and sold there', or that once in Lisbon and Oporto, 'from these two [ports] prodigious quantity of English goods are exported again to the Brasil and also to Goa in the East Indies'.[17]

---

15 Dudley North, *Discourses Upon Trade; Principally Directed to the Cases of the Interest, Coynage, Clipping, Increase of Money* (London: Printed for Tho. Basset, at the George in Fleet Street, 1691).

16 Windham T. Root, "The Lords of Trade and Plantations, 1675–1696," *American Historical Review* 23 (1917): 20–41, quote at 29–30.

17 Daniel Defoe, *The Compleat English Tradesman* (London: Printed for Charles Rivington, 1727) vol 2: 65, 67.

I :Partie vis a vis la pag. I .

Figure 5.1   Frontispiece to Jacques Savary, *Le Parfait Négociant* (1675)

Source: Reproduced with permission from the original held in the W.D. Jordan Special Collections and Music Library, Queen's University at Kingston, Ontario.

Later commercial law dictionaries built on, or borrowed heavily from, the works and words of Malynes, Mun and Dafforne, and in doing so added more geographical information on routes, ports and merchant practices in different regions. Wyndham Beawes, for example, was an influential mid eighteenth-century commercial author. His book's title again signals multiple audiences: *Lex Mercatoria Rediviva; Or, A Complete Code of Commercial Law: Being a General Guide to All Men in Business; Whether as Traders, Remitters, Owners, Freighters, Captains, Insurers, Brokers, Factors, Supercargoes, or Agents.*[18] Beawes's announcement that his work was 'compiled from the works of the most celebrated British and Foreign Commercial Writers', acknowledged amongst others the Frenchman Jacques Savary, whose *Le Parfait Négociant* (1675) had been published in a new edition in 1749. The frontispiece of the Savary book portrays the merchant busy on the quay, ships being unloaded, clerks checking paperwork and the custom house behind (Figure 5.1).

The information Savary provided on specific places was remarkable. In discussions of the Mediterranean trade to Smyrna, for example, there were detailed lists of how many vessels and of what size were in port from Marseilles, Venice, Genoa, and from England and Holland, together with the volume and types of goods each trading company typically brought, and, like Defoe's account, the forwarding roles of Armenians and Jewish brokers.[19] Savary's son, Jacques Savary des Bruslons, inspector-general of the Paris Customs House, composed an alphabetical list of all objects subject to duty, all the words relating to commerce and industry, and the ordinances and rules regarding commerce in France and abroad. This was the starting-point of his *Dictionnaire du Commerce* (1723) and *Universel Dictionnaire du Commerce*, with several editions in the 1730s and 1740s. It was also part of the broader textual project of the Enlightenment when, from the 1720s with Ephraim Chambers's *Cyclopedia* and from the 1750s in France with Diderot and d'Alembert's *Encyclopédie*, encyclopedic texts sought to envision the world anew as they put it to order in print.[20]

In addition to the commercial geographies of the world inventoried by specific merchants, these books and their authors contributed to wider discourses as part of a broad epistemic community that brought together merchants, explorers, political economists, military men and natural philosophers to share and codify their knowledge and advocate policies to further their overlapping interests. As

---

18   Wyndham Beawes, *Lex Mercatoria Rediviva: or, The Merchant's Directory. Being a Compleat Guide to all Men in Business* (Dublin: Printed for Peter Wilson, 1754). The first London edition dates from 1761.

19   Jacques Savary, *Le Parfait Negociant* (Paris: Chez Michel Guignard et Claude Robustel, 1713), 700–704.

20   Robert Darnton, *The Business of the Enlightenment: A Publishing History of the* Encyclopédie, *1775–1800* (Cambridge: Harvard University Press, 1979); Richard Yeo, *Encyclopaedic Visions: Scientific Dictionaries and Enlightenment Culture* (Cambridge: Cambridge University Press, 2001).

William Darity observes, 'Britain was the first to win in pursuit of the grand mercantilist scheme of commercial conquest, naval power, colonialism, slavery and metropolitan industrialization'.[21] For Whiteneck, this success had clear foundations: 'The roots of the *Pax Britannica* of 1815–1873 have their source in the emerging liberal trading community created by the British in the fifty years before the Revolutionary and Napoleonic Wars'.[22] The foundation of this trading community, and the basis of it as an epistemic community, was the forms of knowledge compiled and codified in British works of *ars mercatoria*. At the heart of this, in mid eighteenth-century London, was the prominent eighteenth-century dictionary writer Malachy Postlethwayt. His major work, a 'combined commercial encyclopedia, dictionary and atlas' has been hailed as 'one of the most informative guide-books on the economic facts, thought and policy of the western world in the eighteenth century'.[23]

Malachy Postlethwayt and British Commercial Hegemony

Malachy Postlethwayt's *Universal Dictionary of Trade and Commerce*, in part a translation of and editorial upon the work of Jacques Savary, was the key text of English commercial knowledge from the mid eighteenth century. First published in 1751, it was extended in a second volume five years later, appeared in a second edition in 1757 and its third edition in 1766. A fourth edition of 1774 was based on 'adapting the same to the present state of British affairs in America since the last treaty of peace made in the year 1762'. Postlethwayt's reputation had earlier been associated with his defense of slavery as part of British mercantilist policy in *The African Trade, the Great Pillar and Support of the British Plantation Trade in North America* (1745) and *The National and Private Advantages of the African Trade Considered* (1746), which included a 'new and correct map of the coast of Africa, and all the European settlements'. Over a quarter century later, the map was still being included in Postlethwayt's 1774 edition of the *Universal Dictionary*. Other maps, initially French, were 'improved by Mr. Bolton' for inclusion in Postlethwayt's *Dictionary*.[24]

---

21   William Darity, "British Industry and the West Indies Plantations," in *The Atlantic Slave Trade: Effects on Economies, Societies and Peoples in Africa, the Americas, and Europe*, ed. J.E. Inikori and Stanley L. Engerman (Durham: Duke University Press, 1992), 247–79.

22   Daniel J. Whiteneck, "The Industrial Revolution and the Birth of the Anti-Mercantilist Idea: Epistemic Communities and Global Leadership," *Journal of World Systems Research* 2 (1996), 1.

23   Joseph Dorfman, "Postlethwayt's Pioneer British Commercial Dictionary," in Malachy Postlethwayt's *Universal Dictionary of Trade and Commerce* (New York: August M. Kelley Reprints, 1971), 5, 7.

24   On these works, see Darity, "British Industry and the West Indies Plantations"; Kenneth Morgan, Robin Law, David Ryden and John R. Oldfield, *The British Transatlantic Slave Trade* (London: Pickering and Chatto, 2003).

Figure 5.2   Title page and frontispiece of Malachy Postlethwayt, *The Universal Dictionary of Trade and Commerce* (1751)

Source: Reproduced with the permission of the Trustees of the National Library of Scotland.

The *Dictionary*'s title page acknowledges the 'celebrated Monsieur Savary', but the folio is emphatically centered across the English Channel. The frontispiece is a stunning allegorical representation of British trade (Figure 5.2). Britannia sits at the foot of the classical pillars that fronted the London Custom House; at left, a merchant or customs official stands next to a large barrel on the quayside, beyond which is a forest of masts and sails. Four figures representing the continents bring tribute. Bales and casks and perhaps Neptune (or a figure representing the River Thames) are in the foreground. On the title page is a sketch of sheep being sheared, a reminder that wool and woollen goods were the core export item and key for British mercantile success. Overall, this frontispiece trumps Savary's *Parfait Négociant*, which seems provincial by comparison to London's global reach. Below the main image, in small but legible font, is an extract from a poem by John Gay which also positioned Britain as a trading nation:

> O Britain, chosen Port of Trade,
> May Luxry ne'er thy Sons invade;
> Whenever neighbring States contend,
> 'Tis thine to be the gen'ral Friend.
>
> What is't who rules in other Lands?
> On Trade alone Thy Glory stands.
> That Benefit is unconfin'd
> Diffusing Good among Mankind;
>
> That first gave Lustre to thy Reigns,
> And scatter'd Plenty o'er thy Plains
> 'Tis that alone thy Wealth supplies,
> And draws all Europe's envious Eyes.
>
> By Commerce then thy sole Design.
> Keep that, and all the World is Thine.

This was an extract from a far longer poem, Fable VIII, 'The Man, the Cat, The Dog and the Fly', entitled 'To his Native Country'. Published after Gay's death, an advertisement for these Fables introduced him as 'a man of truly honest heart, and a sincere lover of his country'. For Gay, Britain was, in the four lines omitted from the beginning of the reproduced extract, 'Hail happy land whose fertile ground,/ The liquid fence of Neptune's bounds,/ By bounteous Nature set apart,/ The Seat of Industry and Art'.[25]

Postlethwayt's *Universal Dictionary* brought together phenomenal amounts of material: here, in text and image, was the encyclopedic vision of the eighteenth

---

25   John Underhill, ed., *The Poetic Works of John Gay* (London: Lawrence and Bullen, 1893) vol. 2: 171–6, 138 respectively.

century in commercial form. But his work had more specific purposes too. Part of Postlethwayt's project was to digest recent learned papers on the emerging science of metallurgy. He lived in Pinner Court, off Broad Street near the Pay Office, and several hundred yards from the cluster of coffee houses near the Exchange. Those coffee houses were used for the delivery of mathematical and scientific papers, settings as we have seen for the interaction between traders and financiers on the one hand and scholars and natural philosophers on the other. These were the venues where information for his interest in the new science of metallurgy was acquired: the *Dictionary* cites a dozen English scientists who read and published papers in the Royal Society's *Philosophical Transactions*.

The *Universal Dictionary* came in two volumes, together comprising over one thousand pages of text. It included long entries on the "how-to" aspects of being a merchant (headings on ledgers and book-keeping, for example) and on custom-house regulations for specific commodities. There are numerous entries on places and regions. Typical of such useful commercial place-information is the entry on Oporto, Portugal: 'a famed city and sea-port, on the north side of the Duero about three miles from the sea, is pleasantly built on a rocky ground, that river washing its walls …. The harbour is safe against all winds, but when the floods come down, no anchor can hold the ships, at which time they are forced to squeeze and fasten them to each other along the walls to avoid the fury of the torrent'.[26] Postlethwayt translated a number of entries from Savary, so there are references to French explorers and scientists as well as to France and French territories overseas. Postlethwayt nearly always commented on Savary's entries in essays of his own, headlined as 'Remarks', or 'Observations', or in later editions, 'Further Remarks'. He updated the political situation after the Seven Years War, for example, which diminished France's global holdings, remarking as he did so upon ways in which the British might take advantage of these new economic spaces. The acquisition of former French colonial space was seen as an opportunity for the now larger North American English 'plantation' region. So, in turn, Postlethwayt saw a chance to substitute American lumber for that from Norway, expressed a hope that currants and raisins could be grown in a plantation colony somewhere so as to not rely on supplies from Corfu and the Ionian islands, and, on manufacturers, wrote that 'The great consumptions of German linens, in England and in her Plantations, is one problem in trade balance, and that Irish and Scottish linen production, to substitute and for export to our own plantations, is a positive development'. As we will see later, this advice seems to have been heeded.

His characteristically mid eighteenth-century British antipathy to the French, and its implications for the world of trade, is also evident in the dedication of his second volume to Sir Stephen Theodore Janssen, Baronet, Chamberlain of the City of London. Postlethwayt praises Janssen's role 'as the happy instrument of putting a stop to the general wear of French Cambrics, and instead of this Nation

---

26   Malachy Postlethwayt, *Universal Dictionary of Trade and Commerce*, 2 vols. (London: 1751–56) volume 2, 507.

giving employment to French manufacturers, you promoted that of our English weavers'. He also fought to have the benefits of the Act of 1744–45 extended to ban foreign-wrought silks 'to help give full employ to our industrious Spitafields Manufactures'.

It is clear in many of Postlethwayt's entries that he saw his book working alongside the newspapers and broadsheets which were providing new sorts of spaces for the dissemination of commercial information. Under the entry 'ADVERTISING', he concluded that a new level of normalcy in the provision of information had quickly emerged, as 'great trading and monied corporations', as well as the government, used newspapers to let 'the public know their proceedings'. As he concluded, 'And however mean and disgraceful it was looked on a few years since, by people of reputation in trade, to apply to the public by advertisements in the papers; at present, it seems to be esteemed quite otherwise; persons of great credit in trade experiencing it to be the best, the easiest, and cheapest method of conveying whatever they have to offer to the knowledge of the whole kingdom'.

For Postlethwayt, newspapers were part of the currency in knowledge within coffee houses and counting-houses, alongside his *Universal Dictionary* and works on trade which sought to influence government policy. His work was a crucial part of the mid eighteenth-century making of Britain as a global commercial and imperial power. Postlethwayt's compendium of mercantile knowledge both reflected the rivalry with the French and helped to produce the commercial networks that both underpinned British naval and military victories and benefited from them.[27]

London After the Napoleonic Wars: Anderson's *London Commercial Dictionary*

Over time, commercial dictionaries not only built upon their earlier counterparts but also extended their geographical range bringing new commercial spaces into view. William Anderson's *London Commercial Dictionary and Sea Port Gazetteer* (1819), for example, not only updated Postlethwayt's *Universal Dictionary* and Beauwes's *Lex Commercial*, but shifted its focus to the emerging British Empire. Anderson, the son of an earlier commercial author, cut back Postlethwayt (and thus much of Savary). Some of the North American colonies might have been lost, but trade with America was still critical, and a yet wider world was opening up. For these reasons, Anderson's Preface noted the 'great bulk' and 'high price' of previous dictionaries, especially Postlethwayt, which made them 'inconvenient as books of reference', and argued that 'there is in each of them a considerable mass of materials totally inapplicable to the existing state of trade'.[28] He was keen

---

27   Hancock, *Citizens of the World*.

28   William Anderson, *The London Commercial Dictionary and Sea-Port Gazetteer, Exhibiting a Clear and Comprehensive View of the Productions, Manufactures, and Commerce, of All Nations. The Various Moneys, Weights, and Measures and the Proportions*

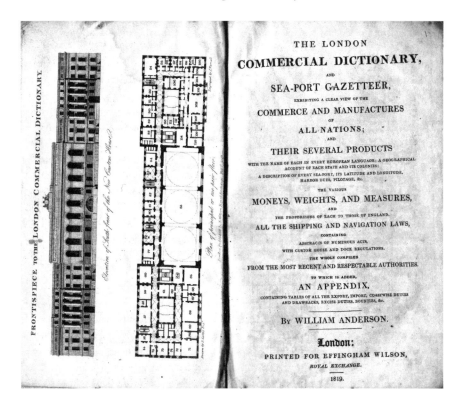

Figure 5.3   Title page and frontispiece of William Anderson, *The London Commercial Dictionary* (1819)

to chronicle all the adjustments in the post-Napoleonic Wars era that prohibited or allowed certain trade goods. Although he reduced the format from folio and adopted a smaller font, it was still 919 pages long (an even smaller font was used for the 1826 second edition, but it was still 888 pages in length).

Postlethwayt had illustrated his *Dictionary* with the London Custom House of old. Anderson's frontispiece was an 'Elevation of the South front of the New Custom House' and 'Plan of principal or one pair floor' (Figure 5.3). The new Custom House, built between 1814 and 1817 after fire had destroyed the earlier buildings, was the jewel in the crown for modern docks that stretched for miles down the Thames. This was the commercial gateway to the City and its business district, and beyond to Britain's empire. David Laing's architectural plan indicated sixty numbered rooms on the principal floor, foregrounding the improved and

---

of Each to Those in England; A Description of All Articles of Merchandize, with Their Marks of Excellency and Names in Every European Language (London: Effingham Wilson, 1819), iii.

enlarged spaces for processing of goods of a now more global trading reach. It also featured the Long Room, 'the most striking object, from its great extent and consequent grandeur of effect, it being nearly the largest room in Europe wherein the roof has no intermediate support: it is 190 feet long by 66 feet wide, and about 55 feet high'. The Long Room saw as much variety each day as the throng of men attending the Exchange, and it was an equally important cog in the circulation of products, with merchants and counting-house clerks seeking the right signatures on paper to get goods released, or duties paid, or fees remitted for re-export. The thickening bureaucracy of global trade is here reflected in this plan and in Anderson's work.

By the early nineteenth century, the complexities of trade across enlarged global networks meant that Anderson's massive compendium contained 475 entries about places, including 379 for the first time containing latitude and longitude locational data, as well as summaries of major imports and exports and specifics of weights and measures associated with that place. Many of those places were in other parts of the British Isles, and in Western Europe. The entry for Oporto now noted:

> OPORTO or PORTO The second commercial city of Portugal, situate about 147 miles N. by E., Lisbon, in W. Lon 8.22. N.Lat. 41.10. The city stands on the right bank of the Douro, and is celebrated for being the port whence the wine to which it gives its name is shipped. The harbour has ample depth of water, but a bar at the mouth prevents loaded vessels from sailing, except at particular periods. Besides the many thousand pipes of wines annually exported, great quantities of cork, lemons, and oranges may be included as chief articles of export.[29]

The entry then summarizes the monies, coins, weights and capacities in use at Oporto compared to Lisbon (for example, 100 alqueries of Oporto = 128 ½ of Lisbon, or = 48 English bushels). This entry, and others like it, is more business-like than that offered by Postlethwayt in the previous century, and provides a broader summary of products and the necessities of trade in order to help a supercargo transact business on behalf of a London merchant.

In contrast to Postlethwayt and Savary, Anderson's gazetteer contained an appreciable Asian presence as a result of the maturing East India trade. He drew on the work of William Milburn, for example, who in 1813 had digested knowledge gleaned from trips to India in the service of the English East India Company into a well-illustrated two-volume treatment of the region, its ports and products.[30] Anderson's *London Commercial Dictionary* also contained over

---

29   Anderson, *London Commercial Dictionary*, vol. 2, 524.

30   William Milburn, *Oriental Commerce; Containing a Geographical Description of the Principal Places in The East Indies, China, and Japan, with Their Produce, Manufactures, and Trade, Including the Coasting or Country Trade from Port to Port; Also the Rise and Progress of the Trade of the Various European Nations with the Eastern World, Particularly That of the English East India Company, from the Discovery of the Passage*

a thousand entries about specific products. For these he drew on Kaufmann's *Dictionary of Merchandize* (1805 and two later editions) and its 402 alphabetical entries, ranging from 'Acacia' to 'Zinc'. Product names were given in the twelve most prevalent languages: French, German, Dutch, Italian, Spanish, Portuguese, Danish, Swedish, Polish, Russian and Latin, as well as English. These multilingual labels became the standard format for product entries in Anderson's *Dictionary*. Anderson's entries on major staples such as tobacco, fish and corn distinguished the best from the ordinary, commented upon what to avoid, and included intriguing detail on esoteric galls, berries and roots used for dyes and in medicine. For dyes, nine different products (brasil-wood, chaya root, cochineal, curcuma, french berries, fustick, logwood, madder and sassafras) and sixteen places are listed. For medicinal drugs, his *Dictionary* contained information on sixteen products sourced and marketed in 47 different places.[31]

Information on the processes of trade took even more space. There were 1246 entries on topics such as the weights, measures and currencies associated with products and regions, as well as with legal and procedural issues. Paralleling the extensive entries found in Postlethwayt, but up-dated with reference to relevant recent acts of Parliament, multi-page entries were devoted to essays on complex topics such as 'Revenue', 'National Debt', 'Quarantine', 'Warehousing', 'Law of Exports' and 'Fisheries'. In the same tradition as Malynes and Mun, some of the extended entries in Anderson's *Dictionary* spoke to contemporary political issues, such as the account of the Ionian Islands, entries on salt and, from the various acts passed under prior monarchs, rules and regulations that prohibited entry or re-export of various items. The tradition of *lex mercatoria* in which Anderson's work stood brought together vast compendia of information to be known and used in metropolitan nodes, distilling as it did so the experience of past times for current use and generating texts which synthesized current knowledge but which always had to be attentive to new knowledge and the changing forms and geographies of trade. The fact that there is an absence of detail on American places in Anderson's work suggests, for example, that a lot of that carrying trade had been conceded to American-based companies in Boston, New York and Philadelphia.[32] They were forging their own spaces of trade and their own epistemic communities.

Round the Cape of Good Hope to the Present Period; with an Account of the Company's Establishments, Revenues, Debts, Assets, &c. at Home and Abroad (London: Black, Parry and Co., 1813).

31   Deryck W. Holdsworth and H.J. Rademacher, "Die worte und welten des handels: waren und die kulturgeographie maritimer räume," in *Das Meer als kulturelle Kontaktzone: Räume, Reisende, Repräsentationen*, ed. Bernhard Klein and Gesa Mackenthun (Konstanz: Universitaetsverlag Konstanz, 2003), 115–42.

32   Robert G. Albion, *Square-Riggers on Schedule: The New York Sailing Packets to England, France, and the Cotton Ports* (Hamden: Archon, 1965); Idem., *The Rise of New York Port, 1815–1860* (New York: Scribner, 1970).

Making American Commercial Knowledge

The mercantile wisdom that had accumulated in London among generations of traders and that was summarized in tomes from Malynes and Mun to Postlethwayt and Anderson was strongly connected with state policy. Science and commerce supported each other, and military and naval success secured individual and national wealth. Constructing a parallel American variant of this nexus would take time. Some efforts came early in the new Republic, including efforts to "reform" and redefine the length of a mile as the basis for the allocation of public lands, and proposals to create a new dictionary with American spelling. As Nathaniel Webster put it, 'as an independent nation, our honor requires us to have a system of our own, in language as well as government'.[33] In these terms, commercial dictionaries in early America sought to distance themselves from their English counterparts as part of the same textual inscription of the new Republic that saw American geographical authors such as Jedidiah Morse produce his own *American Geography* (1789) (See Chapter 9).[34] Nevertheless – and as was also true of English geography books – English commercial culture was still valued and Postlethwayt's book was still important for Americans. It was consulted by Alexander Hamilton as early as 1775, and again for his report on manufactures in 1791. Thomas Jefferson also used the entry on 'CASH' for his discussion of monetary circulation during the war of 1812.[35]

Even so, American editions and updates of European texts soon appeared. Philadelphia-based merchant Stephen Girard's counting-house library contained a copy of one such: W.J. Alldridge's *The Universal Merchant, in Theory and Practice: Improved and Enlarged* (first American edition, Philadelphia, 1797), which included a discussion of coins, rates of exchange in foreign markets and the value of various metals.[36] A second example of the market for American editions of standard European dictionaries and *lex mercatoria* came when Joshua Montefiore's *A Commercial Dictionary: Containing the Present State of Mercantile Law, Practice, and Custom; Intended for the Use of the Cabinet, the Counting-house, and the Library*, which was first published in London in 1803, was modified a

---

33   William D. Pattison, *Beginnings of the American Rectangular Land Survey System, 1784–1800* (Chicago: University of Chicago Department of Geography Research Paper 50, 1957); Andro Linklater, *Measuring America* (New York: Walker, 2002). On Webster and the new language for a new Republic, see Nathaniel Webster, *Dissertations on the English Language with Notes, Historical and Critical, to which is added an essay on a reformed mode of spelling with Dr Franklin's arguments on the subject* (Boston: Isaiah Thomas and Company, 1789).

34   Brückner, *Geographic Revolution in Early America*.

35   Dorfman, "Postlethwayt's Pioneer British Commercial Dictionary", 7.

36   W.J. Alldridge, *The Universal Merchant, in Theory and Practice: Improved and Enlarged* (Philadelphia: Francis and Robert Bailey, 1797). For a note on this in Girard's library, see William F. Zeil, *A Catalogue of the Personal Library of Stephen Girard (1750–1831)* (Philadelphia: American Philosophical Society, 1990), 4.

year later for the American market upon its publication in Philadelphia.[37] Here, too, there are important parallels not just with the world of geography's books but with what Sher has shown of the dissemination and modification of Enlightenment ideas through publishing in Ireland and America, notably in Philadelphia.[38] But how much could change in a year, and what were these new American practices?

One sense of these adjustments comes via the diary of the Philadelphia Quaker merchant Thomas Cope in visiting the writer William Johnson in New York in 1801. Cope's diary entry for 22 July records: 'Wm. is preparing for the press a compendium of the laws of the merchants adapted to the United States. Have no doubt that it will be a valuable work'.[39] The outcome was less an adaptation of English commercial books to American use, but rather a translation of an Italian book, M.D.A. Azuni's *Sistema universale dei Principii del diritto marittimo dell'Europa* (Trieste, 1795, second edition 1796). The American work, translated by Johnson, is still attributed to Azuni but contains an eight-page translator's preface by Johnson, who wanted to educate American merchants on trade in times of war:

> remote from the great theatre of political and military ambition, the United States are happily exempt from that direct pressure which might compel them to take part in the wars of Europe; and they are impelled by interest and policy, to observe a perfect neutrality between the belligerent nations of that continent. To maintain this fortunate and advantageous position, for a long period of time, will be attended with no little difficulty, and may require consummate prudence and address. The surest guide, in this embarrassing situation, is a clear and accurate acquaintance with the principles by which the intercourse between belligerents and neutrals is to be regulated. Whatever relates, therefore to the maritime law of Europe, or what can throw light on the obscure and dubious path of neutral conduct, cannot fail of being interesting to the people of America.[40]

For Johnson, Azuni was the first person to have digested the principles of maritime law into a regular system. This was evident in the 855 pages of the work and its historical reach to Egyptian and Roman practice, through Venetian and Genoan

---

37   Joshua Montefiore, *A Commercial Dictionary: Containing the Present State of Mercantile Law, Practice and Custom. First American Edition With very considerable additions relative to the laws, usages and practice of the United States.* 3 vols. (Philadelphia: James Humphreys, 1804).

38   Richard B. Sher, *The Enlightenment & The Book: Scottish Authors & Their Publishers in Eighteenth-Century Britain, Ireland & America* (Chicago: University of Chicago Press, 2006).

39   Eliza C. Harrison, ed., *Philadelphia Merchant; the Diary of Thomas P. Cope, 1800–1851* (South Bend: Gateway, 1978), 76.

40   M.D.A. Azuni, *The Maritime Law of Europe,* 2 vols. Translated by William Johnson (New York: Riley, 1806), volume I, x.

maritime history and that of the Spanish, Portuguese and Dutch all the way to eighteenth-century conflicts between the English and French. These are case histories of precedent, but they are not yet framed as American examples, except for Johnson's worry about 'the principles by which neutral commerce is to be regulated in time of war', a key issue for American merchants in the first decades of the century.

American merchants in the new Republic inevitably reframed their view of the Atlantic realm in other ways. Once they had to heed British colonial regulations, and their goods flowed to London, Bristol, Liverpool or Glasgow, for use or re-export by British merchants. Now they sought to market their goods directly, both in England and elsewhere in Europe. This required new spaces for gathering and communicating knowledge. There were no longer colonial business sites named The London Coffee House, as had been the case in Philadelphia or in Salem, or the British Crown Coffee House in Boston. They developed instead new Merchant Coffee Houses and grand Merchant Exchanges with libraries of commercial papers and reference material. The merchant's counting-house was nevertheless still important, and in major east coast port cities competing merchants used them to develop trading networks to China, the Caribbean, the Baltic and the Mediterranean.[41]

Stephen Girard, the Philadelphia merchant, owned one such counting-house. He there kept thousands of pages of letters related to his business, political and social life, many of them letters from his supercargoes in places as varied as Cap François in St. Domingue, Montevideo on River de la Plata and Isle de France (Mauritius). Girard also kept a considerable library of maps, charts and books related to places, regions and commodities. Many of those books were lent to his supercargoes for specific voyages. On 4 January 1820, for example, Girard wrote to Joseph Curwin: 'I understand you have a book in England entitled *History of Java* by Raffles Esq; if its price is not extravagant, please send it to me'. Then, in 1822, in a letter to John Garland, Girard stated that 'it is with pleasure that I loan you the *History of Java* by Thomas Stamford Raffles Esq'. On appointing Garland as the supercargo of the ship *Helvetius*, Girard wrote 'I am giving you the opportunity to commence your commercial career'.[42]

An earlier exchange with his supercargoes reveals the problems of knowledge American merchants faced during the embargo days leading up to the War of 1812. In December 1809, with the European ports off limits because of blockades and embargoes, Girard turned to South American and Chinese markets for his fleet of ships. When his ship *Voltaire* arrived in Montevideo, March 1810, one of the

---

41   Susan S. Bean, *Yankee India: American Commercial and Cultural Encounters with India in the Age of Sail, 1784–1860* (Salem: Peabody Essex Museum, 2001); Jonathan Goldstein, *Philadelphia and the China Trade 1682–1846: Commercial, Cultural, and Attitudinal Effects* (University Park: The Pennsylvania State University Press, 1978); Albion, *Square-Riggers on Schedule*.

42   Zeil, *Catalogue of the Personal Library of Stephen Girard*, xx.

supercargoes Edward George sent Girard a precise set of observations on the rules of trade in this Spanish realm, long off-limits to American ships:

> All vessels must be consigned to a Spanish house who must enter the whole of the cargo at the Custom House before the expiration of 24 hours, under severe penalties in case of non-compliance.

> All kinds of goods are admitted except those held as a priviledge of the King (such as Tobacco and quicksilver) and articles of Spanish production (such as wines, brandy, etc) which are strictly prohibited.

> Its productions are confined to a few articles which are Hides, Tallow, different kinds of wool, copper, tin, Jesuit Bark and Spices some of which are brought a great distance from the interior. The quantity of hides on hand is immense.

> The articles of import are principally for wearing apparel, few articles excepted such as Salt, Tin, Steel, Glass and Hollow ware, cutlery, Lumber, Salt Fish Etc.[43]

These are the same kinds of details, of course, that the East India Company employee William Milburn had shared in his book *Oriental Commerce* over how to deal with various Arab traders in Red Sea ports, or what to watch out for in different places in the Indian Ocean. But by their nature, these were new details too, still-private knowledge for the use of Girard alone or for him to share with his other supercargoes heading to the region on other ships in his fleet. The predicament of having arrived with the wrong trade goods is clear in other segments of George's letter, one of which has echoes of Postlethwayt's lobbying for cheap versions of German goods:

> The market is now glutted with all kind of English goods from the Brazils &c and indeed the English imitate the manufactures of every country: Brittaina's and other German goods are brought here from the Looms of Ireland and Scotland extremely well imitated and cotton goods are selling at and under prime cost; therefore few dry goods can be sent here from America to give a profitt.[44]

One variant of linens was called Osnaburgs, a standard element of Scottish mills by this period, though seen as of inferior quality to the real German item. The dilemma for the American traders was obvious:

---

43 John B. McMaster, *The Life and Times of Stephen Girard, Mariner and Merchant* (Philadelphia: Lippincott, 1918). See also Robert D. Schwarz, *The Stephen Girard Collection: A Selective Catalog* (Philadelphia: Girard College, 1980).

44 McMaster, *Life and Times of Stephen Girard*, 150.

we have found a number of English vessels which have glutted this market
with their manufactured goods for which the Spaniards begin to take a fancy on
account of their cheapness, and of the scarcity of the articles which they formerly
received from Spain. The British having no other places now for the vent of their
goods are obliged to sell them at a great loss, and therefore oblige the Americans
to lower the prices of the other European merchandize which they import here.
Several of our countrymen have been lying many months without being able to
obtain the prices which they ask for … we are determined to proceed to Buenos
Ayres that city offering a better prospect of a quick disposal of our cargo.[45]

This is a good example of how specific instructions from a merchant to the
supercargo let alternate decisions be effected, based on how long to wait in a place
in the hope that markets improved, or to go to another place and hope it provided
a market that was profitable. Girard's decision to load opium or specie and go
all the way to China in order to bring back tea and silk was his own decision,
not a policy recommendation laid out in a new American compendium of advice
to all merchants. Yet he must have taken this decision based on the knowledge
accumulated in his counting house, and by working between printed works of *ars
mercatoria* and letters from his agents and supercargoes. This example underlines
American merchants' need for accurate textual information in order to make
commercial decisions connect productively with the changing global geographies
of trade.

Conclusion: Knowledge at the End of the Age of Sail

Although, as we have seen, many Americans first relied on European translations,
they started to craft their own textual and cartographic frames of reference for
understanding regional and global patterns of trade. They began to do so as part
of the emergent sciences of oceanography, charting with others the winds, tides
and currents important to trade in the age of sail.[46] It is an irony, then, that as
with their commercial texts, such endeavours should come at the end of that age.
The impetus for new maritime knowledge was commercial, as more American
merchants sought out South American markets, political, since the American navy
needed to keep a visible presence in those waters to signal their friendship with
new commercial allies, and scientific.[47]

    At the core of this conjunction of new knowledge and new texts was Matthew
Fontaine Maury, the father of American oceanography and a leading figure in the

---

45   Ibid., 148.
46   Michael S. Reidy, *Tides of History: Ocean Science and Her Majesty's Navy*
(Chicago: University of Chicago Press, 2008).
47   Chester G. Hearn, *Tracks in the Sea: Matthew Fontaine Maury and the Mapping
of the Oceans* (Camden: McGraw Hill, 2002).

global science of hydrography.[48] Maury had already updated earlier American maritime works such as Bowditch's *Coastal Navigator*, but from experience as a midshipman and employment at the US Naval Observatory in Washington, he started to compile knowledge of the best routes, drawing also on the experiences of American whalers who had undertaken four-to-five-year voyages into the Pacific.[49] The result was his *New Theoretical and Practical Treatise on Navigation*, published in Philadelphia in 1836. As he updated editions, so Maury added appendices designed to place America more centrally in the networks of global enterprise. As he began to gather such information, Maury imagined it would be quickly available: 'a flight up into the garrets, and a ransacking of time-honored sea-chests in all the maritime communities of the country for old log-books and sea journals'.[50] However, ransacking the past for its textual traditions did not yield the information he hoped for. He turned instead to a more thoroughly military intelligence-gathering exercise that would let him better understand that 'the sea, with its physical geography, becomes as the main spring of a watch; its waters, and its currents, and its salts, and its inhabitants, with their adaptations, as balance-wheels, cogs and pinions, and jewels'.[51] He arrived at this finely-tuned understanding by having American ship captains and navigators take precise measurements on their voyages: 'They were told that if each one would agree to cooperate in a general plan of observations at sea, and would send regularly, at the end of every cruise, an abstract log of their voyage to the National Observatory at Washington, he should, for so doing, be furnished, free of cost, with a copy of the charts and sailing directions that might be founded upon these observations'.[52] The wider use of these maps came in the form of Maury's *The Physical Geography of the Sea* (1855), by which he intended to make merchant shipping more efficient and profitable. He did so by advocating the use of new routes rather than by crossing the Atlantic three times to catch winds as historically dictated in traditional routes. Merchants that followed his suggested "tracks in the sea" arrived in record time; those that didn't, faltered, and sometimes worse. Although he drew on prior scholarship, this was a world being measured by American whalers and clipper captains as they linked the world's oceans. Maury shared the concerns of Postlethwayt, Anderson and

---

48   D. Graham Burnett, "Matthew Fontaine Maury's 'Sea of Fire': Hydrography, Biogeography, and Providence in the Tropics," in *Tropical Visions in an Age of Empire*, ed. Felix Driver and Luciana Martins (Chicago: University of Chicago Press, 2005), 113–35.

49   For a parallel use of tacit expertise and in-the-field study of tides to inform earlier scientific and textual study of the Gulf Stream, see Charles W.J. Withers, "Where Was the Atlantic Enlightenment? – Questions of Geography," in *The Atlantic Enlightenment*, ed. Susan Manning and Francis D. Cogliano (Aldershot: Ashgate, 2008), 37–60.

50   Matthew F. Maury, *Treatise on Navigation* (Philadelphia: Biddle, 1836), v.

51   Maury, *Treatise on Navigation*, 54; idem., *The Physical Geography of the Sea* (New York: Harper and Brothers, 1855).

52   Matthew F. Maury, *Explanations and Sailing Directions to Accompany the Wind and Current Charts* (Philadelphia: Biddle, 1855), xii.

others. But new worlds of commerce and science required new epistemologies, new words and new books.

The nineteenth century witnessed the end of the age of sail and with it the need for counting-house compendia, and, indeed, the very label "counting-house". Shipping companies developed packet lines with scheduled departures between the same ports. Reliability increased as steam supplanted sail. The vagaries of tide and wind diminished. The advent of the transatlantic telegraph in 1866 meant that prices for cotton in New Orleans or New York could be known the same day in Liverpool or in Hamburg. Newspapers routinely printed "prices current", and specialized business magazines emerged such as the *Wall Street Journal* in New York or the *Financial Times* in London. Offices managed flows of paper, not encyclopedic compendia. Merchants and clerks left at night to go home, rather than retire upstairs to rooms above the counting-house. Coffee houses transformed into business clubs, and merchants organized Chambers of Commerce or Boards of Trade to take positions on trade issues and to lobby local and national government. Old merchant exchanges were reorganized into specialized commodity exchanges. In London, in Philadelphia and across the world, merchants no longer needed counting-house libraries nor the same sorts of books of commerce to bring the world to market.

Acknowledgements

I am grateful to Juraj Kittler, Susan Friedman, Charlotte Houghton, Brad Hunter, Hank Rademacher, Fiona Black and the editors for comments on an earlier draft.

Chapter 6

Printing Posterity: Editing Varenius and the Construction of Geography's History

Robert J. Mayhew

Introduction: Editing, the History of the Book and Disciplinary History

Debates amongst geographers about how to write disciplinary histories were catalysed by the appearance of David Livingstone's *The Geographical Tradition* (1992). In this context, geographers questioned Livingstone's attitude to women in the history of geography, whilst further critiques have taken a post-colonial stance and criticised *The Geographical Tradition* as "Eurocentric". Most radical of all have been postmodernist questions about the epistemological possibility of doing disciplinary history at all.[1] In these debates, as in parallel debates in other realms of disciplinary and intellectual history, questions about "editing" taken in a metaphorical sense have been to the fore, with queries about how to write intellectual history, about what is included, excluded and occluded by narrative constructions. In short, disciplinary histories have become highly self-conscious about what is edited "in" and "out" of their narratives.

More recently, and complementing this, historians of geography, in common with historians of science and intellectual historians, have addressed themes from the sociology and history of the book. This has led to a concern with the categories of author, publisher and audience which have preoccupied luminaries of book

---

1   David N. Livingstone, *The Geographical Tradition: Essays in the History of a Contested Enterprise* (Oxford: Blackwells, 1992) and, for commentaries, Felix Driver et al., "Geographical Traditions: Rethinking the History of Geography," *Transactions of the Institute of British Geographers* NS 20 (1995): 403–22 and Felix Driver, Robert J. Mayhew and David N. Livingstone, "Classics in Human Geography Revisited: *The Geographical Tradition*," *Progress in Human Geography* 24 (2004): 227–34. See also Mona Domosh, "Toward a Feminist Historiography of Geography," *Transactions of the Institute of British Geographers* NS 16 (1991): 95–104; Janice Monk, "Women, Gender and the Histories of American Geography," *Annals of the Association of American Geographers* 94 (2004): 1–22; Avril Maddrell, *Complex Locations: Women's Geographical Work in the UK, 1850–1970* (Oxford: Wiley-Blackwell, 2009); James Sidaway, "The (Re)making of the 'Western' Geographical Tradition: Some Missing links," *Area* 29 (1997): 72–80; Clive Barnett, "Deconstructing Derrida: Exposing Context," *Transactions of the Institute of British Geographers* NS 24 (1999): 277–93.

history such as McKenzie, Darnton and Chartier.[2] As a result, studies have looked to the specific conditions of authorship in different times and places, often inspired by Foucault's genealogical approach to authorship;[3] to the economic conditions of print culture over time, bringing ethereal and romantic conceptions of authorship to earth by grounding them in the economic realities of publishing;[4] and to the differential reception of works by different audiences across time and space.[5] Such concerns have culminated in a sophisticated fusion of intellectual, scientific and print history as exemplified in Adrian Johns's *Nature of the Book* (1998) and James Secord's *Victorian Sensation* (2000). The history of geography is also showing interest in the ways in which attending to the cultural history of textuality can enhance understanding. Miles Ogborn, for example, analysing the relationship between geography, penmanship, print and the circulation of power/knowledge, whilst Martin Brückner has interwoven the history of writing and literacy with the history of early American geography, arguing that the didacticism of language learning was transferred into the geographical primers (or "grammars", as they were known) of the time.[6] Questions of the reception of geographical knowledge have

---

    2   These categories structure the only "textbook" in the new history of the book, David Finkelstein and Alistair McCleery, *An Introduction to Book History* (London: Routledge, 2005), chapters 4–6. They are also deployed by book historians Jonathan Topham, "Scientific Publishing and the Reading of Science in Nineteenth-Century Britain: A Historiographical Survey and Guide to Sources," *Studies in the History and Philosophy of Science* 31 (2000): 559–612 and Adrian Johns, "Science and the Book in Modern Cultural Historiography," *Studies in the History and Philosophy of Science* 29 (1998): 167–94. In geography, the same division guides the "textual" approach in Robert J. Mayhew, "The Character of English Geography, c.1660–1800: A Textual Approach," *Journal of Historical Geography* 24 (1998): 385–412.

    3   Michel Foucault, "What is an Author?" in *Essential Works of Foucault, 1954–1984, Volume 2: Aesthetics*, ed. James D. Faubion (Harmondsworth: Penguin Books, 1998), 205–22.

    4   Jerome McGann, *A Critique of Modern Textual Criticism*, 2nd edition (Charlottesville: University of Virginia Press, 1992); Donald McKenzie, *Making Meaning: "Printers of the Mind" and other Essays* (Amherst: University of Massachusetts Press, 2002); James Raven, *The Business of Books: Booksellers and the English Book Trade, 1450–1850* (New Haven: Yale University Press, 2007).

    5   Hans Robert Jauss, *Towards an Aesthetic of Reception*, trans. Timothy Batli (Minneapolis: University of Minnesota Press, 1982); Wolfgang Iser, *The Act of Reading: A Theory of Aesthetic Response* (Baltimore: Johns Hopkins University Press, 1980). A reception history inspired by these theoretical concerns is Nicolaas Rupke, *Alexander von Humboldt: A Metabiography* (Frankfurt: Peter Lang, 2005).

    6   Miles Ogborn, "*Geographia*'s Pen: Writing, Geography and the Arts of Commerce, 1660–1760," *Journal of Historical Geography* 30 (2004): 294–315; Idem., *Indian Ink: Script and Print in the Making of the English East India Company* (Chicago: University of Chicago Press, 2007); Martin Brückner, *The Geographic Revolution in Early America: Maps, Literacy and National Identity* (Chapel Hill: University of North Carolina Press, 2006), esp. chapter 4.

been to the fore in Charles Withers's study of Mungo Park and Innes Keighren's recovery of the reception of the work of Ellen Churchill Semple.[7]

Such studies have been less concerned with issues of editing, perhaps because the "new" approach to book history seems to have paid rather less attention to editing, in general preferring to subsume editing either under the authorship function or putting it under the aegis of the bookseller/publisher. Thus, Iliffe's study of editing in the late seventeenth and early eighteenth centuries looks at editing as an authorship function in itself and as inflecting the idea of the author: 'the manifestation of the "editor" was intimately bound up with the appearance of the "author", and should be taken into account in the history of the latter'.[8] On the other hand, Secord's meticulous work on the history of *Vestiges of the Natural History of Creation* deals with editing only sparsely, consistently seeing it as the product of the demands of John Churchill, Chambers's publisher.[9] Considerable philosophical attention has been paid to paratextual features in the wake of Derridean interest in margins, notably in the work of Girard Genette.[10] One editorial paratext in particular, the footnote, has received extensive recent attention, most notably from Anthony Grafton, but also in works aimed at a popular audience.[11] Neither set of works pays attention to the intentionality of the editor in creating the suite of print features which comprise an edition, in Genette's case due to a poststructural disregard for authorial intention, in Grafton's due to the focus on one type of paratext to the exclusion of others. To the extent that editing has been studied at all as a set of print processes intended and managed by an individual, it has largely been as part of a more traditional conception of bibliography and has

---

7   Charles W.J. Withers, "Memory and the History of Geographical Knowledge: The Commemoration of Mungo Park, Explorer," *Journal of Historical Geography* 30 (2004): 316–39; Innes Keighren, "Reading the Reception of Ellen Churchill Semple's *Influences of Geographic Environment*" (PhD diss., University of Edinburgh, 2007).

8   Rob Iliffe, "Author-Mongering: The 'Editor' Between Producer and Consumer" in *The Consumption of Culture 1600–1800: Image, Object, Text*, ed. Ann Bermingham and John Brewer (London: Routledge, 1995), 167.

9   James Secord, *Victorian Sensation: The Extraordinary Publication, Reception and Secret Authorship of* Vestiges of the Natural History of Creation (Chicago: University of Chicago Press, 2000), 145ff; Secord discusses the editions of George Combe's *Constitution of Man* as publisher-driven at 70ff.

10   Jacques Derrida, *Margins of Philosophy*, trans. Alan Bass (Chicago: University of Chicago Press, 1984); Girard Genette, *Paratexts: Thresholds of Interpretation,* trans. Jane E. Lewin (Cambridge: Cambridge University Press, 1997). On paratexts, see also Kevin Jackson, *Invisible Forms: A Guide to Literary Curiosities* (New York: St Martin's Press, 1999).

11   Anthony Grafton, *The Footnote: A Curious History* (London: Faber & Faber, 1997). See also Chuck Zerby, *The Devil's Details: A History of Footnotes* (New York: Simon & Schuster, 2002).

concerned practical problems in the editing of scientific and geographical texts for modern standard editions.[12]

The intellectual-historical significance of editing remains the preserve of literary criticism, then, with the exception of an important recent contribution from Neil Safier in *Measuring the New World* (2008). Safier looks to the ways in which maps and natural histories relating to South America in general and the La Condamine expedition in particular were edited in the print culture of ancien régime France. Safier's engagement with editing is meticulous in its tracing of the exclusions, inclusions and occlusions performed as works progressed from oral and scribal traces in South America to their printed end products in Enlightenment Paris. But Safier tends to see the editorial process through a (broadly) post-colonial lens critical of the Enlightenment, arguing that editing served to marginalise and silence indigenous and creole voices.[13] This ignores the extent to which editing as a practice in eighteenth-century intellectual life was an interventionist process, one which was far more comfortable excising and reworking original authorial intentions than the era after the romantic adulation of the sovereign author.[14] As this chapter will show, editing in the eighteenth century was quite as happy silencing or reworking the printed and scribal work of the denizens of Europe as those indigenous voices to whom Safier attends. It shows, furthermore, the extent to which, emergent in the mid eighteenth century, was a genuine editorial respect for the views of those from different times and places. The enlightened print culture Safier critiques also forged for the first time the canons of editorial respect for the other, even as it altered their words. It could happily run both projects alongside one another, thereby pointing eloquently to its alterity from our present expectations of the role of the editor.

---

12   See three titles in the "Conference on Editorial Problems" series: Joan Winearls, ed., *Editing Early and Historical Atlases* (Toronto: University of Toronto Press, 1995); Germaine Warkentin, ed., *Critical Issues in Editing Exploration Texts* (Toronto: University of Toronto Press, 1995); Trevor H. Levere, ed., *Editing Texts in the History of Science and Medicine* (Toronto: University of Toronto Press, 1982). See also Michael Hunter, *Editing Early Modern Texts: An Introduction to Principles and Practice* (London: Macmillan, 2006). This is not a blanket statement: Adrian Johns briefly discusses editing in *The Nature of the Book: Print and Knowledge in the Making* (Chicago: University of Chicago Press, 1998), 29–31. In geography's history, editing has received attention in Robert J. Mayhew, "Was William Shakespeare an Eighteenth-Century Geographer? Constructing Histories of Geographical Knowledge," *Transactions of the Institute of British Geographers* NS 23 (1998): 21–37; Idem., "Textual Editing and the Construction of Geography's History: The Vexed Question of the Definitive," *Journal of Historical Geography* 30 (2004): 559–563.

13   Neil Safier, *Measuring the New World: Enlightenment Science in South America* (Chicago: University of Chicago Press, 2008), 123–65 and 200–235.

14   Jerome McGann, *The Romantic Ideology: A Critical Investigation* (Chicago: University of Chicago Press, 1983); William St Clair, *The Reading Nation in the Romantic Age* (Cambridge: Cambridge University Press, 2004).

This essay springs from the desire to show that the fusion of intellectual history and the history of science with the burgeoning awareness of print culture as an historical phenomenon can be furthered by taking editing seriously as a source of insights. Editing matters to historians of science and to intellectual historians at the literal level as well as the metaphorical. Paying attention to editing can deepen the move from a concern with the *history* of the book to the *historical geography* of the book by showing how spatially differentiated editorial practice was, this being a contributor to the geographically differentiated production and reception of knowledge. The case to support these contentions is made with regard to the English editing of a major work in the geographical canon, Bernhard Varenius' *Geographia Generalis* (1650). As we will show, how Varenius was edited varied enormously over time, and resulted in very different positionings of his work on a spectrum between cheap hackwork and revered classic. Editing led to widely varied appearances for the same work from a blank text to a complex page with marginal directions, footnotes, keyed labels and multiple typefaces and sizes. In short, the way Varenius was edited and the contexts in which this occurred had important consequences for the costs of producing the book, for the sense of the authorial status of the Varenian copytext, and for the implied audience at which the *Geographia Generalis* was pitched. The way in which texts like Varenius were edited interacted with and impacted upon each component in the tripartite style of analysis revolving around authorship, production and reception which intellectual historians have taken over from the new history and sociology of the book and furthered the geographical inflection of these processes.

Introducing Varenius

Bernhard Varenius' *Geographia Generalis*, published in Amsterdam by the Elsevier press in the year of his death, 1650, has been accorded an extraordinary status in the history of geographical thought, especially since his work was singled out for praise by Alexander von Humboldt and Paul Vidal de la Blache.[15] The critical reception of Varenius from Humboldt to the present has emphasised his modernity; that his mathematically-grounded form of geographical enquiry leads us into a version of the discipline recognisable to geographers practicing today: 'the intellectual divide separating ancient and medieval geography from modern geography is generally taken to be the publication of the *Geographia Generalis* by Bernhard Varenius'.[16] The fact that Isaac Newton produced two editions of the

---

15 Alexander von Humboldt, *Cosmos: A Sketch of the Physical Description of the Universe: Volume 1*, trans. E.C. Otte (Baltimore: Johns Hopkins University Press, 1997), 66–7; Paul Vidal de la Blache, "Le Principe de la Géographie Générale," *Annales de Géographie* 5 (1895–96): 129–42.

16 Fred Lukermann, "The *Praecognita* of Varenius: Seven Ways of Knowing," in *Text and Image: Social Construction of Regional Knowledges*, ed. Anne Buttimer, Stanley

Geographia Generalis has cemented this depiction of Varenius.[17] Varenius has been assumed, implied or directly asserted to have been widely influential in his day, the suggestion being that *Geographia Generalis* was in the hands of all learned inquirers with an interest in geography. Rebutting the division between general and special geography, for example, Preston James both credits the popularisation of this distinction to Varenius and opines that the *Geographia Generalis* 'remained the standard text for geography in universities for more than a century'.[18] In sum, huge rhetorical weight has been placed upon Varenius; he is seen as ushering in, prefiguring or standing as a symbol for geography as a science, as mathematical, and as numerical, whilst also being the most popular textbook in seventeenth- and eighteenth-century European and North American geographical education. The upshot of this is that the name of Varenius becomes a signifier for the signified of 'scientific geography', a status evidenced in the decision of the US's National Center for Geographic Information and Analysis to name its new project to advance GIS "Project Varenius".[19]

Print History and Varenius

We can scrutinize parts of this edifice of memorial aggrandizement by looking at the print history of the English editions of Varenius. Varenius' *Geographia Generalis* has a long and involved print history in England, spanning from Newton's Latin edition for Cambridge University Press in 1672 to the last impression of the Dugdale-Shaw English translation in 1765. This is clearly a print life far longer than that accorded to most English geography books, although it is by no means unique; in the realm of digests of special geography, for example, both Patrick Gordon's *Geography Anatomiz'd* and William Guthrie's *New Geographical, Historical and*

D. Brunn and Ute Wardenga (Leipzig: Institute für Länderkunde, 1999), 7. See also the influential 1911 edition of the *Encyclopaedia Britannica*, sub. 'History of Geographical Theory' for a similar view. The claim ramifies to the present: see Michael Curry, "Toward a Geography of the World without Maps: Lessons from Ptolemy and Postcodes," *Annals of the Association of American Geographers* 95 (2005): 682.

17   William Warntz, "Newton, the Newtonians and the *Geographia Generalis Varenii*," *Annals of the Association of American Geographers*, 79 (1989): 165–91.

18   Preston E. James, "On the Origin and Persistence of Error in Geography," *Annals of the Association of American Geographers* 57 (1967): 21. For the same argument with respect to early American geographical education, see William Warntz, "*Geographia Generalis* and the Earliest Development of American Academic Geography," in *The Origins of Academic Geography in the United States*, ed. Brian Blouet (Hamden: Archon Books, 1981), 245–63; Geoffrey J. Martin, "The Emergence and Development of Geographic Thought in New England," *Economic Geography*, Special Issue (1998): 1–13.

19   US National Center for Geographic Information and Analysis, "Varenius: NCGIA's Project to advance Geographic Information Science," http://www.ncgia.ucsb.edu/varenius/varenius.html.

Commercial Grammar went through many more new editions and impressions for a similar span of time.[20] If we take a broader definition of geography, other works of geographical knowledge such as Camden's *Britannia* went through editions and abridgements for far longer than did the *Geographia Generalis*.[21]

If we look beyond the fact of the span of years over which Varenius was printed, the case for his pivotal status in later seventeenth- and early eighteenth-century geographical print culture weakens. The two "editions" produced by Cambridge University Press in 1672 and 1681 as edited by Newton have identical pagination and seem to have only changed in their typeface and the correction of a few typographical errors. There is no evidence to suggest that they had exceptional print runs: David McKitterick suggests a run of 1000–1200 copies as a maximum,[22] something which fits with the recorded fact that the third edition to emerge from Cambridge, that annotated by James Jurin in 1712, had a print run of 1000 copies, an entirely standard print run for a mathematical-cum-scientific text with a potential student audience.[23] Despite the fact that Varenius remained recommended reading throughout the century for Cambridge undergraduates, no further Latin edition was called for. Furthermore, we must bear in mind that a reasonable proportion of these three Latin editions would have gone to overseas markets, notably in Holland.[24]

The two major English translations of the *Geographia Generalis*, those by Richard Blome (1682 and 1693) and by Dugdale and Shaw (1733, 1734, 1736 and 1765) reinforce the impression which the Latin editions yield of a work of some but by no means exceptional popularity. This, too, undercuts generalisations about the centrality of Varenius' work as a textbook. Blome entered a copyright claim on a translation of Varenius in the Stationers' Company register in 1663, renewing this in 1668.[25] Yet it was only in 1682 that this translation appeared in print as part 1, "The General and Absolute Part", of *Cosmography and Geography*, conjoined to Nicolas Sanson's *Geographical Description of the*

---

20   For details of their print history, see O.F.G. Sitwell, *Four Centuries of Special Geography* (Vancouver: University of British Columbia Press, 1991).

21   *Britannia* was first published in 1586. The last serious edition was that by Richard Gough in 1806.

22   Personal communication, 3/12/2004.

23   Donald McKenzie, *The Cambridge University Press, 1696–1712: A Bibliographical Survey* 2 vols. (Cambridge: Cambridge University Press, 1966), 1: 340–41; for this as a standard print run, see 1: 100. On estimating edition sizes, see Owen Gingerich, *The Book Nobody Read: One Man's Quest to Visit Every Surviving Copy of One of the World's Great Books* (London: Random House, 2004), chapter 8.

24   David McKitterick, *A History of Cambridge University Press. Volume 1: Printing and the Book Trade in Cambridge, 1534–1698* (Cambridge, Cambridge University Press, 1992), 381–3; Idem., *A History of Cambridge University Press. Volume 2: Scholarship and Commerce, 1698–1872* (Cambridge: Cambridge University Press, 1998), 41, 82 and 86.

25   Henry R. Plomer, *A Dictionary of Printers and Booksellers at Work in England, Scotland and Ireland from 1668 to 1725* (Oxford: Oxford University Press, 1922), 39.

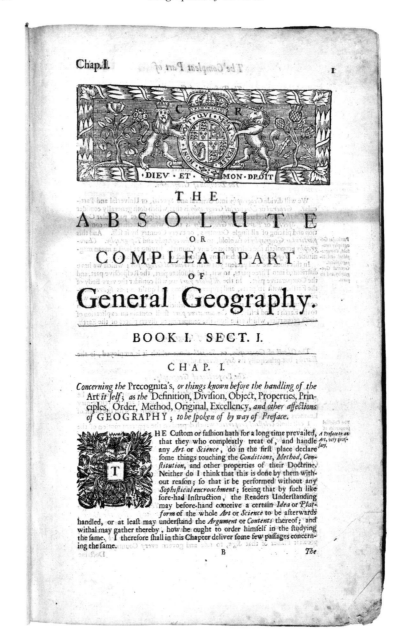

Figure 6.1   First page of Richard Blome, *Cosmography and Geography* (1693) showing Caroline crest

Source: Reproduced by permission of Llyfrgell Genedlaethol Cymru / The National Library of Wales.

World, an English translation of which Blome had first issued in 1670. Blome's 1693 "edition" of Varenius again appears with his Sanson translation and is clearly a reissue of the 1682 project, to which county maps of England derived from John Speed were appended. Whilst the dedication was changed in 1693 – perhaps unsurprisingly given that in 1682 the whole was dedicated to the Duke of Abermarle, who had staunchly supported James II and was now dead – the print ornaments, which still bear the Royal Crest and the initials "CR" (Carolus Rex), point to a product designed to flatter a monarch long since dead from a line deposed five years previously (Figure 6.1). This, of course, was not treasonable Jacobitism but a result of the exigencies of Blome needing to clear his stocks and recoup his costs on a project which had manifestly failed to sell its print run.[26] We can infer, then, that Blome finally published his translation of Varenius to help clear stocks of his Sanson translation, but that this project itself failed to meet his expectations, leading to its reissue in 1693.

A similar print history appears to hold for the Dugdale-Shaw translation. This first appeared in 1733 and is – unlike Blome's project – an edition in its own right, not part of a broader bookseller's strategy. A second edition appeared in 1734, this clearly being a 'real' edition, in that it contains substantial amounts of new material, including new footnotes and major additions to the text such as a collection of geographical paradoxes and a discussion of tidal flows.[27] After this two further "editions" of the Dugdale-Shaw translation emerged, in 1736 and 1765. In both cases, they are reissues of the second edition, presumably designed to shift remaining copies. In both cases the text is identical, and whilst the 1765 "fourth" edition has a new publishing group – L. Hawes, W. Clarke and R. Collins – for the simple reason that Stephen Austen, publisher of the first three editions, died in 1750,[28] it still refers in the text to 1734 as the year of composition and retains a flyleaf advertising Austen's other titles.[29] In short, it would appear that a successful first edition in 1733 led to the rapid development of a second, expanded edition of 1734, this then proving hard to shift and still being cleared when Hawes et al. purchased the property rights on Austen's death.

---

26   For parallel problems for publishers of geographical works caused by the change of regimes, see Edmund Bohun's *Geographical Dictionary* as discussed in Robert J. Mayhew, *Enlightenment Geography: The Political Languages of British Geography, c. 1650–1850* (London: Macmillan, 2000), chapter 5.

27   Bernhard Varenius, *A Complete System of General Geography*, 4th edition, ed. Mr Dugdale and P. Shaw (London: L. Hawes, W. Clarke and R. Collins, 1765), 779ff for paradoxes; and 811ff on motion of tides.

28   Henry R. Plomer, G.H. Bushnell and E.R. McC. Dix, *A Dictionary of Printers and Booksellers who were at Work in England, Scotland and Ireland from 1726 to 1775* (Oxford: Bibliographical Society, 1932), 10; see also University of Birmingham, "The British Book Trade Index," http://www.bbti.bham.ac.uk/Details.htm?TraderID=112714.

29   Varenius, *Complete System*, ed. Dugdale and Shaw, 839 (mispaginated as 339 in the copies I have consulted for 1734 and 1765) which refers to 'the time of writing here of *Anno* 1734'.

The 1765 reissue proved to be the last printed version of Varenius in English. William St Clair has pointed out both that many successful works were 'tranched down' by publishers in abridgements and handy editions to capture an audience in less affluent sections of the market, and that there was a flood of such work after the 1774 decision about property rights.[30] Successful geographical works could be tranched up and down, as Sher shows for the Irish and American editions of Guthrie's *Geographical Grammar*, where publishers competed to make the most authoritative and elegant edition of the work.[31] Varenius' *Geographia Generalis* remained unaffected by either trend: neither Blome in the late seventeenth century nor Austen or Hawes et al. in the mid eighteenth saw enough of a market to create abridgements or digests of Varenius, and the relaxation of copyright after 1774 did not lead to any new reworkings.

Simply put, there were three main incarnations of Varenius in English print culture: the three Latin editions from Cambridge University Press, the two issues of Blome's translation, and the four issues of two editions of the Dugdale-Shaw translation. The Latin editions from Cambridge sold best of all, in good part perhaps because their audience was not restricted to the Anglophone world and because the book was required reading in Cambridge. In each case it would be implausible to suggest there is any evidence of an unusually large demand amongst either students or the learned. General assertions made by historians of geography about Varenius being a standard textbook – if by that is meant a claim that it was unusually widely available or exceptionally profitable as a piece of intellectual property – do not hold good in the English-speaking world. Other geographical texts such as Guthrie's *Geographical Grammar* sold more copies and went through more genuine editions (as well as a goodly number of publisher's impressions), whilst the English editions of Varenius in particular clearly struggled to sell and were not deemed worthy of being reworked for the less affluent.[32] In the cold light of print history, considered in terms of editions and numbers, Varenius emerges as a text more respected than read, and as a text read above all in the Latinate context of university students in later seventeenth- and early eighteenth-century Cambridge.

30   St Clair, *Reading in Romantic Age* and personal communication, 6/05/05; on the 1774 decision, see Mark Rose, *Authors and Owners: The Invention of Copyright* (Cambridge: Harvard University Press, 1993).

31   Richard B. Sher, *The Enlightenment & the Book: Scottish Authors & their Publishers in Eighteenth-Century Britain, Ireland & America* (Chicago: University of Chicago Press, 2006), 487–93 and 573–82.

32   Ibid., 92 and 642–3 lists Guthrie's *Geographical Grammar* as one of the select forty-six "bestsellers" produced by the eighteenth-century Scottish book trade.

From Hackwork to Classic: Editing Varenius as Canonical

Yet there is another element of the print history of English editions of Varenius which might incline us to argue that the *Geographia Generalis* was indeed a pivotal text in the history of geographical thought. To make good on this claim, attention has to turn from *editions* to *editing*, from quantitative questions to the more literary question of the suite of editorial labours which came to be embodied in the English versions of the *Geographia Generalis*. The ways in which Varenius came to be edited in the century over which his text was in circulation in the Anglophone world show that he came to be accorded a special – if not quite unique – place in the geographical culture of the time. Whilst claims about Varenius being the standard textbook of the period hold no weight in terms of print history, and whilst claims about his status in the transition to a modern geography are either beyond the realm of historical substantiation or proleptic and largely meaningless, we can assert Varenius' significance in terms which are robust in terms of intellectual history and print culture. Varenius started as an author whose text was edited and reissued as any other geographical author might be in the Grub Street culture of geographical writing and publishing of the era in England.[33] Over time his work came to be accorded a special status wherein his authorial voice was carefully preserved, even when it was being corrected or questioned in the light of Newtonian natural philosophy. The result of this editing was that the print space of the *Geographia Generalis* came to resemble that developed for classical texts – ancient and vernacular – printed in England at the same time, an honour accorded only to Varenius amongst English geographical textbooks, and perhaps only to Varenius and Camden amongst the broader gamut of works of geographical knowledge at this time.

Varenius as Hackwork: Newton and Blome, 1672–1693

At the start of *Geographia Generalis*'s print history in England, the work was not accorded any special status, something which is apparent in the two Cambridge editions attributed to Newton and especially in Blome's translation of that edition.

Great play has been made of Varenius – and thereby geography – having a direct link with Isaac Newton, but the verifiable facts about Newton's connection with the text are rather less impressive. Whilst Warntz asserts Newton's editorial contributions 'were many and substantial', he only in fact points out that Newton added a series of 31 figures or schemes, promised in but lacking from the Elsevier edition of 1671. It is additionally asserted that Newton corrected grammatical and typographical errors and perhaps reworked a few paragraphs.[34] In fact, from Newton himself we only have the assertion that 'the Book here in Presse is *Varenius*

---

33   On which in general see Mayhew, "Character of English Geography".

34   Warntz, "Newton", 178–81.

his Geography, for which I have described Schemes', something he also asserted to others: 'according to John Conduitt the book's title ... is misleading, for "Sir I.N. told Mr Jones all he did in the edition of Varenius before wch is put Curante Isaaco Newtono was to draw the schemes"'.[35] Newton may well have cleaned up the text to a greater extent than merely adding figures, but it is also possible that a vigilant corrector at the press did much of this work.[36]

Whilst the details of Newton's input to the 1672 edition cannot be ascertained with certainty, it is clear that even on Warntz's reading he only tinkered with elements in the text. Newton chose not to correct Varenius, nor did he take on what he would have seen as manifest errors in Varenius' trigonometry,[37] nor in his Cartesian natural philosophy (although with Newton's Varenius coming over a decade before the *Principia,* we cannot know the full depth of his disagreement with Descartes at this time). Further, whilst Newton aimed the text at students in Cambridge in the 1670s and 1680s, the text retained the original dedication to eight Dutch dignatories, dated 1650, which could hardly have been relevant. Likewise, Varenius' original centring of the text on Amsterdam in terms of its table of longitude and latitude is retained. If Newton hoped this text would appeal to students, he chose not to include even basic finding devices to help them: there is, as in the Elsevier originals, no index and no marginal glosses describing the progress of the argument. In an era of burgeoning methods to aid students in searching and organising data to keep "information overload" at bay,[38] Newton's *Geographia Generalis* was strangely spartan, a text with no editorial signposts to aid the reader. In this, Newton was following his original, but he was hardly pitching the text at its purported audience. As such, Newton's editorial labours were not great and the Varenius edition, whilst it was Newton's first published book, did not involve any great input of mental or editorial labour on his part.[39]

To the extent that Newton or his corrector did update Varenius, they produced as much confusion as they rectified, something Warntz does not note. For example, a number of the branching diagrams which adorn the book do not actually replicate the claims made in the text itself. Looking at Book 1 chapter 8,

---

35   Isaac Newton, *Correspondence. Volume 1: 1661–1675,* ed. H.W. Turnbull (Cambridge: The Royal Society, 1959), 161, Newton to Collins, 25/5/1672; and D.T. Whiteside, ed. *The Mathematical Papers of Isaac Newton. Volume 2: 1667–1670* (Cambridge: Cambridge University Press, 1968), 288–9, n.43.

36   On correctors, see St Clair, *Reading Nation in the Romantic Age.*

37   D.T. Whiteside, ed. *The Mathematical Papers of Isaac Newton. Volume 4: 1674–1684* (Cambridge: Cambridge University Press, 1971), 10 n.28.

38   Ann Blair, "Reading Strategies for Coping with Information Overload, ca. 1550–1700," *Journal of the History of Ideas* 64 (2003): 11–28; Idem., "Note Taking as an Art of Transmission," *Critical Inquiry* 31 (2004): 85–107; Idem., "Annotating and Indexing Natural Philosophy" in *Books and the Sciences in History,* ed. Marina Frasca-Spada and Nick Jardine (Cambridge: Cambridge University Press, 2000), 69–89.

39   A.R. Hall, "Newton's First Book," *Archives Internationales d'Histoire des Sciences,* 13 (1960): 55–61.

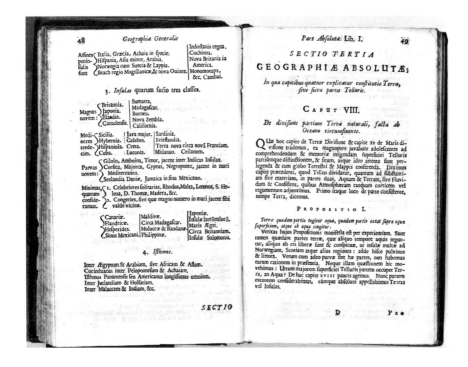

Figure 6.2 Tree diagram of "large islands" from Bernhard Varenius, *Geographia Generalis* (1681) as edited by Isaac Newton

Source: Reproduced by permission of Llyfrgell Genedlaethol Cymru / The National Library of Wales.

the text tells us that the world has ten islands which fit into the category 'large', whilst the supporting tree diagram tells us of 'magnas novem' and duly lists nine (Figure 6.2).[40] Whatever editing Newton did to the text, it was "silent" in that it did not alert readers to the fact that he was departing from the original text, and was partial in that the text and diagrams did not always tally, such alterations not having been worked through the entirety of the text. The result is a text which is in places confusing and contradictory. As a project aimed to update Varenius for a student audience, Newton's edition is really rather deficient: it lacks finding aids for the student, its updating of basic mathematical propositions is partial and its consistency where points are adjusted is uneven. Newton's edition neither cleaves to the authorial original, nor does it thoroughly update it. Above all, it does not make clear which policy is being adopted at any time by way of textual

---

40 Bernhard Varenius, *Geographia Generalis*, 2nd edition, ed. Isaac Newton (Cambridge: Cambridge University Press, 1681), 52–3 for the text and 48 for the supporting diagram.

or paratextual signposts. Whilst one would hesitate to call the result "hackwork", it is clear that the occasional nature of the project – to produce a workable student edition – and Newton's own low key comments about his input to the Cambridge Varenius seem to be borne out by the text itself.

Still more orientated in the direction of hackwork was publisher Richard Blome's English edition of Varenius, published as the first part of *Cosmography and Geography* in 1682. This project looks much grander than did Newton's Cambridge editions, an 18cm octavo being replaced by a 35cm folio. The text also uses more print ornaments such as fancywork and initial drop lettering and contains marginal glosses as a finding aid. Yet, in most respects, Blome's approach to Varenius was that of hack publisher, struggling to sell off copies of a floundering geographical project, his Sanson, first published in 1670 and here repackaged as the second or special part of *Cosmography and Geography*. This creates an unavoidable mishmash, wherein Blome appears to bring together several of his geographical and cartographical publishing ventures in the hope of selling them off by yoking them together.

With respect to Varenius, Blome's translation appears fairly faithful to the Latin text as received from Newton's 1671 edition. The unfortunate result is that Blome also uncritically takes on and perpetuates the sorts of editorial inconsistencies in Newton's edition highlighted above, including for instance the nine grand islands in his diagrams whilst retaining textual reference to ten (indeed eleven, with 'California, if that be an Island').[41] Clearly, the translator (unnamed) was as casual about consistency as the Cambridge edition used as the base text, retaining the single voice and silent alterations which created inconsistency. This inconsistency merely complements the previously-mentioned typographic inconsistencies in the 1693 impression caused by the failure to update the ornamental work to reflect the change of both monarch and line. In text and typography, *Cosmography and Geography* is a hack's palimpsest, differing and inconsistent voices and devices being juxtaposed due to deficient editorial oversight.

Blome's edition was more confusing and contradictory than was Newton's, because of the way in which it spliced together two separate projects, Sanson's and Varenius', a problem only exacerbated by the addition of a third geographical project – Speed's maps – in the 1693 impression. Varenius, of course, was distinguished by his lack of interest in special or descriptive geography in the *Geographia Generalis*, yet Blome then spliced this together with a long special geography of the world by Sanson. Thus we have Varenius stating he only deals with political or special geography because 'we must yield somewhat to Custom and the Profit of Learners', whilst the book itself – in the form of Sanson's *A Geographical Description of the World* – in fact devotes 493 folio pages to this topic.[42] The conjoining of the two texts creates confusion when reference is

---

41   Richard Blome, *Cosmography and Geography* (3rd impression, London: Richard Blome, 1693), 37; cf 34.

42   Ibid., 3.

made in the Varenian sections to "special geography" as to whether this refers the reader to special geography inasmuch as it is treated in the *Geographia Generalis* or to Sanson's *Geographical Description* as appended. Finally, and in a move characteristic of English hack geographies, Blome also expanded the British sections of Sanson's description so as to make the text more appealing to English audiences. Where the other nations of Europe receive 10 to 20 folio pages of description, Britain covers some 108 pages. The British section is clearly taken from Camden's *Britannia*, following his division of Britain according to the kingdoms of the Saxon heptarchy and then into modern counties, hundreds and parishes. This was no doubt itself excerpted from Blome's earlier publication, *Britannia* (1673), which William Nicolson described in his *English Historical Library* as 'a most entire piece of theft out of Camden and Speed'.[43]

By 1693, then, Blome's *Cosmography and Geography* conjoined Varenius with Sanson, despite the contradictory nature of their geographical projects, further failing to correct inconsistencies in Newton's Varenius and inserting material he had already lifted from Camden for other purposes into Sanson's special geographical description of the world. Taken as a whole, Varenius was positioned as just another geographical author ripe for plagiarism, silent emendation and sloppy adjustment in the pursuit of profit or more plausibly in this case in the avoidance of loss. *Cosmography and Geography*, as a polyvocal text whose inconsistencies, typographic and textual, had not been ironed out due to indifference and haste on the part of its anonymous editor(s), places Varenius in a hackwork project *par excellence*. Here, then, there is little of that reverence for Varenius as a canonical author which is enshrined in modern histories of geographical thought. The transition in Varenius' status comes in the annotated editions produced by Jurin and by Dugdale and Shaw.

Varenius as a Latinate Classic: Jurin's 1712 Edition

In many – perhaps in most – respects, James Jurin's 1712 Latin edition of the *Geographia Generalis* for Cambridge University Press continued the editorial indifference of the Newtonian editions which it was designed to update. The text remains in 511 octavo pages and still opens with Varenius' original 1650 dedication. Jurin retains the inconsistencies we have noted previously in Newton's edition. The bulk of the book, in short, is simply a reissue of the earlier work, presumably to meet continuing student demand in Cambridge and beyond for this text.

And yet Jurin's edition also marks a new stage in the English reception of Varenius. More precisely, Jurin's edition begins for the first time to treat Varenius' work as a modern classic of geographical writing, which needs to be respected even as it is emended. Where both Newton and Blome had been happy to adjust

---

43   Cited in the old *Dictionary of National Biography*. Other comments to the same effect from Nicolson are cited with more caution in Stan Mendyk's article on Blome for the *New Oxford Dictionary of National Biography*.

the text silently and had not been too concerned about the textual inconsistencies resulting from such interventions, Jurin is more careful in these respects. In this he reflects a shift in the publishing policy of Cambridge University Press more generally in the years from the 1690s, which aimed to make the press 'contribute to the *respublica litterarum*'.[44] This policy came with the new printing press, whose construction was masterminded by Richard Bentley.[45] By the 1710s, under Cornelius Crownfield's direction, the press was merging this scholarly aim with the need 'to serve the needs of everyday education'.[46] It is precisely this blend of the scholarly and the pedagogic which is reflected in Jurin's edition of Varenius.

Before we look at Jurin's editorial labours, it is worth noting that the 1712 edition, whilst in most respects identical to that of 1682, is reset in a far clearer typeface, part of the Press's policy of using Dutch printers and typefaces.[47] The edition was adorned with ornamental printwork by the Dutch engraver Simon Gribelin, whose skills were also deployed on prestigious projects such as Bentley's edition of Horace.[48] Finally, Newton's figures or "schemes" were redistributed onto eight leaves as opposed to the original five, rendering them easier to consult. As part of the drive to make Cambridge University Press's editions respected in the *respublica litterarum*, there had been a drive to emulate Dutch typography such that the "look" of the print space appeared to be of the highest standard. Jurin's edition clearly benefited from a cleaner and more professional typography than had its Cantabrigian predecessors.

The editorial element in Jurin's work which matches the new ambitions as reflected in the typography of the 1712 edition is contained in his "Appendix". This is a newly paginated 54 page supplement to Newton's edition, the main aim of which is to emend, correct and update statements made in the main text in the light of over half a century of investigations in natural philosophy.[49] The "Appendix" takes the form of a series of notes keyed to chapters or pages in the main text. The appended endnotes are further adorned with footnotes, these being references for claims made in the text of the Appendix, and with some marginal notes directing the reader to three new foldout pages with a total of seven new figures or "schemes" keyed to the text of the appended commentary. The content of the notes is as Jurin's "Praefatio" leads us to expect: he takes the findings of the post-Cartesian generation of natural philosophers, especially English and

---

44   McKitterick, *History*, 2: 43.

45   Ibid., 55–6.

46   Ibid., 78.

47   Ibid., 62. The much-admired double pica roman typeface brought from the Netherlands for the press was used for the new Dedication to Richard Bentley in Jurin's edition: see McKenzie, *Cambridge University Press*, 1: 340.

48   McKenzie, *Cambridge University Press*, 1: 363.

49   Bernhard Varenius, *Geographia Generalis*, ed. James Jurin (Cambridge: Cambridge University Press, 1712); for this ambition, see the "Praefatio, Lectori Jacobus Jurin", 1–2; a workable translation is provided in Warntz, "Newton": 184–5.

French ones, and uses them to update Varenius on basic issues in mathematical geography such as the shape of the earth, tides, the hydrological cycle and global wind patterns. In the course of the Appendix such luminaries of modern science as Boyle, Halley and Huygens are cited and other papers, notably from the Royal Society's *Philosophical Transactions*, are quoted with regularity. Pride of place is given to 'Felicissime *Newtonus*',[50] whose *Principia* forms the backbone to the adjustments made by Jurin.

The content of Jurin's "Appendix" is well known and has been studied for the way in which it brought the *Geographia Generalis* within the ambit of Newtonian science,[51] correcting the Cartesian natural philosophy of Varenius in the light of Newton's *Principia*. What has received less attention is the style of editing Jurin deployed in revivifying Varenius for a new generation of scholars. Jurin's edition not only printed the original dedication (as had its Newtonian predecessor), but also included a new dedication to Richard Bentley, Master of Trinity College, Cambridge where Jurin was a Fellow. This was not just (although it was no doubt in part) a call for patronage as Bentley had suggested the project to Jurin, nor was it simply a sign of his gratitude to Bentley for checking the Latin style of his "Appendix".[52] Instead, it reflected allegiance to Bentley's distinctive style of editing. Bentley was famed for his skills as a classical editor, 'the greatest scholar that England, or indeed Europe, had ever bred',[53] and in particular for his belief in emendatory editing, or the adjustment of established classical texts not in the light of new manuscript evidence but of logic and editorial intuition. In short, Bentley both revered ancient authors and aggressively altered the received text by his own lights: his Cambridge edition of Horace, published the year before Jurin's Varenius, for example, made more than seven hundred alterations 'not controlled by knowledge either of the historical and individual style of the writer or of the history of the text'.[54] Bentley's approach to editing was, as Pfeiffer makes clear, anachronistic by our modern standards of editing, in that he would alter a text in the light of knowledge an author could not in principle have held at the time. Bentley was later to extend the same principle to the realms of natural philosophy and vernacular poetry in his 1732 edition of Milton's *Paradise Lost*, wherein 'the

---

50　Varenius, *Geographia Generalis*, ed. Jurin, Appendix, 13.

51　Warntz, "Newton": 182–8; Andrea Rusnock, ed., *The Correspondence of James Jurin (1684–1750)* (Amsterdam: Rodopi Press, 1996), 10–11.

52　Rusnock, ed., *Correspondence*, 62–3, Bentley to Jurin, 7/6/1711.

53　E.J. Kenney, *The Classical Text: Aspects of Editing in the Age of the Printed Book* (Berkeley: University of California Press, 1974), 114.

54　Rudolf Pfeiffer, *A History of Classical Scholarship: From 1300 to 1850* (Oxford: Clarendon Press, 1976), 154. See also L.D. Reynolds and N.G. Wilson, *Scribes and Scholars: A Guide to the Transmission of Greek and Latin Literature*, 2nd edition (Oxford: Clarendon Press, 1974), 168.

philosophy and natural philosophy ...[are] tested by Bentley against his own standards'.[55]

Importantly, Jurin's approach to editing Varenius clearly bears some parallels to Bentley's then-celebrated efforts. As with Bentley, Jurin does not seek to merely give a text which Varenius as the author penned, instead wanting to use subsequent knowledge to update Varenius. Further, as Bentley was to suggest numerous alterations to Milton in his 1732 edition but only actually inscribed a small number in the text itself,[56] so Jurin makes a clear separation between the text and his additions or corrections, the result being annotations which subvert the text they aim to support.[57] Jurin does not actually annotate Varenius' original Latin text at all, but instead the already-altered text which Newton's edition had established, with all its faults and inconsistencies. In this, Jurin followed the standard editorial practice of the age, wherein new editions were based not on the original text but on the *textus receptus* as handed down by a sequence of editors and printers, 'a procedure so well calculated to foster the survival of old and persistent error'[58] of the sort we have already seen bedevilled Newton's Cambridge editions.

Jurin's editorial method not only carries forward the textual inconsistencies and problems of Newton's received text, but also creates some editorial problems of its own due to the decision to merely add an Appendix. This decision was probably taken on cost grounds: it allowed the Newtonian text to be reset from an earlier printing with no alterations. This decision was not, perhaps, the most satisfactory in scholarly terms, but it reflected the context of the 1710s where the desire for a worthy edition was being balanced with the need to produce affordable texts by Cambridge University Press, especially for a work such as *Geographia Generalis*, whose declared audience was a student one. Yet the decision not to tamper with the received text means that whilst the Appendix refers the reader to which pages of the text it glosses, the main text has no diacritical marks to direct the reader to the Appendix (Figure 6.3). As such, intertextual reference only works from apparatus to text, not both ways round as would normally and preferably be the case. Furthermore, where at the same time Cambridge was receiving praise for its edition of Minucius Felix because its notes were easily perused at the foot of the text,[59] the separation of text and commentary in Jurin's Varenius make cross-

---

55   Marcus Walsh, *Shakespeare, Milton and Eighteenth-Century Literary Editing: The Beginnings of Interpretative Scholarship* (Cambridge: Cambridge University Press, 1997), 73.

56   Ibid., 67.

57   Evelyn Tribble, *Margins and Marginality: The Printed Page in Early Modern England* (Charlottesville: University of Virginia Press, 1993); James McLaverty, *Pope, Print and Meaning* (Oxford: Clarendon Press, 2001); Peter Cosgrove, "Undermining the Text: Edward Gibbon, Alexander Pope and the Anti-Authenticating Footnote," in *Annotation and its Texts*, ed. Stephen Barney (New York: Oxford University Press, 1991), 130–51.

58   Kenney, *Classical Text*, 68.

59   McKitterick, *History*, 2: 85. The edition of Minucius Felix is advertised in the "Catalogus Librorum" after Jurin's Appendix in the 1712 *Geographia Generalis*.

Figure 6.3 The opening page of James Jurin's appended comments to his edition of Bernhard Varenius, *Geographia Generalis* (1712)

Source: Reproduced by permission of Llyfrgell Genedlaethol Cymru / The National Library of Wales.

referencing cumbersome, even if the reader is aware (as they cannot be from the text itself) that a passage has an appended comment. The Appendix adds further confusion by its previously-mentioned addition of three new fold-out leaves with seven new figures, because they were numbered anew, rather than sequentially after Newton's original schemas. As a result, there is more than one figure 1 in Jurin's edition making referencing somewhat confusing. That page 1 of the new figures comes between pages 28 and 29 of the Appendix, whilst page 2 is between pages 4 and 5 compounds the confusion.

Simply put, Jurin's 1712 edition of the *Geographia Generalis* was part of the broad thrust at Cambridge University Press initiated by Richard Bentley to produce scholarly editions aimed at a market which would extend beyond England to the European republic of letters. In the case of Jurin's work, this aim was inflected by the need to keep costs down for a student audience. Its typography reflected this desire, and the way in which it was edited suggested that Varenius was to be taken seriously as an author, that he was a modern classic of geographical writing. For Jurin, as for Bentley, this sense of reverence resulted not in a desire to produce an edition as close as possible to Varenius' *ipsissima verba*, but on the contrary a willingness to carry forward the *textus receptus*, however altered it had been by the labours of subsequent editors and publishers. Coupled with this was a drive to update that received text in the light of new findings in mathematical geography and natural philosophy.

The result is a text which continues some confusions from Newton's edition and generates a few of its own by the lack of integration of the appended material with the Newtonian received text. Despite this, the overall effect of Jurin's edition is to give an unprecedented sense of Varenius as someone worthy of serious editorial labours, as a classic that needs to be transmitted in updated form to a new generation. We also get, for the first time, a clear distinction between the received text and new editorial additions. The flyleaf to Jurin's edition advertises other texts printed by Crownfield for Cambridge including *The Suidas*, and the works of Sallust and Julius Caesar: the company *Geographia Generalis* kept here bespeaks the new light in which Varenius was being edited. With Jurin, the canonisation of the *Geographia Generalis* which has become such a feature of modern writing in the history of geography begins.

Varenius as a Vernacular Classic: The Dugdale-Shaw Editions

The final great English edition of the *Geographia Generalis* first appeared in 1733. Whilst it initially appears to sever the text's link with a Cambridge audience, in fact, as McKitterick notes, numerous Cambridge scientific works came to be printed in London by the 1730s even where they were published for Cambridge University Press by Crownfield. Further, many such texts were sold in London by Stephen Austen, and it is Austen who printed and published the Dugdale-Shaw Varenius.[60] However metrocentric the project was, it still retained connections with the Cambridge nexus which had generated all the English editions of Varenius except Blome's.

Before inspecting the book itself, there are two main reasons to anticipate a text still more confused in its editing than was any previous edition. Once more the base for the translation is the received text, in this case Jurin's 1712 edition,[61] a situation where 'the text selected for editing was the one closest to the editor rather than that closest to the author: the text that had undergone the most rather than the least meditation'.[62] One would anticipate a palimpsest which interweaves Newton's changes, Jurin's adjustments and those of the new translators. Second, and pursuant on the first point, the title page reveals the most complex authorship yet seen for the *Geographia Generalis*, acknowledging the text as 'originally written' by Varenius, as 'improved and illustrated' by Newton and Jurin, 'now

---

60   McKitterick, *History*, 2: 122–3.

61   Acknowledged in Varenius, *Complete System*, ed. Dugdale and Shaw, v.

62   Margreta De Grazia, *Shakespeare Verbatim: The Reproduction of Authenticity and the 1790 Apparatus* (Oxford: Clarendon Press, 1991), 52.

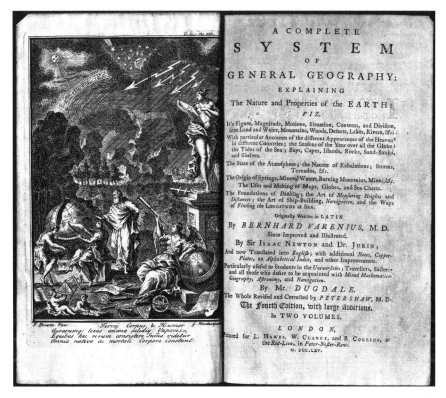

Figure 6.4   The title page to the Dugdale-Shaw translation of Bernhard Varenius, *A Complete System of General Geography* (1765 impression)

translated into *English*; with additional *Notes*, *Copper-Plates*, an *Alphabetical Index* and other Improvements' by Dugdale, 'the Whole Revised and Corrected by PETER SHAW'. By the (spurious) fourth edition, we are told of further 'large additions', although none in fact occur.[63] The Dugdale-Shaw translation takes a polyvocal text, already twice edited and adds at least two further layers, suggesting at least five voices in the authoring of "Varenius'" text (Figure 6.4).

In fact, the Dugdale-Shaw edition is by some margin the most thoroughly edited and meticulously typeset edition of Varenius in the English print history of the work, and can claim furthermore to be one of the most professionally edited texts in the English geographical tradition more generally in this era. This is achieved by the typography and notation being rendered clear, consistent and accurate for the first time in the work's print history, in addition to which annotation is practiced in a new way. The cumulative effect is to create a unique print space for a geography

---

63   Varenius, *Complete System*, ed. Dugdale and Shaw, Title Page.

book which aligns the *Geographia Generalis* with the format of canonical texts, both classical and vernacular. It is only in the last English incarnation of the *Geographia Generalis*, then, that we find its canonization in a way accordant with the reverence that modern histories of geography might lead one to expect.

Looking at the editorial work in more detail reveals the achievement of the Dugdale-Shaw Varenius. This edition was the first to achieve a consistent mode of annotation and cross-referencing, allowing the reader to access information with ease and reliability by means of a series of linked changes. First, the inconsistencies between texts and tables which had ramified down from Newton's editions were corrected. Thus the previously-mentioned inconsistency as to how many "large" islands existed was rectified here for the first time in any edition.[64] Second, the figures or schema which Newton had added to the text and Jurin had expanded were integrated by both marginal references in the text to the figures and by references to the relevant pages in the text inserted next to each figure. This process, it should be said, was not perfect: all four impressions have two figure 32s, whilst the second edition in all three impressions had a set of additional figures which were keyed alphabetically as well as the original 49 images in 12 foldout plates. Yet Dugdale-Shaw was the first edition properly to integrate Newton's schemata with the text to allow for proper cross-referencing. A third major change was the addition of an index of topics and place names, this being the first 'finding aid' in any edition, the marginal glosses of the Blome translation excepted, and coming at a time when a number of classic texts such as Thomas Newton's edition of *Paradise Lost* (1749) began to include extensive indexes for the first time.[65] This move was paralleled by the final great change in the typesetting, namely to move the notes from a collection of unkeyed endnotes in Jurin's Latin edition to footnotes with clear keying to the text in the Dugdale-Shaw translation. This is part of a broader shift at this time to footnotes and away from the marginal gloss, inspired in part by cost factors.[66] In the case of Varenius, the transition cannot be explained merely in these terms, as Jurin's solution of a freestanding Appendix was surely cheaper as it did not involve any resetting of the original text.[67] Most notes are keyed by means of a bracketed letter in the text mirrored at the foot in the notes, although some notes were merely keyed by an asterisk mark.[68] The reason for this inconsistency is not clear: whilst some notes were added to the second edition of 1734, this does not explain the asterisked notes, some of which are present in this form in both editions. It can only be

---

64   Ibid., 101 (table) and 109–11 (text).

65   Walsh, *Shakespeare, Milton*, 107–8.

66   Thomas McFarland, "Which was Benjamin Whichcote? Or, The Myth of Annotation" in *Annotation and its Texts*, ed. Stephen Barney (New York: Oxford University Press, 1991), 165; Lawrence Lipking, "The Marginal Gloss," *Critical Inquiry* 3 (1977): 622.

67   I am grateful to Robin Myers for confirming this point.

68   Varenius, *Complete System*, ed. Dugdale and Shaw, 37–8.

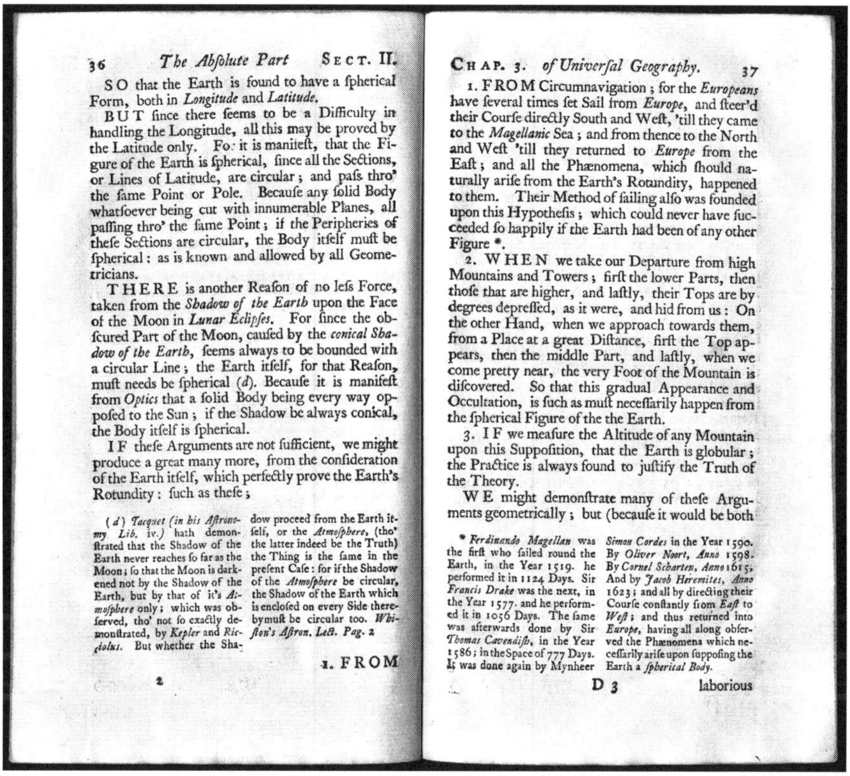

36 *The Abſolute Part* SECT. II.

SO that the Earth is found to have a ſpherical Form, both in *Longitude* and *Latitude*.

BUT ſince there ſeems to be a Difficulty in handling the Longitude, all this may be proved by the Latitude only. For it is manifeſt, that the Figure of the Earth is ſpherical, ſince all the Sections, or Lines of Latitude, are circular; and paſs thro' the ſame Point or Pole. Becauſe any ſolid Body whatſoever being cut with innumerable Planes, all paſſing thro' the ſame Point; if the Peripheries of theſe Sections are circular, the Body itſelf muſt be ſpherical: as is known and allowed by all Geometricians.

THERE is another Reaſon of no leſs Force, taken from the *Shadow of the Earth* upon the Face of the Moon in *Lunar Eclipſes*. For ſince the obſcured Part of the Moon, cauſed by the *conical Shadow of the Earth*, ſeems always to be bounded with a circular Line; the Earth itſelf, for that Reaſon, muſt needs be ſpherical (*d*). Becauſe it is manifeſt from *Optics* that a ſolid Body being every way oppoſed to the Sun; if the Shadow be always conical, the Body itſelf is ſpherical.

IF theſe Arguments are not ſufficient, we might produce a great many more, from the conſideration of the Earth itſelf, which perfectly prove the Earth's Rotundity: ſuch as theſe;

(*d*) *Tacquet (in his Aſtronomy Lib. iv.)* hath demonſtrated that the Shadow of the Earth never reaches ſo far as the Moon; ſo that the Moon is darkened not by the Shadow of the Earth, but by that of it's *Atmoſphere* only; which was obſerved, tho' not ſo exactly demonſtrated, by *Kepler* and *Ricciolus*. But whether the Shadow proceed from the Earth itſelf, or the *Atmoſphere*, (tho' the latter indeed be the Truth) the Thing is the ſame in the preſent Caſe: for if the Shadow of the *Atmoſphere* be circular, the Shadow of the Earth which is encloſed on every Side therebymuſt be circular too. *Whiſton's Aſtron. Lect. Pag.* 2

1. FROM

2

CHAP. 3. *of Univerſal Geography.* 37

1. FROM Circumnavigation; for the *Europeans* have ſeveral times ſet Sail from *Europe*, and ſteer'd their Courſe directly South and Weſt, 'till they came to the *Magellanic* Sea; and from thence to the North and Weſt 'till they returned to *Europe* from the Eaſt; and all the Phænomena, which ſhould naturally ariſe from the Earth's Rotundity, happened to them. Their Method of ſailing alſo was founded upon this Hypotheſis; which could never have ſucceeded ſo happily if the Earth had been of any other Figure *.

2. WHEN we take our Departure from high Mountains and Towers; firſt the lower Parts, then thoſe that are higher, and laſtly, their Tops are by degrees depreſſed, as it were, and hid from us: On the other Hand, when we approach towards them, from a Place at a great Diſtance, firſt the Top appears, then the middle Part, and laſtly, when we come pretty near, the very Foot of the Mountain is diſcovered. So that this gradual Appearance and Occultation, is ſuch as muſt neceſſarily happen from the ſpherical Figure of the the Earth.

3. IF we meaſure the Altitude of any Mountain upon this Suppoſition, that the Earth is globular; the Practice is always found to juſtify the Truth of the Theory.

WE might demonſtrate many of theſe Arguments geometrically; but (becauſe it would be both

* *Ferdinando Magellan* was the firſt who ſailed round the Earth, in the Year 1519. he performed it in 1124 Days. Sir *Francis Drake* was the next, in the Year 1577. and he performed it in 1056 Days. The ſame was afterwards done by Sir *Thomas Cavendiſh*, in the Year 1586; in the Space of 777 Days. It was done again by Mynheer *Simon Cordes* in the Year 1590. By *Oliver Noort*, *Anno* 1598. By *Cornel Scharten*, *Anno* 1615. And by *Jacob Heremites*, *Anno* 1623; and all by directing their Courſe conſtantly from *Eaſt* to *Weſt*; and thus returned into *Europe*, having all along obſerved the Phænomena which neceſſarily ariſe upon ſuppoſing the Earth a *ſpherical Body*.

D 3 laborious

Figure 6.5 The print format and keying to footnotes in Dugdale-Shaw's translation of Bernhard Varenius, *A Complete System of General Geography* (1765 impression)

conjectured that this is a result of the interaction of the two editors, Dugdale and Shaw, not being fully ironed out in the typesetting, or of late notes being added at the printers.[69] Yet the vast majority of the notes are alphabetically keyed to the text. This is an enormous advance on previous editions in terms of ease of reference and one which completely changes the print format of Varenius from one bloc of continuous text to that standard for editions of canonical authors at this time, a large typeface being used for the main text, coupled to a two-column commentary in a smaller typeface at the foot of the page (Figure 6.5).

69 Such processes are described in detail for Boswell's *Life of Johnson* in Bruce Redford, *Designing the Life of Johnson* (Oxford: Oxford University Press, 2002), chapter 1; and for Edward Gibbon's career in print by David Womersley, *Gibbon and the "Watchmen of the Holy City": The Historian and his Reputation 1776–1815* (Oxford: Clarendon Press, 2002).

The change in annotation for this edition is not simply about format – from endnotes to footnotes – but also relates to their nature. Dugdale-Shaw presents notes in different "voices", rather than a univocal or anonymous editorial "voice". More specifically, there are in many footnotes both Jurin's appended comments of 1712 and further updated portions of commentary taking into account new findings in the twenty years intervening between Jurin and this translation or (less frequently) offering differing interpretations. In this, the Dugdale-Shaw Varenius was a parallel project to Peter Shaw's other modernised editions of scientific classics by Bacon and especially Boyle, the aim being 'to reinscribe … [scientific] works within the socio-political and symbolic economies of early eighteenth-century science'.[70] The result is a page which replicates the appearance of the "variorum" edition wherein the opinions of different commentators stood shoulder to shoulder.[71] Unlike the variorum, which only fully emerges for vernacular texts with Johnson's edition of Shakespeare in 1765,[72] what Dugdale-Shaw provides is not a set of competing interpretations of a contested point in a manuscript or print tradition, but a sequence of additions to Varenius' by-now-obsolete geographical and natural-philosophical positions. Textually, the print format which results is identical and bespeaks the degree of reverence in which the author, Varenius, is held by the extent of the editorial labours which have been lavished upon his text at a time when the worth of this editorial structure was still controversial.[73]

One final typographic change needs to be noted, this time in the text itself rather than its many new and reformatted "paratexts". As "The Translator's Preface" notes:

> There is often not the least *Consonance* or *Similitude* between the *Latin* and
> *modern English* [geographical] *Names* … And very often their Names are changed

---

70   Robert Markley, *Fallen Languages: Crises of Representation in Newtonian England, 1660–1740* (Ithaca: Cornell University Press, 1993), 243. On Shaw as an editor see Jan Golinski, "Peter Shaw: Chemistry and Communication in Augustan England," *Ambix* 30 (1983): 19–28; Harriet Knight, "Re-Arranging Natural History Philosophy into Natural Philosophy: Eighteenth-Century Editions of Boyle's Works," in *Science and Beliefs: From Natural Philosophy to Natural Science, 1700–1900*, ed. Matthew Eddy and David Knight (Aldershot: Ashgate, 2005), 31–42.

71   Walsh, *Shakespeare, Milton*; De Grazia, *Shakespeare Verbatim*; Joanna Gondris, "'All this Farrago': The Eighteenth-Century Shakespeare Variorum Page as a Critical Structure" in *Reading Readings: Essays on Shakespeare Editing in the Eighteenth Century*, ed. Joanna Gondris (London: Associated University Press, 1998), 123–39.

72   Walsh, *Shakespeare, Milton*, 170; Simon Jarvis, *Scholars and Gentlemen: Shakespearian Textual Criticism and Representations of Scholarly Labour, 1725–1765* (Oxford: Clarendon Press, 1995), 160.

73   Unease with this structure amongst eighteenth-century critics is canvassed in Marcus Walsh, *Arguments of Wit and Sense: Eighteenth-Century Literary Editing and the Problem of Textual Knowledge* (Birmingham: University of Birmingham School of Humanities, 2000).

by *later Discoverers*, and their *Figures* and *Situations* better discovered since our Author's Time. Wherefore, in such Cases, we have taken the Liberty to *alter their Names, Situations, and Descriptions*, in order to make them conformable to our latest and best *English* Maps; *deviating* as little as possible from our Author's Sense; and making use of the *same Words* as 'tis likely he would have done, had he writ at the *same Time*, and in the *same Language*. We have done this to avoid, in some Measure, *Marginal Notes*, which must necessarily have been inserted to have explained a *strict Translation* ... These Alterations are included in *Brackets*, and for the most part distinguished by a different Character.[74]

Major transformations are announced in this excerpt, which are remarkable in the light of the previous editing of Varenius and in the context of geographical print culture more generally at this time, and bespeak an exceptional editorial reverence for Varenius. There is a clear recognition that Varenius wrote at a different and earlier point in geography's history such that errors made in the light of subsequent knowledge do not impair his authority as a geographer. The historicity of geographical knowledge is acknowledged for the first time, an editorial attitude which was emerging more generally in vernacular editing at this time and was the polar opposite of Bentleian editing. This attitude manifests itself in the way in which this edition is updated in important respects. The old 1650 dedication is omitted for the first time by Dugdale-Shaw, and Varenius' questions for the student about tides, longitudes, and so forth are updated and made to centre on London in the 1730s rather than the Amsterdam of eighty years earlier. Some of this work was done for the first, 1733, edition, whilst more anglicisation of Varenius' text takes place in the second, 1734, edition.[75] Again as a concession to the audience for a vernacular edition of Varenius, the 1734 edition added a set of theorems and 'paradoxes' in mathematical geography, derived from Patrick Gordon's highly popular *Geography Anatomiz'd* on the grounds that

to borrow from an Author, and to improve upon that Author, is not only justifiable, but has been the chief if not the only means by which the present Age has brought their Acquirements in Philosophy, and Mathematics, to the pitch to which they are now arrived; therefore to do justice to Mr *Gordon* in his Grammar, and yet to improve *Varenius* so as to make it the most excellent of it's Kind, the Publisher hath thought proper to add not only some *Theorems* ... but also some of the most useful of *Gordon's Paradoxes*.[76]

---

74   Varenius, *Complete System*, ed. Dugdale and Shaw, vii–viii.

75   Ibid., 266–7 for tide tables centred on London; 692 ff for a table of longitude and latitude centred on London; and 839 for the declination of the compass in 1734. Of these, only the longitude and latitude material was taken over from a previous edition, being in Varenius, *Geographia Generalis*, ed. Jurin, Appendix, 40–45.

76   Varenius, *Complete System*, ed. Dugdale and Shaw, 779. The theorems and paradoxes follow at 780–99.

Varenius is updated in recognition that both his mathematical geography and his pedagogic style may no longer suit an audience in the mid eighteenth century, but all this is done not silently, but with a clear respect for the *ipsissima verba* of the author. Even as the text is adjusted, so this is flagged for the reader typographically and the original is respected, something which marks a clear break with the editorial attitude of both the vast majority of geography books at this time and of previous editions of Varenius.

The overall aim of the changes made in the light of the passage of time is to recast what had always been – in both the Dutch Elsevier and the Cantabridgian editions – a Latinate student text for an erudite audience, as a vernacular work aimed at the burgeoning market for books of popular instruction. This shift helps to explain why, as well as removing Varenius' original dedication, the Dugdale-Shaw translation is the first edition of Varenius to include no dedication, merely containing a "Translator's Preface" to the audience. Instead of working in a system of patronage, be it aristocratic (as in Blome's editions) or academic (as in Jurin's work), Dugdale and Shaw were targeting the more anonymous reader in the public sphere. This new location for Dugdale-Shaw's Varenius is also apparent if one compares the works advertised by Stephen Austen on its flyleaf with those mentioned above on Jurin's flyleaf. Where the Jurin edition linked Varenius to classical authors, Austen chooses to advertise works of popular instruction aimed at an English speaking audience: *The Builder's Dictionary*, Worster's *Principles of Natural Philosophy*, *A New English Dispensary* and the seventh edition of Ogilby and Morgan's *Pocket Book of Roads*. Austen's flyleaf is retained in the 1765 reissue by Lacy Hawes, who himself specialised in similar projects and had a hand in publishing works of useful instruction such as Malachy Postlethwayt's *Universal Dictionary of Trade and Commerce* and Samuel Johnson's *Dictionary*.[77]

Above all, there is in the Dugdale-Shaw edition a demand to respect the authorial intentions of Varenius even as they are superseded, in that all changes made in the text itself as opposed to corrections in footnote commentaries are marked typographically by the use of square brackets and italic typeface. A good example of this is the updated section on 'large islands', a section which, as shown, was erroneous in the Newton, Blome and Jurin editions. Here, the names of the islands are updated to match the then-current English toponyms, but the reader is alerted to this editorial intervention by the use of brackets (Figure 6.6). Occasionally, the text even gives both Varenius' original geographical name and a corrected name in brackets.[78] Any deviation from the copy text is immediately apparent to the reader, unlike all previous editions which had allowed for the silent alterations which had caused so many confusions. It is also worth noting that in this regard the Dugdale-Shaw Varenius is very different from Peter Shaw's other great editorial labour, his 1725 edition of Boyle, wherein Shaw engaged

---

77   Plomer et al., *Dictionary of Printers, 1726–75*, 120.
78   Varenius, *Complete System*, ed. Dugdale and Shaw, 139.

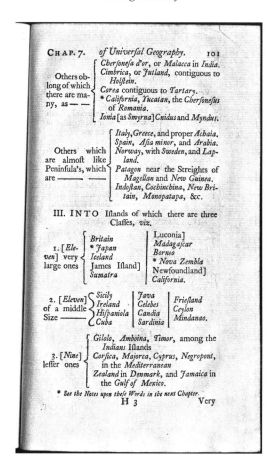

Figure 6.6   **An example of the bracketed alterations in the *textus receptus* of the *Geographia Generalis* made in the Dugdale-Shaw translation (1765 impression)**

in a 'radical, line-by-line rewriting of his predecessor's work'.[79] In the Dugdale-Shaw Varenius, a faithful translation of Varenius dominates the print space, with only very infrequent alterations in the text, the bulk of the "modernising" work being spatially demarcated by its place in the footnotes. In part this simply reflects changing editorial fashions as the silent editor was replaced by the editor as commentator.[80] But it is also tied up with a process by which the author as creator is accorded a new reverence in editorial practice whilst simultaneously

79 Markley, *Fallen Languages*, 217.

80 Jarvis, *Scholars and Gentlemen*, 43; De Grazia, *Shakespeare Verbatim*, 210; Lipking, "Marginal Gloss," 625–7.

being located in the culture from which they came, a move which saw revised theories and practices of translation develop at this time.[81] It should be noted, however, that as in the case of literary editing more generally at this time, the authorial intention which is respected is not that of the original author in their own time and place, in this case Varenius in 1650 Amsterdam, but the composite of intentions as represented in the *textus receptus*, in this case Jurin's 1712 revision of Newton's edition for Cambridge students. As Jarvis notes, 'the pervasiveness of *receptus*-thinking' had yet to be dispelled by a desire to produce fully historical editions and the Dugdale-Shaw edition of Varenius clearly stands on this editorial cusp.[82]

Putting all this together, the Dugdale-Shaw translation of Varenius places the *Geographia Generalis* in two publishing contexts simultaneously. On the one hand, the respect for the intentions of the copy text even as they are altered bespeaks a new reverence for the idea of an original text produced in a certain time and place. Varenius (as updated by Newton and Jurin) is seen as a classic text and is edited as such with an increasingly sophisticated awareness of the passage of time which bespeaks an emergent historical consciousness.[83] The result is a wholly new print space for the geography book in the English publishing tradition which mirrors that for classical texts and comes contemporaneously with the adoption of closely parallel typographical and editorial conventions for the first great editions of vernacular classics such as Theobald's Shakespeare of 1733. On the other hand, in making Varenius speak to an English audience in the mid eighteenth century, the Dugdale-Shaw project also recasts the text as a work of useful knowledge in line with other successful publishing ventures. A Latin text aimed at an erudite audience mutates into a work seeking broader appeal by being made relevant to the times, and by the attempt to make learning mathematical geography more entertaining. To modern eyes, the result is perhaps a bifurcated text which both respects and overrides authorial intention, a work which is at once an edition and a new book, a mishmash of historical respect, *receptus* adjustments and market-driven repackaging. But in this the Dugdale-Shaw Varenius is in fact representative of editorial conventions at this time for classical and vernacular works which moved easily between these different attitudes.[84]

---

81   Stuart Gillespie and David Hopkins, eds, *The Oxford History of Literary Translation in English. Volume 3: 1660–1790* (Oxford: Oxford University Press, 2005), 67.

82   Jarvis, *Scholars and Gentlemen*, 167.

83   J.G.A. Pocock, *Barbarism and Religion. Volume 2: Narratives of Civil Government* (Cambridge: Cambridge University Press, 1999); Mark Sabor Phillips, *Society and Sentiment: Genres of Historical Writing in Britain, 1740–1820* (Princeton: Princeton University Press, 2000).

84   See Walsh, Jarvis and De Grazia for this in eighteenth-century vernacular editing; Reynolds and Wilson and Pfeiffer for the same in classical editing.

Conclusion: Editing, Geography and Book History

This essay points to ways in which editing as an issue fruitfully conjoins the rubrics of geography and the history of the book and does so at two levels. First, an attentiveness to the findings of historians of the book can inflect and deepen our narratives of the history of geography as a discipline. Secondly, editing points to substantive ways in which spaces of book production, reception and dissemination come to exhibit geographical difference.

Taking the first issue, this essay shows that paying attention to the print history in general and the editorial history in particular of geographical texts can challenge received wisdom about the history of geography. In the case of Varenius, English print history initially appears to dissolve many of the platitudes about the importance of the text: in quantitative terms, Varenius was a successful but not exceptional title in comparison with other geographical works of the time. Yet this analysis of the editing of Varenius at least partially reinstates his centrality to the history of geography at the qualitative level: in Jurin's work and especially in the labours of Dugdale and Shaw, we find Varenius being accorded an exceptional status. His text is deemed of sufficient worth to warrant editing as a modern classic: 'to provide vernacular texts with an apparatus ... tacitly claimed a degree of prestige for them (because that apparatus was of the type which conventionally had accompanied the work of the revered *auctores*) ... techniques of exposition traditionally used to interpret "ancient" authorities are being used to indicate and announce the literary authority of a "modern" work'.[85]

To take editing as a topic of interest in its own right is to reverse the interpretive priorities of previous historians of geography with regard to the English editions of Varenius. A major component of Varenius' canonization to date has always been his association with Newton, other editions receiving less attention. The present analysis suggests that Newton's editions are in many ways the least interesting as they essentially reprinted the 1671 Elsevier edition for his Cantabrigian audience with minimal editorial labour on his part. Paradoxically, Newton treated his Varenian labours as mere hackwork, yet it is his connection with Varenius that has in good part led to the canonization of the *Geographia Generalis*. By contrast, the less celebrated efforts of Jurin, Dugdale and Shaw actually bespeak the esteem in which Varenius was held at the time. Clearly, who edits a text and how can play a major part in determining its status and even, in the case of Varenius, its enshrinement in the canon around which disciplinary and intellectual history forms itself.

On the second point, historians have shown increasing interest in the geographies of knowledge production and reception, the ways in which ideas and

---

85   A.J. Minnis, *Medieval Theory of Authorship*, 2nd edition (Aldershot: Wildwood House, 1988), xi and xiii.

books take on different meanings in different spaces.[86] Emergent from the present discussion is the suggestion that one of the reasons for this which has not been canvassed to date is the role of editing: a text like the *Geographia Generalis* took on very different appearances at different places in different times thanks to the labours of different editors. Blome produced a massive folio, where the original Elsevier editions were in the less impressive duodecimo size. The sparse pages of Newton's editions had a different impact on the reader from the complex page of text and commentary which the Dugdale-Shaw edition created out of respect for Varenius' *auctoritas*. A Latin edition *ipso facto* had different implications in terms of authorial reverence and intended readerships from a vernacular translation.

Editing, then, creates two interactions between geography and the book of huge importance to questions concerning the geography of knowledge. First, editing is key to creating the space of the page, the interaction of blank space and print which materialises text into book. If, in line with Jerome McGann and Donald McKenzie, we accept that 'the material forms of books, the non-verbal elements of the typographic notations within them, the very disposition of space itself, have an expressive function in conveying meaning', then to the extent that editing creates this space of the printed page, it also creates meanings which we must examine.[87] The printed page of the different editions of Varenius discussed here varied hugely thanks to editorial work and set different patterns of meaning production and reception. Secondly, how editing is conducted will vary over space, thereby creating a geography to the editing of texts. In the case of Varenius, the original Dutch edition was a dense page of text, thinned out and made easier to the eye in the 1671 edition. The Cambridge editions of Newton and Jurin essentially retained the print appearance of the 1671 Elsevier edition. Blome's folio and the Dugdale-Shaw edition created different appearances for Varenius unique to Britain. It is interesting to note, then, that the Latin editions produced in England and seeking an international audience in the Republic of Letters stayed as close to the appearance of the prestigious Elsevier press as possible, where the vernacular editions developed distinctive print formats. Different editions would have different appearances according to where they were produced, under what stylistic influences and for what audience. As a result, it is possible to envisage

---

86   David N. Livingstone, *Putting Science in its Place: Geographies of Scientific Knowledge* (Chicago: University of Chicago Press, 2003); Charles W.J. Withers, *Geography, Science and National Identity: Scotland since 1520* (Cambridge: Cambridge University Press, 2001); Idem., *Placing the Enlightenment: Thinking Geographically about the Age of Reason* (Chicago: University of Chicago Press, 2007); Secord, *Victorian Sensation*; Simon Naylor, ed., "Historical Geographies of Science – Places, Contexts, Cartographies," *British Journal for the History of Science* 38 (2005): 1–100.

87   Donald F. McKenzie, *Bibliography and the Sociology of Texts,* 2nd edition (Cambridge: Cambridge University Press, 1999), 17. In geography, see Robert J. Mayhew, "Materialist Hermeneutics, Textuality and the History of Geography: Print Spaces in British Geography, *c.* 1500–1900," *Journal of Historical Geography* 33 (2007): 466–88.

complex geographies even for the same title at a set of spatial scales. On the one hand, to the extent that there were distinctive national print cultures, these could generate different formats for the same work. But, as the example of Cambridge University Press's incorporation of Dutch typefaces shows, there were complex networks in the print trade which transcended the national scale and could lead to similarities across borders as in the Latinate editions of Varenius. Equally, the English-language editions of Varenius show that within a nation, complex geographies of the book could emerge thanks to different editorial practices, the format, market and intended readership for Blome's translation, for example, being markedly different from that for the Dugdale-Shaw translation.

Editing should not merely be seen as a function of decisions made by commercially-driven booksellers, nor just as a variant on the author function. This is not because editing is not driven by these two poles: on the contrary it obviously is. It is because there is, equally, intellectual capital to be made out of also treating editing in its own right as we look to the ways of creating intersections between intellectual history and the history (or historical geography) of the book. To take editing as a subject means seeing it as creating a suite of developments, adaptations and adjustments which make a text into a book, or an extant book into a new one, the sum of these decisions powerfully affecting the space of the printed page and, thereby, the projected image of the authorship of and implied audience for the book. It is by attending to these decisions that we can rewrite our disciplinary histories in the light of historical geographies of the book.

Acknowledgements

Thanks for comments and help to: Ann Blair, Michael Hunter, David Livingstone, Fred Lukermann, David McKitterick, Robin Myers, Miles Ogborn, William St Clair and Charles Withers. My thanks also extend to the Leverhulme Trust, whose award of a Philip Leverhulme Prize allowed for the completion of this chapter.

PART III
GEOGRAPHIES OF RECEPTION

Chapter 7

Geography, Enlightenment and the Book: Authorship and Audience in Mungo Park's African Texts

Charles W.J. Withers

Surveying the connections between the Enlightenment, book history and print culture, one leading scholar has recently called for 'a kind of book history that takes seriously and explores fully – in multiple genres and in local, national and international contexts – the values, aspirations, actions, and interactions of eighteenth-century authors and publishers, and that does not seek to restrict one to the realm of the mind and the other to the realm of the purse'.[1] Sher's own intentions to this end in his *The Enlightenment & The Book* centre upon the Scottish Enlightenment, and in showing how eighteenth-century Scottish authors and their books made and promoted the Enlightenment and so served the interests of British publishers. He shows how the re-printing of Enlightenment texts by yet other publishers helped extend publishing and Enlightenment ideas into new realms. Such an interpretation, notes Sher, requires attentiveness to the geographical dimensions of publishing and of authorship: 'the key to understanding the relationship between print and knowledge lies not in any particular local context but rather in the dynamic interplay of authors, publishers, and other members of the book trade in a variety of locations. Enlightenment book history must be viewed through a wide geographical lens'.[2] Sher demonstrates the critical importance of Edinburgh and London as the centres of Scottish Enlightenment book publishing in Britain and shows how re-printing and publishing in Dublin, Boston and Philadelphia helped disseminate Enlightenment ideas further afield.

One Enlightenment author whose books interested audiences in Britain, America and across Europe and whose life, work and publishing history illuminates these issues was the Scots-born African explorer, Mungo Park. His life and geographical achievements have been the focus of considerable attention.[3] Park was the first

---

1   Richard B. Sher, *The Enlightenment & The Book: Scottish Authors & Their Publishers in Eighteenth-Century Britain, Ireland & America* (Chicago: University of Chicago Press, 2006), 11.

2   Ibid., 9.

3   On Park's life and geographical achievements, see Charles W.J. Withers, "Mungo Park (1771–1806)," *Geographers: Biobibliographical Studies* 23 (2004): 105–15; Idem.,

European to return safely having observed for himself the west-to-east course of the River Niger in West Africa. In so doing, he helped solve part of a 2,000-year-old geographical problem concerning the Niger's course. His enduring fame results from that fact, from the widely-acclaimed much reprinted account of his Niger journey, *Travels in the Interior Districts of Africa* (first published in London in 1799), and from his heroic failure and death on a second and unsuccessful Niger expedition which had aimed at solving the remaining problem of where the Niger terminated. Papers gathered together from this second expedition and other sources formed the basis to his second, posthumous, book, *Journal of a Mission in the Interior of Africa in the Year 1805*, published in 1815. The importance of this latter book rests also in the fact that it contains accounts of the circumstances surrounding Park's death and *An Account of the Life of Mr Park (by John Whishaw)* which set in train the varied biographical treatment of Park.[4]

Sher and others have shown that Park's 1799 book was a 'best seller', that Park secured one thousand guineas (£1,050) for the first edition, and that, by January 1800, the profits from three further editions had reached 'about two thousand pounds' with more to come. The success of Park's 1799 *Travels in the Interior Districts of Africa* owed much to Park's personal credibility as an ordinary man made good through his own endeavours, to his claims that his book was 'a plain unvarnished tale; without pretensions of any kind, except that it claims to enlarge, in some degree, the circle of African geography' and to his incorporation of a frontispiece illustration of himself (Figure 7.1) which 'helped establish Park's persona as a credible, even heroic figure whose stories of distant adventures could be believed'.[5]

Such a treatment of Park's 1799 book and of Park's self-representation is largely focused upon questions of production, and, more especially, the final production of the printed text: its publication, price, the use of frontispiece illustrations as indications of authorial credibility, the financial rewards to the author. Such attention is consistent with Sher's survey overall, for while he is aware of the importance of location and the mobility of knowledge over space, and seeks to emphasize the 'dynamic interplay' between authors and publishers, his account is production-oriented. Of course, looking anew at production may allow insight into that other significant feature of book history, namely the reception of texts and, notably, questions of review and reading: 'The process of making and marketing books and the nature of books themselves must be recognized as crucial factors in shaping reading and reception'.[6]

"Park, Mungo," in *The Oxford Companion to World Exploration*, 2 vols, ed. David Buisseret, (Oxford: Oxford University Press, 2007), 2: 141–2. Amongst the numerous biographies of Park, the fullest is Kenneth Lupton, *Mungo Park, The African Traveler* (Oxford: Clarendon, 1979).

4   Charles W.J. Withers, "Memory and the History of Geographical Knowledge: The Commemoration of Mungo Park, African Explorer," *Journal of Historical Geography* 30 (2004): 316–39.

5   These points are from Sher, *The Enlightenment & The Book*, 217, 185–7.

6   Ibid., 32.

Mr. M. Park.

Publish'd April 4.1799, by G. Nicol, Pall Mall.

**Figure 7.1   Park's portrait, from an engraving by Eldredge, which
appears as the frontispiece to his *Travels in the Interior Districts
of Africa* (1799)**

Source: Reproduced with the permission of the Trustees of the National Library of
Scotland.

My concern here is also largely production oriented in focusing on Park's
two African texts before and after they became printed works. Exploring the
relationships between print and knowledge in Park's case involves assessment of
the making of geographical knowledge in the field and its production, in the form
of a best-selling printed book, in London. This geographical displacement was

also an epistemological one. What follows argues that Park's 1799 book was not wholly his – simply, that what he produced as written text and as maps of his explorations, was worked upon by others in a variety of ways. One consequence of this is that, between the surviving manuscript evidence and the printed account, different versions of Africa's geography and of Park's geographical explorations were being produced. More significantly, the story of the making of Park's *Travels in the Interior Districts of Africa* allows us to question notions of authorial credibility of and for Park, to cast light on the idea of the author as an autonomous agent, of authorship as a process and so to consider the geography of the book as a matter of social interaction in and across space and in place. The purposes behind publication of Park's 1815 *Journal of a Mission* were likewise multiple: for some, the book was to provide clues to the circumstances of Park's death; for Whishaw, it was a means to aid Park's widow; for John Murray, the publisher, the book was an opportunity to cash in on tales of hazardous geographical exploration even as he declared philanthropic intentions towards Park's family.

In considering the motivating purposes, cognitive content and the authorial production and redaction of Park's African texts, my purpose is not just critical exegesis with a view to the more complex 'truth' behind Park's 'plain unvarnished tale'. This interpretation of Park and his books is intended to raise questions of wider applicability about the meaning of "production" and "authorship", readily used in book history but perhaps too often as taken-for-granted analytic terms. Similarly, by considering the subscription lists associated with the first London edition of Park's *Travels* and with the first American edition of this book, published in Philadelphia in 1800, I hope to raise similar questions of the terms "reception" and "audience" and about their geographical interpretation and constitution. While it may be possible to see the publication of Park's 1799 book in America as one expression of a wider geographical Enlightenment and so test Sher's general thesis about the wider geographies necessary to comprehend book histories, my concern is less with the geographical reach of Park's book. It is more with the different geographies *in* the book and with the complex social and intellectual processes that lie behind its production in print and reception in space. The geography of Park's books – the geographical making of Park's books – is, I contend, to be understood differently in relation to questions of production and reception. For the first, Park's 1799 work was made in Africa and, chiefly, in London, but only partly by Park. Of the second, what we know of Park's purchasers suggests that his reception in the later Enlightenment on both sides of the Atlantic was, largely, a metropolitan affair. Park's book geographies were dispersed complex affairs, not the simple locational expression of simple categories of author, publisher and audience.

Mungo Park and the Niger Problem

Mungo Park was born in Foulshiels in the parish of Selkirk on or about 10 September 1771. His father, also Mungo, was a tenant farmer, his mother, Elspeth

Hislop, the daughter of a tenant farmer. Park was educated at home and at Selkirk Grammar School. His father intended Park for the ministry but, in 1786, he was apprenticed to Thomas Anderson, a surgeon in Selkirk. In 1789, Park was admitted to the University of Edinburgh to study medicine. He there came under the instruction of leading Enlightenment men including Joseph Black the chemist, Alexander Monro *secundus* for anatomy and surgery, Daniel Rutherford for botany and John Walker in natural history. By 1791, Park had gained his surgical diploma but was without gainful employment. In association with his brother-in-law, James Dickson, a botanist, Park undertook botanical excursions throughout the Highlands of Scotland in the summer of 1792. Dickson had earlier secured the patronage of Joseph Banks, and Dickson introduced Park to Banks. Through Banks' influence, Park was appointed to the position of Assistant Surgeon on the East India Company's ship *Worcester* which sailed for Sumatra in February 1793. Park's stay at Benkulen, the East India Company's station in Sumatra, provided him with time to sketch and to collect botanical and zoological specimens, some of which Park presented to Banks upon his return to London in May 1794.

Strictly, it is inaccurate to see Park's Sumatran experience as an apprenticeship for his African travels. Yet Park's work as a physician and natural historian and his experience of foreign travel as well as his declaration of interest in related work certainly commended him to Banks. In May 1794, Banks advanced Park's name to the African Association, and, in July that year, interviewed him with a view to West African exploration on behalf of the Association. Park was formally appointed from 1 August 1794. The African Association had been founded in June 1788. Its intentions were plain: 'That as no species of information is more ardently desired, or more generally useful, than that which improves the science of Geography; and as the vast Continent of Africa, notwithstanding the efforts of the Antients [sic], and the wishes of the Moderns, is still in a great measure unexplored, the Members of this Club do form themselves into an Association for Promoting the Discovery of the Inland Parts of that Quarter of the World'.[7] Enlightenment concerns to extend geographical knowledge of Africa were paralleled by interest in the continent's commercial possibilities: both matters came together in attention to the River Niger.

The Niger Problem and Park's First Niger Journey

By the later eighteenth century, the Niger was a 2,000-year-old two-part geographical problem. The first part concerned the course of the river: did it flow east-west as some hypothesized, or west-to-east as others argued? The second part concerned where the river ended. Herodotus in the fifth century BC believed that

---

7   *Proceedings of the Association for Promoting the Discovery of the Interior Parts of Africa. In Two Volumes* (London: Bulmer, 1810), and also edited, in two volumes, by Robin Hallett (London: Dawson Facsimile Series, 1967), I, 9. Hereafter referenced as Hallett, *Proceedings*.

it flowed west-east and that it eventually joined the headwaters of the Nile. In the second century AD, Ptolemy subscribed to a similar view, but held that the river terminated in a great inland lake. Islamic geographers had also written upon the subject. The first, Sharif al-Idrisi, accepted Ptolemy's argument for the existence of a central lake from which the Nile flowed northwards. But he also asserted that the Niger, what he termed the *Nil as-Sudan* – the Nile of the Sudan – flowed *westwards* out of this lake. Leo Africanus, writing in 1526, confirmed al-Idrisi's view that the Niger flowed from east to west, but denied the existence of the inland lake. In the early and mid eighteenth century, French geographers such as DeLisle and Buache published maps derived from officials in the Senegambia region which gave the Niger a west-east flow but showed the Niger ending in an inland lake.

In short, no one could agree which way the Niger went and where it ended. The work of Arab geographers contradicted classical authorities. European map makers were not agreed. But because one of the several theories about the Niger was that it flowed across Africa and joined up with the headwaters of the Nile, solving the Niger problem was of considerable commercial interest to the British and to the French and would add to the sum of late Enlightenment geographical knowledge.[8]

With the backing of government, notably of Banks, with some scientific training and an enduring geographical question at the heart of his concerns, Park sailed for West Africa on 22 May 1795. He landed in the Gambia on 21 June 1795. His local contact was Dr John Laidley, honorary consul in Pisania, a slave and trading centre, to which Park travelled in July 1795. Park was ill with fever during the first two months of his stay. During his convalescence, he engaged in botanical study, learnt Mandingo, the principal language of the local peoples, and studied Islamic customs and laws.

Park's expedition to determine the course of the Niger and to know more of the region's lands and peoples began on 2 December 1795. It lasted until May 1797. In truth, it was not much of an expedition. Park was accompanied by only two other persons: Johnson, a freed slave, who acted as interpreter, and Demba, a servant. Park's records of his travels are a mixture of personal adventure, commentaries upon the trading networks of the interior and ethnographic descriptions of the locals, notably of the Mandingo people. These were, he wrote, 'of a mild, sociable, and obliging disposition'.[9] He comments upon the slave trade. He was himself an object of ethnographic wonder. He is quizzed by the ladies of one court over his whiteness – which they attributed to his being dipped in milk as a child – and over 'the prominency of my nose [which] ... they insisted had been pinched every day, till it had acquired its

---

8   Charles W.J. Withers, "Mapping the Niger 1798–1832: Trust, Testimony and 'Ocular Demonstration' in the Late Enlightenment," *Imago Mundi* 56 (2004): 170–93.

9   Mungo Park, *Travels in the Interior Districts of Africa* (London: W. Bulmer for G. Nicol, 1799), 21.

own present unsightly and unnatural conformation'.[10] He was also threatened with being inspected to see if he was circumcised. Park's response was 'that it was not customary in my country to give ocular demonstration in such cases, before so many beautiful women; but that if all of them would retire, except the young lady to whom I pointed [selecting the youngest and handsomest], I would satisfy her curiousity'.[11] Not everyone treated him kindly. He was regarded with suspicion – by the Moors because he was a Christian and by others who simply refused to believe that his justification for travelling was to observe the course of a river – and he was on more than one occasion imprisoned and threatened with death. Park reserved his criticism mainly for the Moors: 'at once the vainest and proudest, and, perhaps, the most bigoted, ferocious and intolerant of all the nations on the earth: combining in their character, the blind superstition of the Negro, with the savage cruelty and treachery of the Arab'.[12]

Park accomplished the principal aim of his travels on 21 July 1796. As he wrote, 'I saw with infinite pleasure the great object of my mission – the long sought for majestic Niger, glittering in the morning sun, as broad as the Thames at Westminster, and flowing slowly *to the eastward*' [original emphasis].[13] Following this, he travelled further east to the Moorish kingdom of Timbuktu (but not to the town of that name) and to the settlement of Sansanding on the Niger. In late July 1796, near the town of Silla, Park turned westwards, initially intending to explore the southern bank of the Niger. In late August 1796, he was again beaten and robbed. Park's account often describes his frailty in the face of an adversarial nature and his good fortune when threatened. At one point – in which he recalls how his manuscript notes survived only because they were in his hat band – he was robbed and stripped by Foulah bandits.

> Humanity at last prevailed: they returned me the worst of my two shirts, and a pair of trowsers; and as they went away, one of them gave back my hat, in the crown of which I kept my memorandums; and this was probably the reason they did not wish to keep it. After they were gone, I sat for some time looking around me with amazement and terror. Which ever way I turned, nothing appeared but danger and difficulty. I saw myself in the midst of a vast wilderness, in the depth of the rainy season; naked and alone; surrounded by savage animals, and men still more savage. I was five hundred miles from the nearest European settlement. All these circumstances crowded at once on my recollection; and I confess that my spirits began to fail me. I considered my fate as certain, and that I had no alternative, but to lie down and perish. The influence of religion, however, aided and supported me. I reflected that no human prudence or foresight, could possibly have averted my present sufferings. I was indeed a stranger in a strange land, yet

---

10   Ibid., 56.
11   Ibid., 132.
12   Ibid., 160.
13   Ibid., 194.

I was still under the protecting eye of that providence who has condescended to call himself the stranger's friend.

Reflections like these, would not allow me to despair. I started up, and disregarding both hunger and fatigue, travelled forwards, assured that relief was at hand: and I was not disappointed.[14]

Park completed his return journey on foot. He survived largely due to the kindness of an Arabic slave trader, Karfa Taura, who fed and sheltered Park in return for the fee of one slave which Park would provide in the event of his safe return to the Gambia. Park, in effect, enslaved himself to ensure his own safe passage. Park was resident in Taura's care in the town of Kamalia for seven months from mid September 1796, leaving there on 19 April 1797. Immobility in Kamalia afforded Park further opportunity for reflection. 'Being now left alone and at leisure to indulge my own reflections, it was an opportunity not to be neglected of augmenting and extending the observations I had already made, on the climate and productions of the country, and of acquiring a more perfect knowledge of the natives, than it was possible for me to obtain in the course of a transient and perilous journey through the country.'[15] Chapters on the geography of the region were thus put together not as a result of Park's movement through the country, but from reflection and accumulation whilst immobile. This is a common motif of Enlightenment geographical exploration and a problem of science's making more generally: observation borne of in-the-field mobility lends immediacy and the benefits of direct encounter and personal credibility to reported facts – even allowing that the sense of immediacy is often subsequently and elsewhere manufactured through the craft of writing – but the observer is always compromised by the wealth of data and by the frailties of memory. Sedentary reflection upon accumulated evidence affords time for measured thought but raises doubts over the truth claims of what is reported upon and by whom.[16]

Park reached the Gambia River on 31 May 1797, and, on 7 June 1797, returned to Pisania, where, on 12 June, he was reunited with Laidley who greeted Park 'with great joy and satisfaction, as one risen from the dead'.[17] Park returned to London via the West Indies and Falmouth and, on the morning of Christmas Day in 1797, met up (by chance) with his brother-in-law James Dickson as the two walked in the gardens of the British Museum.

------

14   Ibid., 243–4.

15   Ibid., 257.

16   Dorinda Outram, "On Being Perseus: New Knowledge, Dislocation, and Enlightenment Exploration," in *Geography and Enlightenment*, ed. David N. Livingstone and Charles W.J. Withers (Chicago: University of Chicago Press, 1999), 281–94.

17   Park, *Travels in the Interior Districts of Africa*, 358.

Making Enlightenment Travel: Park's *Travels in the Interior Districts of Africa*

Questions of Authorship and Authenticity

Park's account of his Niger work, in full *Travels in the Interior Districts of Africa, performed under the Direction and Patronage of the African Association in the Years 1795, 1796, and 1797*, was printed and published by the London fine printer William Bulmer in April 1799. All 1500 copies of the first edition sold out within a month. The principal selling agent in London was the firm of G. and W. Nicol in London's Pall Mall. In addition to being the King's bookseller, Nicol was a close friend to Sir Joseph Banks and bookseller to the African Association: Bulmer was printer to that body. Since Park's book was well received, Nicol, more so than Bulmer, recognized the importance of keeping the book in print with further editions: writing to Banks on 7 May 1799, he observed 'If we allow it [Park's *Travels*] to be long out of print, some fresh Tub will be thrown to the Whale, and poor Mungo will be forgotten'.[18] Banks, reporting to the African Association on 25 May, in turn noted that 'Mr. Nicol, of Pall Mall, the publisher [sic], in consequence of the rapid sale which the book had experienced, with a liberality which ever marks his character, had undertaken to publish a second edition for Mr Park, without retaining any part of the proceeds of the first, for the purpose of defraying the further expense'.[19] Two more London editions appeared that year. French, German, Dutch and Italian translations followed. The first American edition, printed from the first London quarto edition, was published in Philadelphia in 1800 by James Humphreys. An edition was also published in New York that year. Editions were published throughout the nineteenth century: in Dublin, in Huddersfield and Newcastle and in other towns in Britain and later editions appeared in Philadelphia and in New York.[20] The fact that Park's 1799 *Travels* is still in print is testimony to an enduring interest in the work, its author and the geographical exploration of Africa in the late Enlightenment.

Rather than read Park's success then and *Travels'* later print history as a chronology of production of a successful exploration narrative, I want to interrogate the initial making of the book and the meanings ascribed to it by contemporaries. It is clear that the book's initial publication and its further editions owed much to enduring geographical concerns and particular social networks which centred upon the African Association. Contemporary reviewers thought well of the book because it addressed geographical questions. The *Gentlemen's Magazine* stressed Park's contribution to geographical knowledge – for, in effect, answering the first

---

18 This is from Peter Isaac, *William Bulmer The Fine Printer, 1757–1830* (London, Bain and Williams, 1993): the letter from Nicol to Banks is cited on page 47.

19 Hallett, *Proceedings*, II, 3.

20 K.G. Saur, *Bibliography of American Imprints to 1901* (New York: American Antiquarian Society, 1993), 227.

part of the Niger problem. So, too, the *Edinburgh Magazine* which remarked how 'a considerable portion of Africa is now known, which hitherto has been impervious to every traveller, and to no one has the world been so much obliged as to this gentleman'.[21] One modern commentator has considered the text, with its dramatic scenes and unassuming style, a touchstone for European traveller-writers for decades to follow; others take Park to be an epitome of the simply 'curious' or a Romantic traveller.[22]

We need to be careful, however, in considering the production and the reception of Park's 1799 work solely in regard to its edition and reprint history for the former and in terms of reviews for the latter. To understand why this is so, it is important to note the essential elements that together constituted Park's 1799 *Travels*. The book is made up of text, a frontispiece portrait of Park, five engraved plates, a song from 'Mr Park's Travel' [with words later added by the Duchess of Devonshire], three folding maps and a lengthy appendix, entitled 'Geographical Illustrations of Mr. Park's Journey By Major Rennell'. Three of the plates were topographic scenes from Park's journey (the other two are of plant specimens), and were clearly drawn after his travels and are described in the book as 'J.C. Barrow del. From a Sketch by Mr. Park'. No manuscript version of these sketches is known to exist. These materials are not always positioned consistently in relation one to another. In different later editions of the *Travels*, and even in some copies of the first 1799 edition, the positioning of the maps is varied, a reflection probably of different binding instructions from different customers.

The text by Park, two of the maps and the geographical appendix by Rennell are the most important features here. The first map, commonly folded to the front of the text, shows the route of Park's travels. The second, commonly placed after Park's narrative and before Rennell's geographical appendix, was of North Africa and was entitled 'A Map shewing the Progress of Discovery & Improvement in the Geography of North Africa'. It was published for Rennell by act of Parliament on 25 May 1798. This map and the route map of Park's travels help illuminate the complex authorship of Park's 1799 book. (The third map, of magnetic variations in the Atlantic, also by Rennell and used by him to elaborate a point in his 'Geographical Illustrations', is unimportant in this context).

Let me turn first to the text. Initial reviews of Park's 1799 work were based not on the printed book but on a published *Abstract* of it prepared, using Park's notes in summary form, by Bryan Edwards, Secretary to the African Association, an appointment he owed to Banks' influence. Edwards, then MP for Grampound in

---

21   Lupton, *Mungo Park*, passim.

22   Mary Louise Pratt, *Imperial Eyes: Travel Writing and Transculturation* (Routledge: London, 1992), 74–82. On Park in the context of travel writing and curiosity, see Nigel Leask, *Curiosity and the Aesthetics of Travel Writing, 1770–1840* (Cambridge: Cambridge University Press, 2002). On Park the Romantic traveller, see Richard Holmes, *The Age of Wonder: How the Romantic Generation Discovered the Beauty and Terror of Science* (London: Harper Press, 2008), 211–34.

Cornwall, had been a colonist and plantation owner in Jamaica for most of his life. An historical author in his own right,[23] Edwards was, broadly, pro-slavery although in more emancipated forms than advocated by contemporaries. Nevertheless, the connection between the Association's at least tacit support for slavery and Edwards' direct involvement in the question was to be a factor in shaping Park's 1815 book *Journal of a Mission*.

In respect of Park's first book, Edwards occupied a key role between Park's notes and the finished volume. Under Banks' directions – and because he, Edwards, was Secretary to the Association rather than from any particular literary or editorial qualities – Edwards assisted Park, then resident in London, in revising his notes and improving his style for what would become *Travels*. This role changed the better Park's writing became. 'Park goes on triumphantly' [Edwards wrote to Banks]. 'He improves in his style so much by practice, that his journal now requires but little correction; and some parts, which he has lately sent to me, are equal to any thing in the English language'.[24]

Park's *Travels* thus emerged in part from the preliminary Abstract and from Edwards' close tutelage of Park as author. Whether this fact was widely known and if it was, whether it raised doubts as to the nature of Park's authorship and the content of his work is not clear. Such evidence as there is suggests this matter of authorial credibility was of some concern. At the meeting of the African Association on 25 May 1799, it was noted that 'the public opinion on the merits of the work [Park's *Travels*], called for acknowledgement, that it had been accomplished in a manner creditable to the author, and honourable to his employers'.[25] Elaborating upon this question, Sir Joseph Banks observed:

> that the recital of discoveries and events throughout, bore every mark of veracity; that the remarks and observations interspersed, were intelligent and unassuming; and that the whole was conveyed in a stile unaffected, clear, and such as might distinctively and properly be termed "the language of truth". In such language, Mr Park's character bore a representation which did honour to him, and reflected credit on the choice of the Society [the African Association]. Strength to make exertions; constitution to endure fatigue; temper to conciliate; patience under insult; courage to undertake hazardous enterprise, when practicable; and judgement to set limits to his adventure, when difficulties were likely to become insurmountable, were every where exemplified in his book.[26]

Banks offered additional observations regarding Park and Edwards:

---

23   Edwards's principal works were *History, Civil and Commercial, of the British Colonies in the West Indies* 2 vols. (London: John Stockdale, 1793) and *An Historical Survey of the French Colony in the Island of St Domingo* (London: John Stockdale, 1797).

24   Hallett, *Proceedings*, II, 8.

25   Ibid., II, 1.

26   Ibid., II, 2.

In arrangement of the materials of this Journal [Park's book], and in the manner
and stile with which it has been introduced to public notice and approbation,
Mr. Park professed his acknowledgements, to the advice and critical taste of the
worthy Secretary of the Society.

The literary character of Mr. Edwards, stood too high in the learned world, to
require testimony; but the candour of the avowal was honourable to Mr Park,
and vouched his general accuracy, and laid claim to further confidence.[27]

Banks is here attributing epistemological certainty to the content and to the
literary style of Park's work – and, thus, imparting credibility to the book – by
appeal to the moral quality of its author, Park, even as he recognizes Edwards'
intervention. He is doing so by an appeal to the virtue of Park's conduct, in Africa
for his geographical fortitude and in London for his gentlemanly preparedness in
acknowledging Edwards' assistance. In thus bestowing upon Park the man and
Park the author the appropriate qualities of what Shapin, with reference to an
earlier period, has termed 'epistemological decorum'[28] – the associations between
social status, the cognitive content and plausibility of recounted tales and, thus,
the moral basis to the epistemology of authorship – Banks is also establishing the
credibility of himself and of the African Association. Authorship and authenticity
are here being attested to in terms of moral qualities: in the man, in the field, in
London literary circles *and* by virtue of association with a particular institution
and individuals of influence.

These claims are substantiated when we examine the maps within Park's
Travels. For at the same time as Edwards was acting to assist Park's words
concerning his African travels, Park's maps were put to order by James Rennell.
A leading geographer, map maker and hydrographic surveyor, Rennell had earlier
established his credentials in leading mapping work in Bengal but, following injury,
he lived in Britain, mainly in London, from 1778 until his death in 1830.[29] He
was, in effect, resident geographical consultant to the African Association. Given
that body's interests and his own capacities, it is natural that Rennell should have

---

27   Ibid., II, 2.

28   Steven Shapin, *A Social History of Truth: Civility and Science in Seventeenth-
Century England* (Chicago: University of Chicago Press, 1994), chapter 5.

29   James Rennell (1742–1830) was a leading late Enlightenment map maker and
geographical author. His principal works focused on Bengal but he was also interested in
oceanography and in the historical and political geography of Asia and the ancient world,
of which his *Geographical System of Herodotus* (London, W. Bulmer, 1800) was one
part of a never-realised larger project. See Andrew S. Cook, "Major James Rennell and
A Bengal Atlas (1780 and 1781)," *India Office Library and Records Report for the Year
1976* (London: British Library, 1978), 5–42; Matthew Edney, *Mapping an Empire: The
Geographical Construction of British India, 1765–1843* (Chicago: University of Chicago
Press, 1997); Clements R. Markham, *Major James Rennell and the Rise of Modern English
Geography* (London: Century Science, 1895).

Figure 7.2   Park's route as illustrated on the map in his 1799 *Travels*. From *Travels in the Interior Districts of Africa*, facing page 1

Source: With the permission of the Trustees of the National Library of Scotland.

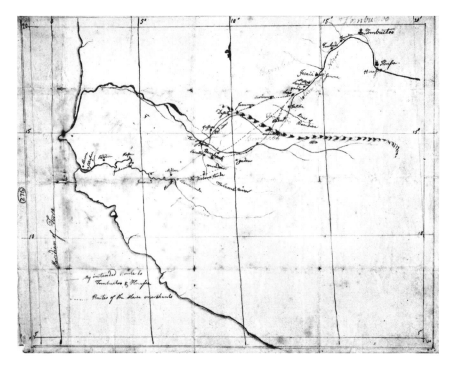

Figure 7.3   One of two manuscript maps of western Africa and the Niger River, drawn by Mungo Park at an unknown date and location: note his writing of the words 'My intended route to Tombuctoo [sic] & Housa'

Source: Reproduced with the permission of the Trustees of the National Library of Scotland.

turned to Park's findings. As Park put it: 'Major Rennell was pleased also to add, not only a Map of my Route, constructed in conformity to my own observations and sketches (when freed from those errors, which the Major's superior knowledge and distinguished accuracy in geographical researches, enabled him to discover and correct), but also a General Map, shewing the progress of discovery, and improvement in the geography of North Africa'.[30] Park is here referring to the map of his travels and to Rennell's more general map of 1798.

Park's route between December 1795 and May 1797 is clearly shown on the map that appears in his 1799 *Travels* (Figure 7.2). What is not clear from this evidence, however, is either what the 'errors' were to which Park refers, exactly what Rennell used as evidence for the amended version other than Park's direct testimony and his own 'superior knowledge and distinguished accuracy', and

---

30   Park, *Travels in the Interior Districts of Africa*, viii–ix.

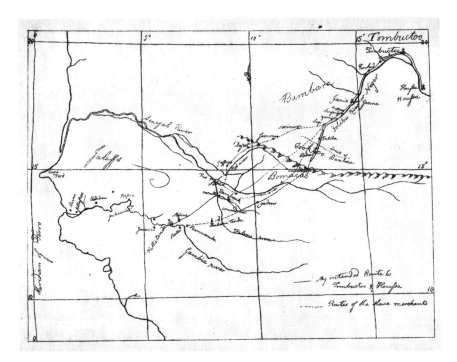

Figure 7.4   The second of the two manuscript maps of western Africa and the Niger River, drawn by Mungo Park. The form of the lettering and the amount of detail suggests, perhaps, that this is a later refinement of the partner map (Figure 7.3)

Source: Reproduced with the permission of the Trustees of the National Library of Scotland.

how and when Rennell went about such cartographic adjustment. Rennell was certainly fulsome in his praise of Park's achievement: 'The late journey of Mr. PARK, into the interior of WESTERN AFRICA, has brought to our knowledge more important facts respecting its Geography (both *moral* and *physical*), than have been collected by any former traveller'.[31] But he is silent on the matter and manner of his correcting Park. Banks tells us only that 'Major Rennell, who has never ceased to promote with diligence, and serve with extraordinary sagacity and learning, the views of the Association, had generously bestowed on Mr. Park, the property of his able Memoir on the Geography of Africa, compiled for the information of the Society, and annexed to the secretary's last report; and it now formed a valuable part of Mr Park's publication'.[32]

---

31   Ibid., viii.
32   Hallett, *Proceedings*, II, 3.

Two manuscript maps by Park do survive although it is not known when Park drew them (Figures 7.3 and 7.4).[33] It is possible, but unlikely, that these maps were amongst the 'memorandums' that Park kept in his hat and which survived his being robbed. But neither can we assume they were drawn by Park in London as sketches to assist Rennell. What is clear is that the map evidence varies and that, to judge from the manuscript maps' relative states – the detail and the lettering of the two rough copies (*cf.* Figure 7.3 and Figure 7.4) – Park refined the cartographic evidence for his Niger researches at least once before publication and that Rennell amended those results still further prior to their being engraved and printed.

With respect to his 1798 map and his 'Geographical Illustrations' in which the map featured, Rennell was working on both map and text before 1798. The fact that Rennell's 1798 map of North Africa presents an erroneous geography of the course of the River Niger – a reflection of Rennell's incorrect views of where the Niger terminated – has been discussed elsewhere.[34] What is important is the fact that Rennell was working on versions of this map as early as 1790. He was doing so on behalf of the African Association in order to place the routes of, and the geographical evidence arising from, the several explorers who had preceded Park in search of the solution to the Niger problem. On 27 March 1790, Rennell had by act of Parliament published a map entitled 'Sketch of the Northern Part of Africa: Exhibiting the Geographical Information Collected by The African Association'.[35] This map was intended as an accompaniment to Rennell's written text of March 1790, entitled 'Construction of the Map of Africa' which appeared in print later in 1790 in the first volume of the published proceedings of the African Association.[36]

Rennell continued to work on the geography of African exploration in the years immediately following his map and text of March 1790. This is clear from his 'Elucidations of the African Geography, from the Communications of Major Houghton, and Mr. Magra; 1791' which was compiled in 1793 and published in 1799.[37] And it is evident from manuscript additions he made to his 1790 map. Rennell noted how 'Since the publication of the Proceedings of the Society for promoting the discovery of the interior regions of Africa, two years have elapsed [Rennell is making reference to the first published volume of the Association's proceedings, in 1790], and in that period their Committee has received, thro' the medium of distinct and unconnected channels, new and interesting intelligence'. Rennell is here alluding to the geographical explorations of Major Houghton, who had earlier set out to solve the Niger problem on behalf of the African Association,

---

33   National Library of Scotland [NLS], MS 10290, ff. 3 and 4. There is no other evidence in the NLS Park material to allow more accurate dating of these maps.

34   Withers, "Mapping the Niger".

35   This map is held in the Archives of the Royal Geographical Society (with the Institute of British Geographers) as "Africa Div. 274" and in the Map Library of the British Library [BL] as Maps.14.dd.40.

36   Hallett, *Proceedings*, I, 209–36.

37   Ibid., I, 263–92.

and from whose surviving papers Rennell and others learned that the Niger ran west to east: Houghton's death before he could return, however, meant his claim could not be verified. Rennell further noted:

> But tho' we have now an assurance that the Niger has its Rise in the chain of Mountains which bound the Eastern side of the Kingdom of Bambouk, and that it takes its course in a contrary direction from that of the Senegal and the Gambia, which flows on the opposite side of the same Ridge, yet the place of its final destination is still unknown: for whether it reaches the Ocean, or is lost, as many of the Rivers of Mount Atlas are, in the immensity of the desert; or whether, like the Streams of the Caspian, it terminates in a vast inland Sea, are questions on which there still hangs an unpenetrated Cloud.[38]

Sometime in early 1792, Rennell added Houghton's information to a copy of his 1790 map. We know this from a letter to the King's Librarian, Frederick Augusta Barnard, from Henry Beaufoy, MP, member of the African Association and earlier, like Park, a student at the University of Edinburgh. Beaufoy recorded his pleasure at sending Barnard 'a Map by Major Rennell of the late discoveries of the African Association. Those discoveries are distinguished by Red Lines from the content of their knowledge in 1790, when they published, or rather printed for the use of their Members a narrative of their Proceedings'.[39]

Taken together, the evidence suggests that the text of Park's *Travels* was amended at least twice between its making in the field and its publication: from Park's own notes to draughts worked on by Park and Edwards to its final publication. The final map of Park's African travels – the adjusted route of his geographical exploration – was prepared in London, being revised by Rennell from manuscript sketches by Park. Rennell was, with others in the African Association, gathering information on African exploration well before Park set out. The 1798 map and the 'Geographical Illustrations', in preparation from at least 1790, were intended as supplements to Park's narrative in order to correct his account, and to provide credibility for Park through association with Rennell and the African Association. It is highly probable, then, since Rennell knew of Houghton's claims and had inscribed them in manuscript on a working copy of his 1790 map, that the west-to-east course of the Niger was 'known' before Park left Britain, and known by Park (see below). Park's endeavours were about confirmation and not 'discovery'. His acclaim was to do with the credibility of direct observation, his survival and his conduct in London rather more than with the facts he secured. Park had to acknowledge the authorial involvement of others or face criticisms from his public and from influential individuals within the organization that sponsored his work:

---

38   BL, Map Library, K.Top.117.241.d, f.11v.

39   BL, Map Library, K.Top.117.24.1.g, f.1. The important fact to note is that the "Red Lines" referred to are in Rennell's hand.

Availing myself therefore, on the present occasion, of assistance like this, it is impossible that I can present myself before the Public, without expressing how deeply and gratefully sensible I am of the honour and advantage which I derive from the labours of those Gentlemen; for Mr. Edwards has kindly permitted me to incorporate, as occasion offered, the whole of his narrative into different parts of my work; and Major Rennell, with equal good will, allows me not only to embellish and elucidate my Travels, with the Maps before mentioned, but also to subjoin his Geographical Illustrations *entire*.[40]

For these reasons, *Travels in the Interior Districts of Africa* is not wholly Park's book. And there is further evidence to suggest that what was published in London in 1799 was not a full account of the author's role in the geographical exploration in west Africa, several years before. This is Park's own admission regarding his endeavours. Recall Park's words in the Preface to his *Travels*: 'As a composition, it has nothing to recommend it but *truth*. It is a plain unvarnished tale, without pretensions of any kind, except that it claims to enlarge, in some degree, the circle of African geography'. In respect of his claims to enlarging contemporary understanding of African geography, Park's words strike true.

Yet what Park claimed in writing to be 'truth', a 'plain unvarnished tale' may not have been so. For in private conversation with Sir Walter Scott, Adam Ferguson and Dugald Stewart in 1804, Park admitted that his published work was only partial. Later commentary makes this plain: 'Mr Park himself was prepared, and with a judicious caution ... [to omit] the relation of many real incidents and adventures, which he feared might shake the probability of his narrative in the public estimation. This fact has been proved beyond doubt, by the testimony of many of his intimate friends and relatives, to who ... he freely mentioned many singular anecdotes and particulars, which he scrupled to admit to the jealous eye of the critical public'. Not unnaturally, Sir Walter Scott asked why: 'His answer was to this effect: "Whatever I have had to tell where the information contained appeared of importance to the public, I have told it boldly; leaving it to my readers to give it such faith as they may be disposed to bestow. But I will not shock their credulity, or render my travels more marvellous, by producing anecdotes which however true, can only relate to my own personal escapes and adventures"'.[41] Scott concluded at the end of this testimony: 'I wish I could remember the anecdotes to which the above declaration alluded – But I feel no confidence at this distance of time, that I could relate them with any accuracy & will not do my deceased friend

---

40   Park, *Travels in the Interior Districts of Africa*, ix.

41   This observation by Scott appears in a manuscript letter, from Scott via John Murray the publisher to John Whishaw, in or about 1814. The letter is affixed in the copy of Mungo Park's 1799 *Travels in the Interior District of Africa* held by Selkirk Library Archives (Borders Council) as SC/5/56 [Letter 10 in a set of interleaved letters]. The quote comes from f.2 of this letter. The copy of Park's book bears the bookplate of one Patrick Muir of Dalswinton.

the injustice to produce them imperfectly'.[42] One reason for Scott's forgetfulness may have been that he, Park, Ferguson and Stewart were the worse for drink: 'I have often endeavour'd to recollect the passages you mention' [Scott wrote to John Whishaw], 'but they were communicated near the close of an evening of conviviality & although I am positively certain of the scope of the conversation I cannot at this distance of time rely on my memory as to the particular narrative which led to it'.[43]

Park's private admission and Scott's testimony has significant implications. The text that is central to Park's reputation is, by its author's own admission, partial. This admission certainly qualifies later interpretations of Park's work as a narrative 'of personal experience and adventure'[44] since the author chose, or so he tells us, to diminish just those elements in his published account. More importantly, it qualifies contemporary reaction to the book and Park's achievements. Park's authenticity in terms of the geographical claims in his *Travels* ought not, perhaps, to be cast in doubt. His role as the sole author of that book, and his claim that 'As a composition, it has nothing to recommend it but *truth*' must certainly be.

Audience Figures: Readers and Subscribers to Park's 1799 Travels

Outside the circle of the African Association, it is unlikely that Park's audience knew of these matters relating to the authorship of the book. In ascertaining exactly what Park's readers would know of the wider context to his work, we face several difficulties. Moving from questions of production to matters of reception in the history and geography of books and print culture is more easily stated than accomplished. Questions concerning the reception of geographical and scientific texts and the geographies of reading and of reception have been to the fore in recent research.[45] Yet as these and other studies have shown, books are read in different ways: marginalia, for example, indicates active engagement with the text, with the reader as author; translations alter content in order to serve different agenda; the circulation of printed works in different social settings – what Secord

---

42   Ibid., SC/5/56, f.2v.

43   H.J.C. Grieve, ed., *The Letters of Sir Walter Scott*, 12 vols (London: Constable, 1932–37), 4, 52.

44   Pratt, *Imperial Eyes*, 75.

45   See, for example, Owen Gingerich, *The Book Nobody Read: In Pursuit of the Revolutions of Nicolaus Copernicus* (London: Heinemann, 2004); Innes Keighren, "Bringing Geography to the Book: Charting the Reception of *Influences of Geographic Environment*," *Transactions of the Institute of British Geographers* 31 (2006): 525–40; Nicolaas Rupke, "A Geography of Enlightenment: The Critical Reception of Alexander von Humboldt's Mexico Work," in *Geography and Enlightenment*, ed. Livingstone and Withers, 319–39; James Secord, *Victorian Sensation: The Extraordinary Publication, Reception, and Secret Authorship of* Vestiges of the Natural History of Creation (Chicago: University of Chicago Press, 2000).

has called the problem of 'knowledge in transit' – means that we cannot assume that local reading practices reflect more general patterns of textual consumption, and vice versa.[46]

In understanding reception as it relates to the book-buying audiences for Park's *Travels*, we are fortunate that the second London edition of this book was published with a list of subscribers and that a comparable list appears in the first American edition. Subscription lists are valuable more as indicators of literary patronage and of purchasing capacity than as guides to collective practices or the hermeneutics of individual reading.[47] They nonetheless offer one measure of who in contemporary society was supporting which books and, often, where the book was being engaged with, if less evidently how. Here, we can see where in Britain and in America Park's *Travels* was being subscribed to and so discern in outline the geography of Enlightenment publishing and the reception of Enlightenment geography beyond the place of publication.[48]

The London edition lists 336 subscribers, the majority London-based and individual men of letters and political influence: Sir Joseph Banks, John Barrow, Henry Cavendish, Alexander Dalrymple the hydrographer, and James Rennell are each listed, as is William Wilberforce, the abolitionist MP, and members of the nobility such as the Duke of Buccleuch, the principal landowner in Park's home district in the Scottish borders. Subscription or local libraries and book societies are also listed, nineteen in total: Bristol, Brompton, Bury St Edmunds, the county library in Cornwall, Deal, Exeter, Halifax, Kendal, two different libraries in Leeds, Macclesfield, Stamford, Trowbridge and Warminster Book Society, Repton Book Society, Rochester Book Society, Sittingbourne, and, in Yarmouth, the Old Book Society and the Monthly Book Society. The Literary Society of Newcastle upon Tyne subscribed as did, in London, the Society for the Encouragement of the Arts.

The metropolitan social status of many of Park's subscribers is unsurprising given the networks associated with the African Association, and the geography of purchase and presumed readership supports what others have disclosed of the making of literate Enlightenment culture in provincial England.[49] Yet such facts

---

46 Lorraine Daston, "Taking Note(s)," *Isis* 95 (2004): 443–8; Nicolaas Rupke, "Translation Studies in the History of Science: The Example of *Vestiges*," *The British Journal for the History of Science* 33 (2000): 209–22; James Secord, "Knowledge in Transit," *Isis* 95 (2004): 654–72.

47 Francis J.G. Robinson and Peter J. Wallis, *Book Subscription Lists: A Revised Guide* (Newcastle upon Tyne: Hill, 1975).

48 There are no subscription lists in any of the non-English language first editions of Park's 1799 *Travels* that I have examined: in French (Paris 1799), Danish (Copenhagen 1800), German (Berlin 1799), Dutch (Amsterdam 1801) or Italian (Naples 1833).

49 For example, Peter Clark, *British Clubs and Societies, 1580–1800: The Origins of an Associational World* (Oxford: Clarendon Press, 2000); Paul Elliott, "The Birth of Public Science in the English Provinces: Natural Philosophy in Derby, *c.* 1690–1760," *Annals of Science* 57 (2000): 61–100; Mary K. Flavell, "The Enlightened Reader and the

about the geographies of the book are important not just for what they show about the mobility of print culture in general and Park's exploration narrative in particular, but for what they suggest about the nature of the Enlightenment as a whole. Rather than think in simple terms of national Enlightenments, or of one essential and supra-national Enlightenment, a picture emerges of altogether 'messier' Enlightenments, of locally situated practices and values, of social worlds overlapping *and* of connections across geographical and cultural space apparent in the shared purchase of a book that engaged different reading publics.[50]

A similar picture emerges of the localities making up Enlightenment America from study of Park's American subscribers.[51] Of 137 listed names, many were prominent figures in the early Republic: Benjamin Rush, Professor of Medicine in the University of Pennsylvania (who, like Park, had trained in medicine at Edinburgh) and Treasurer to the US Mint, Jedidiah Morse, author of texts on America's geography, Samuel Stanhope Smith, the geographical and anthropological author and president of Princeton, and numerous congressmen. In contrast to the listed subscribers in Britain, however, the Philadelphia edition shows that several American purchasers of Park's *Travels*, some of them booksellers but by no means all, bought multiple copies: Thomas Dobson with thirty copies, for instance, Messrs Thomas and James Swords of New York, 100 copies, William Wilkinson of Rhode Island with a dozen copies. While we may never know what these individual subscribers did with these stocks, such evidence is suggestive of the local social and perhaps familial networks through which Enlightenment print culture moved – and of the difficulties faced in capturing fully the geographies of knowledge in transit.

If Park's readers may be presumed to be many, knowing how his book was read, by subscribers and others – and, even perhaps, why – is likewise difficult to discern. We know his book was read by members of that self-improving philosophical body in Enlightenment London the Spitalfields Mathematical Society. Bryan Edwards' copy survives but, as with the copies of other contemporaries, including Banks, Dugald Stewart, Sir Walter Scott and the Duke of Buccleuch, there are no marginalia or other reader's marks to allow speculation upon their motivation for turning to the text or their reaction after having done so. To judge from a law student's copy in the University of

---

New Industrial Towns: A Study of the Liverpool Library, 1758–1790," *British Journal for Eighteenth-Century Studies* 8 (1995): 17–35; Roy Porter, "Science, Provincial Culture and Public Opinion in Enlightenment England," *British Journal for Eighteenth-Century Studies* 3 (1980): 20–46.

50   On this point at greater length, see my *Placing the Enlightenment: Thinking Geographically About the Age of Reason* (Chicago: University of Chicago Press, 2007), especially chapter 4.

51   Mungo Park, *Travels in the Interior Districts of Africa* (Philadelphia: Humphreys, 1800): the list of subscribers' names is given between pages ix–xv of this edition.

Edinburgh Library, Park was read in about 1829 not with an interest in the Niger but for what Park had to say about Mandingo native law, slavery and systems of magistracy and civil governance.

James Rennell, by contrast, read Park with a view to correcting himself. Rennell made several amendments to Park's *Travels* and geographical observations in his copy, mainly in annotations to his own 'Geographical Illustrations'. In addition to notes on the accuracy of Ptolemaic placenames, Rennell offered further adjustment to Park's record of compass bearings and, thus, Park's latitudinal positioning and, by further implication, the accuracy of Park's route map. One amendment is made by Rennell using notes from Park's second Niger journeys – 'Mr Park actually found 17° 2/3 W variation near Sego, in 1805'. Quite when Rennell penned these observations is not known, but here is evidence that Park's published geographical claims were being added to, after his death, by the person who had first helped produce and amend them in life.[52]

Geography, Exploration and Textual Commemoration: Park's *Journal of a Mission in the Interior of Africa*

The Niger (Problem) Re-Visited

Park solved – in effect, confirmed – only part of the Niger problem: the direction of flow of the river. He did not solve where the river ended: if it entered the sea, or (if it did) where it met the supposed great lake in Africa's interior. Because there were other theories around, because Park had left one part of the Niger problem unanswered and because commercial-political interests were still prevalent, the Niger remained high on the geographical agenda. That is what prompted Park's second trip which he undertook from January 1805.

By the time he set out to solve the second part, five theories were extant concerning what contemporaries regarded as 'this most doubtful and obscure problem in modern geography' (i.e., where the river ended).[53] The first (and most widely-held) was that it ran into an inland lake or simply dissipated into the sands

---

52   These remarks are based on a book census survey of copies of Park's *Travels* held in UK university libraries and from examination of copies in other libraries. The Spitalfields Society copy is held in the Fisher Rare Book Library, University of Toronto; Bryan Edwards's is in the University of Birmingham Library, Banks's in the British Library, Dugald Stewart's in the University of Edinburgh Library and Sir Walter Scott's in his home at Abbotsford. The Edinburgh University student is, probably, John Brown Innes, who was in the "Senior Law Class" in 1829–30. Rennell's annotated copy is held in the Archives of the Royal Geographical Society (with the Institute of British Geographers) in London, the latitudinal amendment here quoted, in Rennell's hand, appearing on page xxviii of the "Geographical Illustrations".

53   Park, *Travels in the Interior Districts of Africa*, ix.

of central Africa. A second held that it terminated in the Nile. A third argued on linguistic grounds that the Niger flowed under the Sahara to emerge on Libya's coast, that is, that the Niger emptied into the Mediterranean. A fourth argued that the Niger flowed into the River Congo. Park knew of this idea. He had it from correspondence in 1804 with a Mr George Maxwell, a trader in Africa. To judge from remarks in his *Journal*, Park seems to have been reconciled even before he set off to a longer second expedition in consequence. It is clear that Park held to certain views but that he could (and did) change them. Of his first trip, for example, he remarked:

> The circumstances of the Niger's flowing towards the east … did not surprise me; for although I had left Europe in great hesitation on this subject and rather believed that it ran in the contrary direction, I had made such frequent enquiries during my progress concerning this river; and received from Negroes of different nations, such clear and decisive assurances that the general course was *towards the rising sun*, as scarce left any doubt on my mind: and more especially as I knew that Major Houghton, had collected similar information in the same manner.[54]

Had Park known of the fifth theory, however, he might never have gone at all. This was the view advanced by the German geographical commentator, Christian Gottlieb Reichard, and published in Germany in 1802 in the *Monatliche Correspondenz*, and in 1803 in the *Allgemeine Geographische Ephemeriden*. Arguing by analogy from the disposition of deltas in comparable rivers of the world (the Ganges, the Nile and the Congo), from ancients' geographical accounts and the reports of traders on the Guinea Coast, Reichard argued that the Niger curved south to empty into the Atlantic, into the Bight of Benin. He was right: Reichard, a town clerk to trade who never saw Africa solved the Niger problem before Park set out on his second expedition and died in the attempt and almost thirty years before the Niger's end point was identified in the field by Richard and John Lander in 1830.[55]

This is, in an important sense, a problem of the mobility of print culture, of the geography of knowledge in transit. One could argue that had the exchange of printed material been better between Germany and Britain in 1802–1804, Park need never have gone. A sedentary and speculative theorist got right what numerous field workers had not been able to determine. But that argument will not hold. For even if Rennell, Banks or Park had known of Reichard's claim, it is likely they would have rejected it given Rennell's adherence to the notion that the Niger flowed eastwards into an inland lake and, possibly, given Reichard's status. The production of print culture is no guarantee of the truth claims of its cognitive content. As with Houghton's claims in respect of the flow of the Niger, Park would

---

54   Ibid., 195.
55   Withers, "Mapping the Niger," 183–8.

in all likelihood have returned to the Niger had he known of Reichard's claims because exploratory truth depended upon direct encounter as a basis to textual authority and authorial credibility.

Park's Second Niger Expedition

Between his two Niger trips, Park prepared his notes for publication as we have noted. His fame did not bring security. Surviving letters show him uncertain how to proceed. From 1799, Banks had been urging the government to appoint a consul in the Senegambia region with a view to further Niger exploration. Park was not immediately involved in this renewed attention to West Africa. In January 1800, he hinted at an appointment in China, but was dissuaded by Banks. Park thought about farming, but as he wrote to Banks, the 'high price of cattle' and 'the enormous rents' made this 'a dangerous speculation': Park noted the 'painful but not degrading reflexion that I must henceforth eat my bread by the sweat of my brow'.[56] He toyed with plans to settle in Australia. He entertained the possibility of a regimental surgeonship in the West Indies. Early in 1801, he took the examination to qualify for membership of the Royal College of Surgeons of London and from September 1801, he set up practice as a country doctor in Peebles. It was there that he befriended the Enlightenment thinkers Adam Ferguson and Dugald Stewart, and, from 1804, Sir Walter Scott.

In September 1803, he received an invitation from Lord Hobart, Secretary of State for War and the Colonies, with a view to a further African expedition. Park sailed again for Africa on 31 January 1805 with a temporary commission as Captain and with his brother-in-law, Alexander Anderson, who accompanied him, as Lieutenant. In contrast to his first expedition, Park was accompanied on his second trip by a military detachment from the Royal African Corps garrison at Gorée in West Africa. One officer, Lieutenant John Martyn, and 35 soldiers enlisted for Park's expedition: with others, the party totalled 44 persons.

The record of the expedition – which we know from Park's surviving letters and notebooks as they were brought together to become his 1815 *Journal of A Mission* – is one of struggle against the depredations of Moors and hostile locals, of sickness and the difficulties faced by travelling in the wet season. By 19 August 1805, when the expedition reached Bamako on the Niger, only twelve men remained alive. Those who had survived to this point seemed to have owed much to the efforts of Park, Anderson and to the expedition's native guide, Amadi Fatouma. Despite these difficulties, Park kept up his geographical measurements. In a letter of 26 April 1805 to Banks, Park wrote how he 'was going forward with as much success as I could reasonably expect' and, later in the same letter, asked that Banks 'give my compts [compliments] to Major Rennell and tell him that I hope to be able to

---

56   Mungo Park to Joseph Banks, 13 October 1801, in Warren R. Dawson, *The Banks Letters: A Calendar of the Manuscript Correspondence* (London: British Museum, 1958), Volume 12, ff.265–66. [hereafter BMNH, DTC].

correct my former Errors. The course of the Gambia is certainly not so long as is laid down on the Charts. The watch [Park was making use of a chronometer] goes so correctly that I will measure Africa by feet and inches'.[57] In a letter to Banks of 26 May 1805, Park returned to this theme of his own earlier errors, remarking 'I find that my former journeys on foot were underrated – some of them surprise myself when I trace the same road on horseback'.[58] On 26 September 1805, Park was resident at Sansanding preparing for the final stage of the journey to determine the mouth of the Niger by building an expedition canoe, '*H. M. S. Joliba*' as he named it. The expedition did not leave Sansanding until about 20 November 1805. By then, in addition to Amadi Fatouma and two slaves, it numbered only five men: Park, Lieutenant Martyn (who was in all probability a mentally unstable alcoholic), and three private soldiers, each of whom was ill, one having lost his mind from the rigours of the journey.

On 17 November 1805, Park wrote to Lord Camden that 'I shall set sail to the east with the fixed resolution to discover the termination of the Niger or perish in the attempt'. He also wrote that 'I have heard nothing that I can depend on respecting the remote course of this mighty stream, but I am more inclined to think that it can end nowhere but in the sea'.[59] In a letter of the previous day to Banks, Park recorded how he had hired a guide and that it was his intention, thus guided, 'to keep the middle of the river, and make the best use I can of winds & currents till I reach the termination of this mysterious stream'. Further, 'He [the guide] says the Niger after it passes Kashna runs directly to the right hand or the South: he never heard of any person who had seen its termination, and is certain that it does not end anywhere in the Vicinity of Kashna or Bornou, having resided some time in both these Kingdoms'.[60]

These are Park's last known letters. His death meant he never confirmed his suspicions. Park and his party sailed downstream as far as the town of Boussa. At Boussa, sometime early in 1806, at a narrow gorge where the river ran as rapids, natives attacked the party from the shore. All save one slave were drowned. The precise reasons for the attack upon Park are unclear. What is true is that Park's party, thinking they were being attacked, had earlier shot and killed several natives. It is likely that Park, by then short of tribute gifts for local leaders, had unwittingly transgressed customary expectations. He had long been regarded as a spy by the Moors. Park's body was never recovered.

Rumours of Park's death reached the coast later that year. His fate was not confirmed until 1811. Only by 1812 were the notes of his second expedition brought safely to British officials in the Gambia through the efforts of one Isaaco, who interviewed (amongst others) Park's guide, Amadi Fatouma. Park's expedition

57   BMNH, DTC, Vol. 15, f.356, Mungo Park to Joseph Banks, 26 April 1805.
58   Borders Council Archives, Selkirk Library MS SC/S/56, f.6, Mungo Park to Joseph Banks, 28 May 1805.
59   NLS, MS TD 3005, Acc. XIX, f.1.
60   BMNH, DTC, Vol. 16, f.159, Mungo Park to Joseph Banks, 16 November 1805.

notes and Isaaco's testimony, together with Park's memoir of 1804 to the Colonial Office and some private letters, were brought together by the London barrister, John Whishaw, and became *Journal of a Mission*.

Producing Geographical Authority: Publishing Journal of a Mission

Journal of a Mission to the Interior of Africa was published by John Murray in 1815. Once again, Park's geographies were amended between the field and publication in print. As the 'respectable artist, employed by the publisher to construct the map intended to illustrate the present work' put it in the Advertisement to the *Journal*: 'In compiling the map of Mr Park's route in 1805, much difficulty has arisen from the bearing of places not being mentioned in the Journal .... Considerable pains have been taken to reconcile these differences; but the general result has been, that it was found necessary in adhering to the astronomical observations, to carry Mr. Park's former route in 1796 farther north, and to place it in a higher latitude than that in which it appears in Major Rennell's map annexed to the former volume of Travels' (Figure 7.5).[61] It is likely that Rennell was behind these alterations as he had been earlier. In writing to John Murray in March 1814, John Whishaw noted 'I have already acquainted you that Major Rennell is willing to look over the Papers, in order that he may make any Geographical memoranda which may be useful for the publication', noting too how 'the obtaining of such observations from Major Rennell may be attended with some delay'.[62] By late April 1814, Whishaw was still agitating to Murray about the material to come from Rennell: 'Major Rennell's papers will be material for the purpose of the intended Map, which ought to be put in hand immediately [original emphasis], & about which I am particularly desirous of talking with you.'[63]

Although *Journal of a Mission* was published in order to cast light both upon the circumstances of Park's death using surviving papers and Isaaco's testimony, and upon geographical questions whose cartographic expression changed in transit from field to print, two other matters must not be neglected. One concerns the fact that Whishaw was Secretary to the abolitionist African Institution. Whishaw sought to use that body to distance Park, through his posthumous publication, from his earlier connections with the African Association. In death, Park was thus made to serve an agenda he never much commented upon in life. The second is that the book was used by Whishaw not to serve geographical matters but to

---

61   [Mungo Park], *Journal of a Mission to the Interior of Africa* (London: Murray, 1815), i.

62   John Murray Archives, NLS, MS 41908, John Whishaw to John Murray, 26 March 1814.

63   John Murray Archives, NLS, MS 41908, John Whishaw to John Murray, 21 April 1814. In material on the costings of the 1815 *Journal*, the words "Mr Neale Map" appears in the ledgers with a cost for engraving of £52, 10s, 6d. Neale was the engraver employed by John Murray for the engraving of maps and prints to be included in Murray's books.

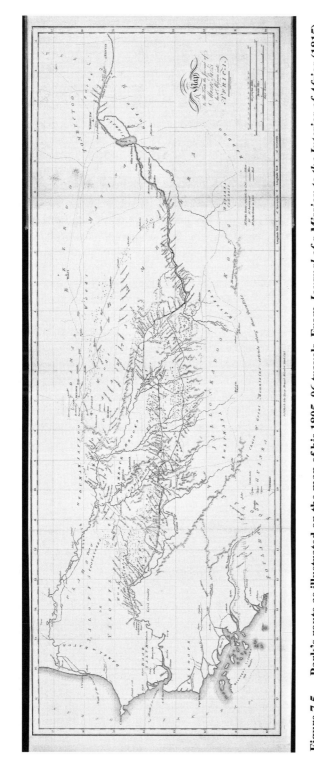

Figure 7.5 Park's route as illustrated on the map of his 1805–06 travels. From *Journal of a Mission to the Interior of Africa* (1815)

Source: Reproduced with the permission of the Trustees of the National Library of Scotland.

promote himself and the African Institution even as he also worked to secure the returns for the publication for the benefit of Park's heirs.

In both respects, Whishaw had to confront the status and influence of Sir Joseph Banks and the African Association and Banks' view of the later publishing enterprise. Writing to an unknown recipient in about 1814, Whishaw reported how a request had earlier been made to Banks over access to sources and letters from Park in his care, and that 'Sir Joseph felt some difficulty about communicating these papers, in consequence of a doubt whether the publication was to be for the benefit of Mr Park's family'. Whishaw continued that since this doubt had now been removed 'and Mr Park's family being entirely satisfied with the conduct of the African Institution (as can be most decisively shewn by letters which I have received from Mr Park's Brothers), I am induced to hope that Sir Joseph Banks may perhaps re-consider his former determination, and allow me to inspect the papers in his possession; with a view to the question whether any of them may be fit for publication or not'.[64] And as Banks held close, at least at first, to material on the geographical content of Park's second trip, Murray proved, initially, an awkward publisher, not making clear the terms of his contract until he could be assured that the work would be to the benefit of Park's family. In a letter to Banks of 26 November 1814, Whishaw reported how, after several conversations with Murray, 'he seems now inclined to be much more reasonable at first'.[65] Even so, Murray delayed the date at which he would pay to Park's executors the first instalment of £1,000 from the total contract of £1,200 – and held back the remainder until it was clear that his profits by the publication would allow it. Upon publication, Banks was, reported Whishaw 'no longer hostile, a circumstance which must be attributed to the success of the work, so far exceeding all national expectation'.[66]

Conclusion

This account of Mungo Park's African texts supports, but also complicates Sher's view of the geography of Enlightenment book history as matters of production and dissemination through reprinting. Park's 1799 work was a successful Enlightenment book, even if its role as a record of successful geographical exploration must be tempered by Park's failure fully to answer the Niger problem and, in effect, his confirmation of an already-circulating view about the course of the river. The fact that Park's book was widely reprinted confirms the view that the Enlightenment's geographical spread may be traced by looking at local patterns of book reprinting.

---

64   Borders Council Archives, Selkirk Library, MS SC/S/56, Item 7, f.2. [It is possible that Whishaw's letter – undated and addressed only to "My Lord" – is to Lord Camden].

65   Borders Council Archives, Selkirk Library, MS SC/S/56, Item 8, John Whishaw to Joseph Banks, 26 November 1814.

66   John Murray Archives, NLS, MS 41908, John Whishaw to John Murray, 18 May 1815.

Yet Park's 1799 *Travels* was not wholly Park's book. Its authorship was the result of work in Africa, mainly by Park but also by Houghton, of close textual supervision in London and cartographic amendment on more than one occasion. These facts of varied authorship – as well as the facts of geography to which Park was acting as credible reporter – were sufficient to prompt contemporaries to ask questions about Park's authenticity: over his work in Africa and his word at home.

Production understood in terms of the location of publication is one means to bring a geographical dimension to book history. Production understood as a drawn-out process of textual refinement that involves the safe passage over space and time of different claims to truth and, as here for Park's 1799 work, that embraces questions of style, geographical accuracy, the agency of other actors and admissions from the author over his work's completeness, is quite another. Similarly, the story of Park's 1815 work is that it was the product of more than one mind to more than one purpose: to use Isaaco's testimony as a warrant of Park's geographical and moral credibility; to promote the African Institution; and to support Park's widow and family.

The implications of this for thinking about the geographies of the book are several-fold. Production is less a location than a process distributed across space and different persons. However much it may be looked at as an end in view, production may cover more than authorial credibility. At issue – at least in Park's 1799 *Travels* – were matters to do with the nature of African exploration and its articulation beyond Africa, with the accessibility of the explorer's prose to diverse audiences and the relationship between text and map in establishing credibility in the tale and in the teller. Authorship as a far from singular element in the process of Park's production – of his book and of him as credible author and resourceful traveller – was influenced by the places in which he and others worked.

Reception and its sub-themes of reviewing and reading are no easier as terms. What did the one hundred New York purchasers of Park's 1799 *Travels*, or the perhaps more distant readers who ordered through the East India Company 'five dozen copies' of his 1815 *Journal* make of the works? Yet reception is perhaps more important in understanding the geographical dimensions to book history for audiences: sites of reading and of review are usually spread further afield than the sites of a book's production or purchase. Thinking about the geography in book history is crucial in addressing such issues. Yet geography is more than a matter of distribution and of location. The story of Park's books about African geography is a story of overcoming the hardships presented by geography in the books' doing and the constraints offered by geography in the books' making in order to recount narratives about geography. It is also a story of the social geographies of particular institutions, of the intellectual geographies that exercised Enlightenment minds and of the many places and spaces of production, reception, readership and debate that need to be recognized if we are, in other contexts and authorial settings, to be able to chart the geographies of (Enlightenment) books.

Acknowledgements

I am grateful to the British Library, the National Library of Scotland, and to those many university library staff who returned information on the copies of Park's 1799 *Travels* in their care and who provided assistance with this research, and to Miles Ogborn, Innes Keighren, Bill Bell, David Livingstone, Diarmid Finnegan, Rick Sher and anonymous reviewers for comments on an earlier draft. The financial assistance of the British Academy and of the Arts and Humanities Research Council is gratefully recognized.

Chapter 8

Books, Geography and Denmark's Colonial Undertaking in West Africa, 1790–1850

Daniel Hopkins

In a quietly dramatic passage in *Heart of Darkness*, Conrad's protagonist is astonished to come upon a book, *An Inquiry into some Points of Seamanship*, in an abandoned hut in the African forest. 'Fancy a man lugging with him a book of that description into this nowhere and studying it – and making notes', he thought. 'The simple old sailor, with his talk of chains and purchases, made me forget the jungle … in a delicious sensation of having come upon something unmistakably real'.[1] It is with somewhat the same surprise that one comes upon the traces of Milton and Shakespeare in the archives of the old Danish slaving forts on the West African coast in what is now southeastern Ghana.[2] Indeed, substantial numbers of books can be followed through these archives into the hands of colonial officers and merchants in Denmark's African enclave in the late eighteenth and early nineteenth centuries. To do so is to throw light on the unmistakably real geographical perspectives that underlay Danish colonial enterprise on the African coast.

Geography, Colonialism and Books

Far-flung commercial and colonial ventures were inseparably connected to the apprehension and understanding of geography. Geographical knowledge – of routes and risks, climates and customs, manufactures and markets, languages and law – has always been as important as capital, credit and cash (if less easily converted and tallied). The indispensable agents and instruments of the colonial powers' expansion of their geographical horizons for four centuries were the navigators, governors, garrison doctors, merchants, book-keepers and warehousemen of Europe's outposts across the world. If remotely situated, these men were by no means isolated from the currents of global colonial thought. On the contrary, their operations in these extraordinary new circumstances generated and directed these

---

1 Joseph Conrad, *Heart of Darkness*, in *A Conrad Argosy* (Garden City: Doubleday, Doran & Company, Inc., 1942), 27–77, on 51.

2 All archival material cited is held in the Danish National Archives (*Rigsarkivet*), Copenhagen.

flows. Situated on the periphery of the European world – or, more properly, on its active economic frontiers – these individuals' ideas and ambitions shaped policies at the metropolitan core: it could be no other way, for European colonial rulers had no direct experience of their colonial territories. Colonialism depended utterly upon the transmission of large and complex bodies of information back and forth across the seas.

The bulk of this communication was, necessarily, in writing, and much of it was in print. The literature of colonialism – travel narratives, geographical and historical accounts, natural histories, handbooks on the management of plantations, exhortatory colonial tracts, atlases, pilots, all manner of books and pamphlets – rapidly became a vital and inseparable element in Europe's expansion into the world. Colonialism was as much an intellectual construction as a tangible affair of territories, guns and ships. The consumption of this literature was by no means limited to the metropolis. In the case of Danish colonialists on the Guinea coast of Africa in the late eighteenth and early nineteenth centuries, the colonial thinking and aspirations of individuals utterly unknown in the established narratives of Denmark's colonial history inspired their government's policies.[3] These men's ideas were framed and shaped by published opinion from elsewhere in the colonial world. What was being read by individuals in colonial settings was a matter of considerable moment.

The value of the library catalogues of some of the great thinkers at the turn of the nineteenth century – Thomas Jefferson, Alexander von Humboldt or Joseph Banks – is well appreciated.[4] More closely focused lists of books, such as those Banks carried with him when he travelled with Cook into the Pacific, reflect the more immediate concerns and intellectual ambitions of their owners.[5] The import of such lists of books is not only in the interest crucially-placed individuals may have had in them, but in the evidence they present of sometimes far-flung intellectual

---

3   See Paul Erdmann Isert, *Letters on West Africa and the Slave Trade. Paul Erdmann Isert's* Journey to Guinea and the Caribbean Islands in Columbia *(1788)*, trans. Selena Axelrod Winsnes (Oxford: British Academy, Oxford University Press, 1992), 227–39; Daniel Hopkins, "Danish Natural History and African Colonialism at the Close of the Eighteenth Century: Peter Thonning's 'Scientific Journey' to the Guinea Coast, 1799–1803," *Archives of Natural History* 26:3 (1999): 369–418; Idem., "Peter Thonning's Map of Danish Guinea and its Use in Colonial Administration and Atlantic Diplomacy, 1801–1890," *Cartographica* 35:3–4 (1998): 99–122; Idem., "The Danish Ban on the Slave Trade and Denmark's African Colonial Ambitions, 1787–1807," *Itinerario* 25:3–4 (2001): 154–84.

4   E. Millicent Sowerby, comp., *Catalogue of the Library of Thomas Jefferson* (Washington: Library of Congress, 1952–59); Jonas Dryander, comp., *The Joseph Banks Bibliography of Natural History* (Bristol: Thoemmes Press, 2003 [1796–1800]); Alexander von Humboldt, *The Humboldt Library; a Catalogue of the Library of Alexander von Humboldt; with a Bibliographical and Biographical Memoir by Henry Stevens* (London: Henry Stevens, American Agency, 1863), lists more than eleven thousand items.

5   Bernard Smith, *Imagining the Pacific* (New Haven: Yale University Press, 1992), 43–5, 242–3.

networks. These books, in these hands, help define times, places and cultural and political contexts. Most of the books known to have belonged to Danish colonialists in West Africa served mainly to connect these readers back to their cultural origins in Europe, but because colonialism was such a broad cultural undertaking, even an ABC in the home of a Danish merchant at Accra, not to mention several dozen of them in a fort's official inventory, is not innocent of colonial freight. Of particular interest here, however, are books that help elicit the geographical thought of the colonists in Africa and their correspondents in Denmark. Some of the books were of concrete colonial utility – guides and handbooks and the like – and reflect something of European imperial economics at this period; others contributed more subtly to the culture of colonialism.

Early Danish Colonial Undertakings on the Guinea Coast

The Danes established themselves in the seventeenth century in a string of forts and lodges on a hundred-mile stretch of the Guinea coast between Accra and the Volta River, in what is today Ghana. Many thousands of slaves were sent to the Americas through these "establishments" (as the Danes termed their armed outposts) over the ensuing century and a half.[6] In 1788, Poul Isert, a physician of German extraction, published a memoir of his five years' service in the Danish forts on the coast and of his observations of the plantation societies of the West Indies. An admirer of Rousseau and an abolitionist (although he participated in the slave trade), Isert inquired why European plantations of tropical products could not be established in Africa, thereby eliminating the need to transport African laborers to the New World.[7] His book was widely translated and appears to have inspired a new national policy.[8] In 1788, Isert was commissioned by the Danish government to return to the Guinea coast to initiate efforts to introduce plantation agriculture there. He died of fever within a few weeks of his arrival, however, and for a time the project lapsed into disarray. Momentum against the slave trade and in favor of an African colonial undertaking had nonetheless been established: Denmark legislated against the slave trade as early as 1792 (effective from 1803),

---

6   Georg Nørregård, *Danish Settlements in West Africa, 1658–1850*, trans. Sigurd Mammen (Boston: Boston University Press, 1966); Per O. Hernæs, "The Volume of the Danish Transatlantic Slave Trade 1660–1806," in *Slaves, Danes, and African Coast Society* (Trondheim: Department of History, University of Trondheim, 1995), 129–303.

7   Paul Erdmann Isert, *Reise nach Guinea und den Caribäischen Inseln in Columbien* (Copenhagen: J.F. Morthorst, 1788); the book appeared in Danish in 1789: *Letters on West Africa*, 3–4, 10–11, 190, 196.

8   By 1797, the book had been published in German, Danish, Swedish, French and Dutch: *Letters on West Africa*, 10–12; for rather indifferent contemporary Danish reviews, see *Kritik og Antikritik* 23 (13 May 1788): 365–6 and *Nyeste Kjøbenhavnske Efterretninger om lærde Sager* 39 (1788): 621–3.

even as the government continued to explore its African territory's suitability for plantation crops.[9]

All along the West African coast from the late eighteenth century enclaves of Europeans, West Indians and black North Americans established new commercial and social footholds and experimented, if tentatively, with the tropical export crops that had already proved so important in the Americas.[10] Indeed, the introduction of export agriculture was generally regarded as the only way to get to the roots of the Atlantic slave trade, which continued for decades after the Danish, English and French bans were enacted.[11] These plantation undertakings were strongly formative steps in the evolution of Europe's colonial interest in Africa. Although Denmark's African ambitions did not survive its own domestic political upheavals in the 1830s and 1840s, the foundation laid down on the eastern Gold Coast in the Danish period was built upon in the ensuing century of British rule.[12] These structures in part rested on the direct experience of Danish officers, traders and gentleman farmers on the ground but also on the cultural, economic and technical notions they drew from their reading about the whole of the colonial world. Printed knowledge of the geography of places, including soils, climates and growing conditions and techniques, was of great significance, for this was an era in which many important plants – cinchona, rubber, breadfruit, tea, coffee, spices, sugar cane, wheat itself – to say nothing of people and domestic animals, were being moved between hemispheres in search of more economically favorable zones of

---

9   Hopkins, "The Danish Ban on the Slave trade"; Idem., "Peter Thonning's 'Scientific Journey'": 406–7.

10   Hopkins, "The Danish Ban on the Slave Trade", 167; Ray A. Kea, "Plantations and Labour in the South-East Gold Coast from the Late Eighteenth to the Mid Nineteenth Century," in *From Slave Trade to Legitimate Commerce*, ed. Robin Law (Cambridge: Cambridge University Press, 1995), 119–43; Edward Reynolds, *Trade and Economic Change on the Gold Coast, 1807–1874* (Harlow: Longman, 1974), 63–9; H.W. Daendels, *Journal and Correspondence of H. W. Daendels, Governor-General of the Netherlands Settlements on the Coast of Guinea*, Part 1, *November 1815 to January 1817* (Legon: Institute of African Studies, University of Ghana, 1964); Eveline C. Martin, *The British West African Settlements, 1750–1821* (London: Longmans, Green and Co., 1927; reprint, New York: Negro Universities Press, 1970), 153–4; Christopher Fyfe, *A History of Sierra Leone* (London: Oxford University Press, 1962), 46, 72–3, 94; A.G. Hopkins, *An Economic History of West Africa* (New York: Columbia University Press, 1973), 137; Kwamina B. Dickson, *A Historical Geography of Ghana* (Cambridge: Cambridge University Press, 1969), 120–32; Philip D. Curtin, *Economic Change in Precolonial Africa* (Madison: University of Wisconsin Press, 1975), 215.

11   Thomas Fowell Buxton, *The African Slave Trade and its Remedy* (London: John Murray, 1840); Hopkins, "The Danish Ban on the Slave Trade"; David Lambert, "'Taken Captive by the Mystery of the Great River': Towards an Historical Geography of British Geography and Atlantic Slavery," *Journal of Historical Geography* 35 (2009): 44–65.

12   Nørregård, *Danish Settlements in West Africa*, 222–9; E.A. Boateng, *A Geography of Ghana* (Cambridge: Cambridge University Press, 1959), 159.

cultivation. This was one of the main concerns of Linnaeus himself. Denmark's African venture after the abolition of the slave trade was motivated, at base, by a need to transplant the plantation system from politically unsustainable settings in the West Indies to what was hoped would prove fertile new territory on the Guinea coast.[13] Books were essential guides in this new colonial commercial undertaking.

Books in Circulation in the Danish Colonial Establishments

On the Guinea coast, as everywhere, books were borrowed and lent and otherwise passed from hand to hand. No one's collection here was really very large. One chaplain in the Danish forts is said to have 'founded a little lending library', but a list of his effects made after his death mentions only books of sermons, a bible concordance, and a small number of other books of a pious nature besides a few bottles of rum.[14] There were certainly no booksellers here: these people dealt in slaves, rum and guns, not in books, and the largest part of the garrison was doubtless illiterate.

The population of Danes in the enclave was never very large, but because death rates were high there was a brisk succession of estate auctions. The auction lists of these men's effects were sent home to Denmark by the colonial government to document the proper settlement of their estates and so came to be preserved in the colonial archives in Copenhagen. The auctions were popular gatherings in this circumscribed expatriate society: simple domestic goods tended to be endlessly recycled in such places, where a new corkscrew, or a set of dishes or a pair of boots might otherwise be difficult to obtain. A Dr Clexton, who died in 1800 at Fort Christiansborg, the main Danish fort at Accra, left, besides twenty volumes of a 'Universal History', a box of watercolors, which was bought for ten times the auctioneer's estimate; a clarinet; a thermometer; a gold-dust scale; a small mahogany table; and a secretary with a lock and key.[15]

Having banned his subjects' involvement in the slave trade in March 1792, the regent, Crown Prince Frederik (Frederik VI), took over the administration of the

---

13   See, for example, Alfred W. Crosby, Jr., *The Columbian Exchange: Biological and Cultural Consequences of 1492* (Westport: Greenwood Publishing Company, 1972), especially his chapter "Old World Plants and Animals in the New World," 64–121; Lisbet Koerner, *Linnaeus: Nature and Nation* (Cambridge: Harvard University Press, 1999), esp. 113–39.

14   Hans W. Debrunner, *A History of Christianity in Ghana* (Accra: Waterville Publishing House, 1967), 91; auction of the estate of Pastor Grundtvig, 8 September 1800, protocol-book opened 29 January 1800, Auktions- og Skifteprotokol, 1800–12, bundle no. 63, Diverse arkivalier fra Guinea (hereinafter cited as DAFG no. 63).

15   Estate of John Clexton, who died in February 1800, DAFG no. 63.

African forts from a private consortium of slave traders.[16] The auction list of his first official representative on the Coast, Governor Bendt Olrick, who died within six weeks of his arrival, gives an idea of how some of the forms and mechanisms of state were to be transferred from Copenhagen to the coast of Africa.[17] Olrick had with him an edition of Christian V's fundamental Law of Denmark and Norway, promulgated in 1683; enormous bound sets of royal ordinances; a guide to juridical practice; and a book of legal formulas – standard deeds and contracts and such.[18] It is not known if all these books were Olrick's own, or whether they should properly have been regarded as government property: they were at any rate auctioned off to the highest bidder.

Olrick had clearly not intended to be isolated there. His effects included a copy of *Kiøbenhavns vejviser* (meaning 'Copenhagen way-finder'), an indispensable tool, with the help of which he could address his correspondence; this book (or other exemplars of it published under various titles between 1770 and 1815) shows up repeatedly in the auction lists. He owned a set of *Danmarks og Norges oeconomiske Magazin*,[19] whose subtitle is a short catalogue of the official and intellectual concerns, both domestic and colonial, of the eighteenth century, 'containing a mix of articles sent in by various well-disposed patriots regarding possible improvement in farm and garden cultivation, forestry, mineral-works, building, cattle-breeding, fisheries, manufacturing, [and the like]'. His copy of Isert's book was snapped up. Olrick, an old Greenland hand, also had with him a book about Iceland.[20]

Olrick brought English, French and German grammars and readers with him. Foreign-language dictionaries appear repeatedly in the auction lists. International

---

16   Erik Gøbel, *Det Danske Slavehandelsforbud 1792* (Odense: Syddansk Universitetsforlag, 2008).

17   Estate of Bendt Olrick, auction register, no. 1a, 8 August 1793, Skiftebreve, indsendte af det kgl Gouvernement paa Guineakysten (Orig. og Afskr.) No. 1–17, 1793–99, General Chamber of Customs, Vestindiske og Guineiske Sager (hereinafter cited as GTK).

18   Jacob Henric Schou, *Chronologisk Register over de Kongelige Forordninger og Aabne Breve, som fra Aar 1670 til 1775 Aars Udgang ere Udkomne*, 10 vols. (Copenhagen, 1777–93); Caspar Peter Rothe, *Casper Peter Rothes Fuldstændige Udtog og Samling af Hidindtil Utrykte Kongelige Rescripter, Resolutioner, Cancellie-Breve og Cammer-Ordres*, 3 vols. (Copenhagen, 1754–69); Ped. Kofod Ancher, *Anviisning for en Dansk Jurist ang. Lovkyndigheds adskillige Deele, Nytte og Hielpemidler* (Copenhagen: 1777); and, from the auctioneer's brief title, J.H. Bärens, *Formularer til Kontrakter, Skiøder, Obligationer, Vexel-Obligationer, Vexler, Opsigelser etc. samt Formularer til Bryllupsbreve, Fadderbreve, Sørgebreve etc.* (Copenhagen, 1775).

19   *Danmarks og Norges oeconomiske Magazin befattende en Blanding af adskillige velsindede Patrioters indsendte smaae Skrifter angaaende den muelige Forbedring i Ager- og Have-dyrkning, Skov-plantning, Mineral-brug, Huus-bygning, Fæe-avling, Fiskerie, Fabrik-væsen o. desl.*, published in Copenhagen, 1757–64.

20   This was presumably Niels Horrebow, *Tilforladelige Efterretninger om Island* (Copenhagen, 1752).

commerce and diplomacy demanded competence in foreign languages, especially in a place like Accra, where Danish, Dutch and English forts stood within a gunshot of one another and the ships of many nations passed down the coast trading from their decks. In the 1830s, in the midst of a dispute with the British colonial administration over access to commercial territories on the Guinea coast, the Danish government in Copenhagen saw to it that a two-volume Danish-English dictionary was sent out to Africa, with orders it be incorporated in the standing inventory at Fort Christiansborg.[21] A German-speaking botanist ordered to the Coast in 1791 provided himself with a Latin-German dictionary, as well as Latin and Greek grammars.[22]

Probate and auction lists record that other men on the Coast owned works of history and fine literature, not merely utilitarian books: they read the Arabian nights, Don Quixote, and works on the lives and times of Luther, Catherine the Great, Napoleon, and other eminent historical figures.[23] Shakespeare was there, and so was the great Danish dramatist and historian Ludvig Holberg.[24] There were Juvenal's satires, Ovid's *Metamorphosis*, Petrarch, Walter Scott in German translation, the works of Byron[25] and of Francisco de Quevedo Villegas, the seventeenth-century Spanish satirist,[26] and Plato's *Republic* in French.[27] There

---

21   The Chamber of Customs to the Government on the Coast, 4 September 1834, in the Guinea correspondence copy-book (hereinafter Guin. kopibog), GTK, sending C.F. Bay, *Fuldstændig Engelsk og Dansk Ordbog* (Copenhagen, 1806), and the third edition of Idem., *Fuldstændigt Dansk og Engelsk Haand-lexicon* (Copenhagen, 1824). Carl Mariboe, *Engelsk Sproglære* (Copenhagen, 1829), was also sent on this occasion.

22   Wrisberg, Christiansborg, 4 May 1798, a list of Julius von Rohr's books, *ad* [i.e., attached to] Wrisberg's letter, recorded at number 22/1799 in the Guinea Journal (the Chamber of Customs colonial office's journal of incoming Guinea correspondence, hereinafter abbreviated GJ), GTK. Wherever reference is only to the journaled files of letters themselves (*journalsager*), rather than to the journal, the abbreviation GJS is employed; if to both the journal and the files, GJ(S); Wrisberg's list is in GJS 35/1799. The Latin-German dictionary appears to have been Benjamin Hederich, *Beniamini Hederici Lexicon Manuale Latino-Germanicum* … [and vice versa], 2 vols. (Leipzig, 1739 or 1766).

23   Peter Meyer's skiftebrev, Xborg [Christiansborg], 26 May 1824, *ad* GJS 604/1825; Ole Justesen, "Henrich Richter 1785–1849: Trader and Politician in the Danish Settlements on the Gold Coast," *Transactions of the Historical Society of Ghana*, New Series, 7 (2003): 187.

24   "Shakespears Comoedier": probate documents of Christian Fleischer, died July 6, 1794, Skiftebreve, indsendte af det kgl Gouvernement paa Guineakysten. Olrick owned Holberg's *Niels Klims Underjordiske Reise,* perhaps the 1789 Copenhagen edition, and the merchant Peder Meyer owned Holberg's *Dannemarks Riges Historie* (Copenhagen, 1762–64).

25   All in the estate of a Doctor Mundt: Registrerings og Vurderings-Forretning i afdøde Regimentschirurg C. Mundts Stervboe, 13–14 February 1828, with Hein, Copenhagen, August 9, 1828, GJ(S) 108/1828.

26   In the estate of Tønne Block Ramus, dead 5 April 1812, DAFG no. 63.

27   *La Republique de Platon*, in two volumes, owned by Olrick.

was Erik Pontoppidan's *Danske Atlas* (1763–81) (a reference work of minute local geographical description, rather than a collection of maps).[28] The Danish expatriate geographer Conrad Malte Brun was represented,[29] and an amateur Danish mapmaker on the Guinea coast refers, rather disparagingly, to the local information in an 'African pilot'.[30] A merchant and planter who operated a small coastal vessel owned a 'skipper's handbook' and a work on navigation.[31] There was a well-known Danish work of national statistics[32] (dealing not with numeric data but with the condition of the state) and an important study of Danish national economy that, briefly, discussed the Guinea trade.[33] There was an edition of William Guthrie's *Geographical, Historical, and Commercial Grammar*, and a Danish geography for young readers.[34] The Danish astronomer Thomas Bugge's *Astronomie* came to the Coast,[35] and one of his practical treatises on land surveying was in the library at Fort Christiansborg, probably brought there by Governor Steffens, a military surveyor. The royal historian and government minister Ove Malling's patriotic semi-official paeon to the lives and deeds of good and great Danes was there, too.[36]

Natural History and Colonialism

The lists of books elucidate the important connection, characteristic of the late Enlightenment, between Denmark's colonial undertakings and the highly influential

---

28   Peter Meyer's skiftebrev, *ad* GJS 604/1825.

29   Dr Mundt's books included "Malte Bruns Atlas": Conrad Malte Brun, *Atlas complet du précis de la géographie universelle*.

30   See *The African Pilot: Being a Collection of New and Accurate Charts, on a Large Scale, of the Coasts, Islands, and Harbours of Africa, from the Straits of Gibraltar to Cape Negro* (London: Robert Laurie and James Whittle, 1799); Hopkins, "Peter Thonning's Map of Danish Guinea," 107–8.

31   Peter Meyer's skiftebrev; see Wrisberg, 24 August 1796, GJS 216/1796.

32   Ramus owned one of Friderik Thaarup's statistical treatises, most likely *Archiv for Statistik, Politik og Huusholdnings Videnskaber*, 4 vols. (Copenhagen: published by Professor Thaarup, 1795–1800).

33   "Fortegnelse over de fra Guinea hjemsendte Sager," GJS 161/1851, actually filed after GJS 311/1891; M.L. Nathanson, *Historisk Statistisk Fremstilling af Danmarks National og Stats Huusholdning* (Copenhagen: Faaes hos C.A. Reitzel, 1836), 398, 503–4, 508.

34   Glad's estate; Peter Meyer's skiftebrev; one of the editions of Chr. Sommerfeldt's *Geographie til Ungdommens Brug*, published from the 1770s into the nineteenth century.

35   In Ramus's effects. The book referred to may have been Thomas Bugge, *De Første Grunde til den Sphæriske og Theoretiske Astronomie, samt den Mathematiske Geographie* (Copenhagen, 1796).

36   Ramus's effects; Ove Malling, *Store og Gode Handlinger af Danske, Norske og Holstenere*, published in many editions.

ideals and institutions of natural history.[37] Natural history at that period was a very broad intellectual edifice, encompassing the study not only of fauna and flora, but of geology, climatology, agronomy, economic production of every description, and ethnography. All were matters of great interest to educated Europeans and their governments: they were central to the colonial enterprise. Most specifically, the assessment of the suitability of the Danish African enclave's locally varying soils and climates for the introduction of tropical plantation crops from around the world was one of the main concerns of the Danish state as it approached the agricultural colonization of West Africa. It is not for nothing that the botanical dissertations of Linnaeus and other authorities appear again and again in Danish records from the Guinea coast. The naturalist Peter Thonning, in a letter from the Coast in 1800, availed himself of the authority of particular descriptions and illustrations in the 15th edition of Linnaeus' *Species plantarum*.[38] Passages in Isert's book suggest that he had with him a treatise on natural history by Michel Adanson, a French naturalist active in Senegal in the 1740s and 1750s. Peter Thonning used Adanson's work in a passage on natural fibres in his own ethnographic notes on the plants he collected.[39] Writing to Sir Joseph Banks from Africa, Isert also cited illustrations of botanical specimens from Jamaica in a work we may now identify as Patrick Browne's *Civil and Natural History of Jamaica* (London, 1756). Apart from their scientific utility, such books of natural history embodied vivid colonial

---

37   See Jorge Cañizares-Esguerra, "Iberian Colonial Science," and Michael A. Osborne, "Science and the French Empire," *Isis* 96:1 (2005): 64–70 and 80–7; Richard Drayton, *Nature's Government: Science, Imperial Britain, and the "Improvement" of the World* (New Haven: Yale University Press, 2000).

38   Thonning also referred to Charles Louis de l'Heritier de Brutelle, *Stirpes Novæ aut Minus Cognitæ, quas Descriptionibus et Iconibus Illustravit Carolus-Ludovicus L'Heritier, Dom. de Brutelle* (Paris: ex typographia Philippi-Dionysii Pierres, 1784–85). Ramus's effects included Linnaeus' *Systema Vegetabilium* and K.P. Funcke's *Naturgeschichte und Technologie für Lehrer in Schulen und für Liebhaber dieser Wissenschaften*, in three volumes. The great Danish entomologist J.C. Fabricius' *Philosophia Entomologica, Sistens Scientiæ Fundamenta, Adiectis Definitionibus, Exemplis, Observationibus, Adumbrationibus* (Hamburg: Impensis Carol. Ernest. Bohnii, 1778), was to be found there, and so was Martin Vahl's *Symbolae Botanicae* (Copenhagen, 1790–94), or at least the first couple of volumes: Wrisberg's list of Julius von Rohr's books.

39   Isert, *Letters on West Africa*, 164, 212; Michel Adanson, *Histoire Naturelle de Sénégal* (Paris: C.-J.-B. Bauche, 1757), might be the book in question. Isert's book is in the form of letters home from the Coast and from the West Indies. It was not published until after his return to Denmark, where he had access to the libraries of the metropolis, so mentions of books in his memoir cannot be taken as proof that the books were in his possession in Africa. However, even if Isert's letters were only a literary device (see Isert, *Letters on West Africa*, 8–9), it is likely that the book was written in Africa and the West Indies and on the voyage home: see F.C. Schumacher, *Beskrivelse af Guineiske Planter som er Fundne af Danske Botanikere, især af Etatsraad Thonning* [Særskilt aftrykt af det kongelige danske Videnskabers Selskabs Skrifter] (Copenhagen: trykt i Hartv. Frid. Popps Bogtrykkerie, 1827), 174–5.

depictions: they can be taken to have had a profound effect on these men's colonial views and sensibilities.

One of the most extensive collections of books in the Danish enclave on the Guinea coast was assembled by the botanist Julius von Rohr. Von Rohr, of German extraction, had served the Danish monarchy in its West Indian colony in the capacity of land surveyor and municipal buildings inspector for thirty years.[40] His appointment in 1757 had also directed him to study the islands' natural history, then largely unknown.[41] His name became well known in natural historical circles in Copenhagen, and, in 1782, he was given leave from his regular duties and ordered to travel throughout the Lesser Antilles to acquaint himself with the techniques of cotton cultivation in the other European colonies there, with an eye to improving the quality and quantity of production in the Danish West Indies.[42] His report was well received by the government, and in 1791 von Rohr was permitted to retire to devote himself to the study of natural history.[43] He had little time to do so, however. Later that year, as the Danish government was formulating its ban on the slave trade, the regent ordered von Rohr to the Guinea coast to attempt to revive Isert's colony and to investigate the suitability of the surrounding territory for plantation agriculture.[44] Von Rohr had a substantial collection of books sent ahead. Some are known to have been shipped by Sir Joseph Banks directly to Accra from London, some appear to have been sent to him from Denmark, while others are likely to have come from his own library on St. Croix.[45] Their titles provide priceless insight into the wide-ranging interests of a well-read and intellectually well-connected colonialist.[46]

Von Rohr regarded his African errand as an expedition into the unknown. He asked Banks for the *Proceedings* of the Association for Promoting the Discovery

---

40   Carl F. Christensen, *Den danske botaniks historie* (Copenhagen: H. Hagerups Forlag, 1924–26), vol. 1, 116.

41   19 April 1757, Amerikansk og Afrikansk Kopibog, 2249.10, Vestindisk-guineisk renteskriverkontor 1754–60, Rentekammeret.

42   The Commerce Collegium, 14 December 1782, to von Rohr, letter copy-book, Kommercekollegiet; Hopkins, "Peter Thonning's 'Scientific Journey'," 375–8.

43   Von Rohr, undated, correspondence journal no. 854/1786, Kommercekollegiet; the Commerce Collegium to von Rohr, 16 December 1786, letter copy-book; royal resolution of 13 April 1791, Vestindiske forestillinger og resolutioner, 1791 and 1792, GTK.

44   Royal rescript, 28 October 1791, to Governor Bjørn, Vestindiske forestillinger og resolutioner, 1791 and 1792, GTK.

45   Wolffs and Dorville, 7 August 1794, and an invoice for goods shipped, 4 August 1794, Rohr vedk., Schimmelmannske papirer vedk. Kommissionerne betræffende Guinea og Negerhandelen, samt forskellige Vestindiske papirer, 1778–1809, GTK.

46   Wrisberg's list of von Rohr's books, *ad* GJ 22/1799, filed at GJS 35/1799; the bookseller Malling, Copenhagen, 10 May 1803, GJS 382/1803, in a printed auction catalogue itemizes von Rohr's books on pp. 225–8; Governor v. Hager, Christiansborg, 6 March 1795, to Waltersdorph, St. Croix, Journaliserede og ikke-journaliserede skrivelser fra Vestindien, 1795, General Gouvernementet, 1760–1854, Vestindiske lokalarkiver. I am grateful to George Tyson for bringing this document to my attention.

of the Interior Parts of Africa, which Banks had founded a few years before: von Rohr supposed this publication 'could perhaps be of greater Use in general, then I my self can guess at, for the Present'.[47] Three numbers of the *Proceedings* were listed among the books auctioned for his estate ten years later.[48] Von Rohr's library also included Jean Baptiste Labat's *Voyage aux isles de l'Amerique* (1742), Andreas Sparrman's 1784 account of his journey into the country of the Hottentots, the Forsters' 1784 account of their voyage around the world with Captain Cook in the *Resolution*, and a German edition of James Bruce's *Travels to Discover the Source of the Nile*.[49] Von Rohr had several of Linnaeus' works and the first six numbers of the *Skrivter* of the Danish Natural History Society, formed in 1789, in which some of his own West Indian botanical work was published.[50] There were also William Mason's 1772 treatise on English gardening style, in the form of a long poem; Jean-Jacques Rousseau's 'Botany for Ladies', in German; and a *Flora Indica*, which von Rohr wanted for comparative purposes.[51] In his last letter to Banks, von Rohr had begged him to send him Drury's book on exotic insects: it is listed among his effects after his death.[52]

Because the success or failure of European colonial settlement in Africa could depend on the colony's ability to come to grips with tropical disease, von Rohr, a licensed practitioner of medicine, supplied himself with a number of books and journals of medicine.[53] His library also included books on mining and metallurgy

---

47   Von Rohr, New York, 10 August 1793, to Joseph Banks, Add. MS 8098 fol. 280–1, Department of Manuscripts, British Library.

48   Wrisberg's list of von Rohr's books; printed auction catalogue.

49   Jean Baptiste Labat, *Nouveau voyage aux isles de l'Amerique: contenant l'histoire naturelle de ces pays, l'origine, les mœurs, le religion & le gouvernement des habitans anciens & modernes: les guerres & les evenemens singuliers qui y sont arrivez pendant le séjour que l'auteur y a fait* (Paris: Chez Theodore le Gras, 1742); Andreas Sparrman, *Reise nach dem Vorgebirge der Guten Hoffnung, den südlichen polarländern und um die welt, hauptsachlich aber in den ländern der Hottentotten und Kaffern* (Berlin, 1784); Georg Forster and Johann Reinhold Forster, *Johann Reinhold Forster's ... reise um die welt, während den jahren 1772 bis 1775 in dem von Sr. itztregierenden grosbrittannischen Majestät auf entdeckungen ausgeschickten und durch den capitain Cook geführten schiffe the resolution unternommen*, 3 vols. (Berlin: Bey Haude und Spener, 1784); James Bruce, *Reisen zur Entdeckung der Quellen des Nils in den Jahren 1768, 1769, 1770, 1771, 1772 und 1773*, 5 vols. (Leipzig, 1790–91).

50   These included Linnaeus' *Amoenitates academicae*, his *Flora Zeylanica*, and *Genera plantarum*.

51   J.-J. Rousseau, *Botanik für Frauenzimmer, in Briefen an die Frau von L*** [i.e., Mme Delessert] (Frankfurt and Leipzig, 1781); Nicolaas Laurens Burman, *Flora Indica* (London and Amsterdam: 1768).

52   Dru Drury, *Illustrations of Natural History; wherein are Exhibited upwards of two hundred and forty Figures of Exotic Insects, according to their different Genera*, 3 vols. (London: B. White, 1770–82).

53   These included Linnaeus' *Materia medica*, Johann Peter Eberhard's *Conspectus medicinae theoreticae* (Halae Magdeburgicae, 1757–61), and the Swede Peter Jonas

(which might be of great use there on the 'Gold Coast') and on engineering, fortification and the military arts.[54] The four Danish forts on the Guinea coast were in constant need of repair, and it was possible that new defensive works might be required for an inland agricultural colony and its communications. Von Rohr, already old when called to his new duty, had indicated to Banks that he expected to end his days on the deadly African coast.[55] At least he had plenty to read, including many such works of general interest as Edward Gibbon's *Decline and Fall*, in German, in fourteen volumes, and a Danish translation of Voltaire's history of Russia under Peter the Great.[56] Time could hang heavy in the forts, especially between visiting ships, and for this reason alone we may surmise that many of these books listed in the establishments were actually read there. '[T]he whole day I spend on the sofa reading', one officer wrote home.[57] Another read in the evenings after dinner with his friend the fort doctor and his wife.[58] Governor Carstensen, in the 1840s, recorded in his diary one day that he was bored to distraction and much plagued by the drunken company in the fort; he was reading Goethe.[59] Yet another diarist, a non-commissioned officer, wrote that 'There could be no question of reading for my part, for when I sat myself down to read or write, I soon became weary in the eyes'.[60]

Bergius's *Materia medica e regno vegetabili, sistens simplicia officinalia, pariter atque culinaria* (Stockholm: typis Petri Hesselberg, 1778). The surgeon H.C. Frahm, who died in 1807, left a "Grundrisse der Experimental-Pharmacie," DAFG, no. 63; Tønne Ramus left a "Materia Medica". Dr Mundt left a *Pharmacopoea Danica, Regia autoritate a Colleg. sanitatis regio medico-chirurgico Hafniensi conscripta* (Copenhagen, 1805), a work on inoculation, a *Handbuch der Materia medica*, translated from the French of Henri Milne-Edwards and Pierre Vavasseur (Weimar: Verl. des Landes-Industri-Comptoirs, 1827) and an edition of Jean Louis Baudelocque's *L'Art des accouchemens* (Paris: Méquignon, 1822).

54   Christoph Traugott Delius, *Anleitung zu der bergbaukunst* (Wien: J.T. Edlen u. Trattnern, 1773); a seventeenth-century edition (according to the printed auction catalog) of Georg Conrad Stahl [Martius], *Europäischer Ingenieur*; and Karl August von Struensee, *Anfangsgründe der Artillerie* (Leipzig: Bey David Siegert, 1788).

55   Von Rohr, New York, 10 August 1793, to Banks, B.L. Add. MS 8098 fols. 280–1.

56   F.-M.-A. de Voltaire, *Keiser Peter den Stores Liv og Levnet* (Copenhagen, 1766–69).

57   *Da Guinea var dansk*, ed. Carl Behrens (Copenhagen: Nyt Nordisk Forlag, 1917), 105. For an English translation, see Wulff Joseph Wulff, *A Danish Jew in West Africa, Wulff Joseph Wulff, Biography and Letters 1836–1842*, ed. Selena Axelrod Winsnes (Trondheim: Department of History, Norwegian University of Science and Technology, 2004).

58   Entry for 31 August 1831, Balthazar Christensens Dagbog fra 1830 31/7 – 1831 26/10, C. Sager vedr. Balth. Christensens Ophold i Guinea 1830–31, box 2 (C–F), Privatarkiv no. 5262, Balthazar Christensen [hereafter cited as Balthazar Christensen's Diary].

59   "Noter mit liv angaaende," duplicated typescript, described on an archival label as a transcription of the diaries in the box, p. 108, Privatarkiv 5258, Edward J.A. Carstensen. I am very grateful to Selena Winsnes for making available to me a copy of the typescript.

60   "Kort Fremstilling af Johan Vilhelm Svedstrups Tilværelse her paa Jorden fra den 12 August 1819, til den …" (a blank was left for the closing date, which Svedstrup prayed

While von Rohr undoubtedly regarded the expedition as a great scientific and colonial opportunity, it may also be that the undertaking appealed to him for religious reasons. The importance of carrying Christian civilization to the Africans through commercial relations and colonies figured centrally in the rhetoric deployed in favor of the abolition of the slave trade. For von Rohr, it may have been a matter of pious faith more than of enlightened Christian sentiment: besides four Danish Bibles, von Rohr's books included two copies of each of the philosophical and theological writings of the Swedish polymath Emanuel Swedenborg, who, when he was not studying the physiology of the brain and theorizing on molecules and magnetism, conversed with angels. Swedenborg speculated that a new religious movement, the Church of New Jerusalem, would soon be founded by God in some 'previously heathen nation', most likely in the interior of Africa, and his teachings were surprisingly influential as European colonial plans for West Africa were being laid in the 1780s and 1790s.[61]

African Colonial Perspectives

The great public issue of the day, the abolition and suppression of the slave trade, scarcely registers in the lists that have survived in the archives. The most substantial commentary, which explicitly linked African colonialism with the abolition of the trade, appeared in the periodical *Minerva*: this has not been traced to the establishments on the Guinea coast.[62] When the Danish prohibition of the trade took effect on the first day of 1803, this event appears not even to have been publicly proclaimed in the Danish establishments on the Coast (or so it was later claimed); the law's applicability to the Africans in the enclave was still a matter of official concern at least as late as 1821.[63]

Commerce, either in slaves or of legitimate commodities, bound everyone, officer and soldier, black, white and mulatto, to the economy, the culture and to the literatures of the colonial Atlantic. Although the Danish establishments might be

---

his sons would insert when the time came, in which filial duty they failed him), Privatarkiv 6411, Johann Vilhelm Svedstrup.

61   Philip D. Curtin, *The Image of Africa* (Madison: University of Wisconsin Press 1964), vol. 1, 26.

62   See, for example, the translation of "Substance of the Report of the Court of Directors of the Sierra Leone Company to the general court, held at London on Wednesday the 19th of Octob. 1791," *Minerva* 2 (May 1792), 216, and, in the same number, [Hans Christian Sneedorf's] "Breve fra en dansk Reisende," 29, dated London, 4 May 1792: 257–8; Hopkins, "The Danish Ban on the Slave Trade and Denmark's African Colonial Ambitions", 164–7; Henning Højlund Knap, "Danskerne og slaveriet. Negerslavedebatten i Danmark indtil 1792," in *Dansk kolonihistorie. Indføring og studier*, ed. Peter Hoxcer Jensen et al. (Århus: Forlaget Historia, 1983), 153–74.

63   Frederik VI, 28 June 1821, GJ 34/1821; the Chamber of Customs's "Allerunderdanigste Betænkning," 4 September 1821, GJS 226/1836.

thought remotely situated, they were nevertheless well within the trade circuits of the western seas and apprehension of their place in the colonial world was a matter of great interest to readers there. A local literature even sprang up in this narrow circle of consumers of books, and people's experience on the Guinea coast was clearly coloured by what they read of it in these literary mirrors. Isert's book was sold repeatedly at auction.[64] So were two books published three or four decades earlier by Ludewig Rømer, a German-born slave trader who spent many years on the Coast in Denmark's service.[65] The Lutheran pastor H.C. Monrad, who for five years early in the nineteenth century ministered (largely in vain) to the spiritual needs of the Danes in the forts, published an account of the establishments in 1822.[66] The Chamber of Customs ordered twenty copies of Monrad's book; two were sent to the Coast with orders that one was to be placed in the official working archives at Fort Christiansborg to be consulted in the event that administrators in Copenhagen might refer to it in future communications.[67]

Books arising directly out of the colonial experience on the Guinea coast were thus of administrative use there. In 1837, the government on the Coast asked for 'a botany'. Ten years earlier, the Royal Danish Academy of Sciences and Letters had published Thonning's African flora. Thonning's work was in Latin, save for the extensive ethnobotanical notes, which were in Danish (to the chagrin of a Danish reviewer who feared that this would limit the work's usefulness to the international scientific community).[68] Thonning, who had gone on to a career in the central colonial administration in Copenhagen, saw to it that at least one

64   Johann D. Westermann, surgeon, died with a copy of Isert on his shelves: Skiftebreve, indsendte af det kgl Gouvernement paa Guineakysten, GTK; so did Peter Meyer.

65   Ludewig Ferdinand Rømer, *Tilforladelig Efterretning om Negotien paa Kysten Guinea* (Copenhagen: Trykt og bekostet af Ludolph Henrich Lillie, 1756), and *Tilforladelig Efterretning om Kysten Guinea* (Copenhagen: Trykt hos Ludolf Henrich Lillies Enke, 1760). See Selena A. Winsnes in *A Reliable Account of the Coast of Guinea (1760) / by Ludewig Ferdinand Rømer* (Oxford: Oxford University Press, 2000); Olrick, Ramus and Meyer all had one or the other of these. I am very grateful to Ole Justesen for information regarding Rømer.

66   H.C. Monrad, *Bidrag til en Skildring af Guinea-Kysten og dens Indbyggere* (Copenhagen: printed by Andreas Seidelin, 1822). Monrad also translated extracts from a publication on Houghton and Mungo Park put out by the African Association: *Efterretninger om det indre Africa, uddragne af de af det i London oprettede afrikanske Selskab udgivne Oplysninger om Houghtons og Parks Reiser, som Foreløber for denne sidstes større Reisebeskrivelse* (Copenhagen, 1799). It seems likely that Monrad had submitted this publication as a credential when he sought the position and that he had it – and the English original – with him in Africa.

67   Molbech, 19 April 1822, GJ 130/1822; the Chamber of Customs to the Government on the Coast, 20 April 1822, Guin. kopibog.

68   Schumacher, *Beskrivelse af Guineiske Planter*, see also J.W. Hornemann's review in *Maanedskrift for Litteratur*, vol. 1 (1829): 318.

copy of the book was sent to the Coast.[69] Governor Schiønning, a former slave trader who acted as governor of the Danish enclave during the Napoleonic Wars, contributed to this production of a Danish colonial literature with a translation into the Accra language of the Ten Commandments and other holy texts. The Bishop of Sjælland (whose see extended to the coast of Africa) had this printed in a hundred copies, but it was never put into the trade: we may assume a few copies were sent to Schiønning on the Coast.[70] The Danish linguist Rasmus Rask, apparently on the basis of conversations with one of the king of Osu's sons who had been sent to Denmark for an education, published a guide to the Accra language in 1828.[71] One of the Danish officers at Fort Christiansborg reported a couple of years later that he had found this book useful.[72]

Schiønning was committed to what he called 'the noble idea that a fertile Africa can in time make up the loss of the West Indian colonies'.[73] (An English force occupied the Danish West Indies beginning in 1807, and, for all Schiønning knew, writing in 1809, the islands were lost to Denmark forever). He had planted coffee in Africa with great success, and he urged his government to support this new enterprise. He had never been to the West Indies, he wrote, but his young trees bore fruit a year before they could be expected to do so in the Caribbean. This he knew from 'various writings ... in various languages, of which I have translated fragments that I still possess and make profitable use of'. Schiønning did not identify these sources, but there can be little doubt that his own agricultural undertakings and his colonial ambitions for his country depended on published descriptions of plantation agriculture on the other side of the Atlantic.

Schiønning's opinions regarding the agricultural potential of this stretch of the West African coast were in turn cited by his English contemporary there, Henry Meredith, whose *Account of the Gold Coast of Africa* was published in London in 1812. Meredith, commandant at Fort Annamaboe up the Coast from Fort Christiansborg, had not visited Schiønning's plantation, but he had spoken with men who had, and they 'agreed in all their reports as to the fertility of the soil, and the vigorous condition of its productions'.[74] He quoted Schiønning: '"I have been

---

69   The government on the Coast, 18 November 1837, GJ(S) 585/1838 (duplicate at GJ 664/1838).

70   C. Schiønning, *De ti Bud, det apostoliske Symbolum og Fader Vor, oversatte i det Accraiske Sprog* (Copenhagen: Trykt i det det Kongl. Vaisenhuses Bogtrykkerie af C. F. Schubart, 1805), 6 and 7.

71   Rasmus Rask, *Vejledning til Akra-Sproget pa Kysten Ginea, med et Tillæg om Akvambusk* (Copenhagen: Trykt i S. L. Møllers Bogtrykkeri, 1828); Poul Erik Olsen, "Supplement," in *Scandinavians in Africa*, comp. J. Reindorf, ed. J. Simensen (Oslo: Universitetsforlaget, 1980), 127.

72   [Balthazar Christensen], "Breve fra og om Guinea," 3rd letter, Christiansborg [30 April] 1831, *Valkyrien* 3 (1831): 57.

73   Schiønning, 4 February 1809, GJ(S) 1081/1809.

74   Henry Meredith, *An Acount of the Gold Coast of Africa* (London: Printed for Longman, Hurst, Rees, Orme, and Brown, 1812), 211–2.

almost over the whole country," says he; "and as to beauty of prospect, pleasing variety, and local advantages, I never saw anything equal; … in short, you may compare it to what poets describe when they depict nature in all her elegance"'. Meredith accordingly quoted Milton: '"Nature here/ Wanton'd as in her prime, and play'd at will,/ her virgin fancies pouring forth more sweet,/ Wild above rule or art, enormous bliss"'.[75] A Danish diarist at Fort Christiansborg in the 1830s recorded that he had read Meredith's book twice in the year since he had arrived on the Guinea coast.[76]

Denmark was thrown into deep economic depression following its involvement in the Napoleonic wars after 1807. The government was not able to pay renewed attention to its African possessions until the 1820s. A new governor, Peter Steffens (the younger brother of the philosopher Henrik Steffens), travelled to the Coast in 1820 and besides surveying instruments and other scientific apparatus took with him several technical treatises provided, at the crown's behest, by the professor of astronomy at the University of Copenhagen.[77] Among these books were a French manual of topography and military science, a French work on mechanics, an astronomical atlas and tables of logarithms.[78] The director of Copenhagen's Royal Natural History Museum supplied him with specimen-bottles, corks, tweezers and the like; a long hand-written set of instructions for collecting zoological specimens; a book on taxidermy; and a German guide to zoology, which, as the museum director said, 'because of the number of clear engravings it contains, is particularly useful for those that have not before occupied themselves systematically with zoology'.[79] Some of the administration's colonial concerns were thus purely scholarly, but its scientific agents were necessarily often amateurs and other men pressed into the service of scholarship, and books – lavishly illustrated books – to

---

75   Ibid., 226–7; see John Milton, *Paradise Lost*, ed. Richard Bentley (New York: Georg Olms Verlag, 1976 [1732]), 157–8 (book 5, lines 294–7).

76   Entry for 17 July 1831, Balthazar Christensen's Diary.

77   The government on the Coast, 20 January 1821, GJ 13/1821; *Fonden ad Usus Publicos, Aktmæssig Bidrag til Belysning af dens Virksomhed*, vol. 2, 1765–1800 (Copenhagen: Rigsarkivet, C.A. Reitzel, 1902), 323.

78   Steffens, Copenhagen, 14 June 1820, *ad* GJ(S) 143/1822, receipt for items received from Professor Schumacher; *Memorial Topographique et Militaire*, published in Paris by the Dépôt de la Guerre, beginning in *An XI* [1802]; François Callet's tables of logarithms ("à 7 décimales"); Siméon-Denis Poisson, *Traité de mécanique*, 2 vols (Paris: Courcier, 1811); "Bodes Atlas mit Stern-Catalog," presumably Johan Elert Bode, *J.E. Bode's Sternatlas*, or *Vorstellung der Gestirne auf XXXIV Kupfertafeln nach der Pariser Ausgabe des Flamsteadschen Himmelsatlas* (Berlin: G.A. Lange, 1782).

79   Reinhardt, Copenhagen, 4 March 1821, journaled correspondence files no. 150, Kommissionen for Museet for Naturvidenskaberne (1796–1829); Johann Friedrich Naumann, *Taxidermie, oder, Die Lehre Thiere aller Klassen: am einfachsten und zweckmäßigsten für Kabinette auszustopfen und aufzubewahren* (Halle: Bei Hemmerde und Schwetschke, 1815); Lorenz Oken, *Okens Lehrbuch der Zoologie* (Jena: A. Schmid und Comp., 1815–16).

guide and instruct them were essential to communications between the colonial and the metropolitan scientific spheres.

Books, Transoceanic Administrative Dialogue and the Wider Colonial World

Like so many, Governor Steffens did not long survive the fevers and fluxes of the Coast, but his administration re-established a certain amount of colonizing momentum.[80] In the late 1820s, the Danish government undertook to overhaul its African colonial government. A two-man commission was dispatched to look into administrative practices in the forts and to investigate the feasibility of a scheme for a penal colony submitted by a junior scion of a Danish family long established on the Coast.[81] In preparing for their journey, the commissioners applied for royal funds with which to purchase books 'dealing with plantations in general and with the Coast of Guinea and Africa in particular'. The king approved the expenditure, and twenty-one books were purchased from various dealers in Copenhagen and sent to be incorporated in the library at Fort Christiansborg.[82]

These books included Monrad's book about the Danish establishments, C.B. Wadström's book on Sierra Leone, Archibald Dalzel's book on Dahomey, two French accounts of explorations in Senegal, Thomas Winterbottom's 'account of the native Africans in the neighbourhood of Sierra Leone', and Denham's, Clapperton's and Oudney's account of their travels in North and Central Africa – all in German editions.[83] There were also Bowdich's *Mission from Cape Coast Castle to Ashantee* (1819), Captain Tuckey's *Narrative of an Expedition to Explore the River Zaire, usually Called the Congo* (1818), both in English, Hugh Murray's and John Leyden's *Historical Account of Discoveries and Travels in Africa, from the Earliest Ages to the Present Time* (1818), an edition of Humboldt's *Reise in die Aequinoctial-Gegenden des neuen Continents* (1815), a French work on Saint

---

80   Forestilling, 24 September 1822, Resolution, 15 October 1822, no. 110, Vestindiske Forestillinger og Resolutioner, 1822, GTK.

81   Philip Wrisberg, 19 November 1825, GJ 630/1826.

82   Frederik VI, April 14, 1827, GJ 912/1827; Hein, Copenhagen, April 10, 1827, *ad* GJ 912/1827; Hein, Copenhagen, 9 July 1827, GJ(S) 987/1827, including list of books purchased.

83   C.B. Wadström, *Über die Gründung der Colonier zu Sierra-Leona und Boulama* (Schnepfenthal, 1799); Archibald Dalzel, *Geschichte von Dahomy* (Leipzig: Schwickert, 1799); J.B.L. Durand's work, translated as *Nachrichten von des Senegal Ländern* (Weimar, 1803), and A.C. La Jaille, *Reise nach Senegal in den Jahren 1784 und 1787,* trans. M.C. Sprengel (Weimar: Landes-Industrie-Comptoir, 1802); Thomas M. Winterbottom, *Nachrichten von der Sierra-Leona-Küste und ihren Bewohnern,* ed. T.F. Ehrmann (Weimar, 1805); Dixon Denham, Hugh Clapperton and Walter Oudney, *Beschreibung der Reisen und entdeckungen im Nördlichen und Mittlern Africa in den Jahren 1822 bis 1824* (Weimar, Im Verlage des Landes-Industrie-Comptoirs, 1827).

Domingue, and a German description of British, Dutch and French Guyana.[84] The list also mentions an account of the English colony in New South Wales,[85] four volumes in German of Baron von Cuvier's natural history of the animal kingdom,[86] and an early edition of Adolf Stieler's *Hand-Atlas*. Administration of Denmark's establishments depended upon acquaintance with an international literature of geography and colonialism.

One of the most ambitious libraries in the Danish establishments, although it was apparently not there for long, was that of a military surgeon sent out in 1827.[87] Dr Mundt appears not to have lived out his first year on the Coast and on his deathbed asked that his books be sent home to his brother.[88] Besides works on medicine and pharmacy, the doctor had brought with him the botanical works of Linnaeus and Wildenow,[89] books of chemistry and of philosophy, Gibbon, an *Histoire de la Revolution Française*, the Holy Bible in Arabic[90] and the Koran in English,[91] a Hindustani dictionary, and the Danish natural scientist and linguist Jakob Bredsdorff's newly-published *Geognosie*.[92] Mundt owned one of the works of Conrad Schmidt-Phiseldek, then director of the Danish colonial office. This book is likely to have been his widely-translated *Europa und Amerika* – which, in English, was subtitled 'the relative state of the civilized world at a future period' – in which Schmidt-Phiseldek predicted the rise of the Americas to predominance in the affairs of the world.[93]

---

84   The last mentioned were S.J. Ducoeur-Joly and Jean Baptiste Poupée Desportes, *Manuel des habitans de Saint-Dominique* (Paris: Lenoir, 1803) and Johann Christoph Friedrich Guts Muths, *Vollständige und neueste erdbeschreibung des britischen, niederländischen und französischen Guyana's und des kaiserthums Brasilien mit einer einleitung zu Südamerica* (Weimar: Verlag des Geographischen Instituts, 1827).

85   David Collins and Philip Gidley King, *An Account of the English Colony in New South Wales: with Remarks on the Dispositions, Customs, Manners, &c., of the Native Inhabitants of that Country* (London: Printed for T. Cadell and W. Davies, 1798).

86   Georges Cuvier, *Das Thierreich eingetheilt nach dem Bau der Thiere als Grundlage ihrer Naturgeschichte und der vergleichenden Anatomie*, trans. H.R. Schinz (Stuttgart: J.S. Cotta, 1821–25).

87   Royal rescript, 10 March 1827, GJ 860/1827.

88   Hein, Copenhagen, 9 August 1828, with a list of Mundt's effects, GJ(S) 108/1828. The brother was presumably the mathemetician Carl Emil Mundt, of the Sorø Academy.

89   Karl Ludwig Wildenow, *Species Plantarum* (n. p., n.d.).

90   This was doubtless *The Holy Bible, containing the Old and New Testaments, in the Arabic Language*, ed. J.D. Carlyle and Henry Ford (Newcastle-upon Tyne: Printed by Sarah Hodgson, 1811).

91   This was an edition of George Sale's translation, old by this time, *The Koran, commonly called the Alkoran of Mohammed*.

92   Jakob H. Bredsdorff, *Begyndelsesgrunde af geognosien* (Copenhagen: A. Seidelin, 1827).

93   C.F. von Schmidt-Phiseldek, *Europe and America, or the Relative State of the Civilized World at a Future Period*, trans. Joseph Owen, facsimile edition (Copenhagen: The Royal Danish Ministry of Foreign Affairs, Rhodos, 1976 [Copenhagen: Printed

In late 1828, the Danish government sent another man of science to the Coast to replace the unfortunate Mundt. This was J.J. Trentepohl, who had lately returned to Copenhagen after a voyage to China as ship's surgeon in a Danish Asiatic Company ship.[94] Early in 1828, his application for support of zoological research on the collections he had brought home from the East was laid before the king. For good measure, Trentepohl had also applied for the post of doctor in the Guinea establishments in case such a position should become vacant. His application was supported by Professor Reinhardt of Copenhagen's Natural History Museum, and Professor J.F. Schouw, the plant geographer, who mentioned the names of Cuvier and Humboldt in praising Trentepohl's promising oceanographic work.[95] The king appointed Trentepohl surgeon on the Coast from October 1828 with a grant to cover four years, provided he remained on the Coast that long.[96] He arrived at Accra in January 1829, and by March had purchased a plantation in the hills of Akuapem.[97] The government reported that he was particularly interested in African colonization.[98] He soon wrote to the colonial office to suggest, since such an excellent foundation had recently been laid for a small but select library at Fort Christiansborg, that a German economic encyclopedia and the complete run of the proceedings of the Royal Danish Academy of Sciences and Letters would be useful additions: he regretted that such books were beyond his own means.[99]

Trentepohl's proposal helps explain the context in which he thought the books might be used. 'These two works would not only be interesting reading for a man of science here on the Coast', he wrote, 'but particularly appropriate and useful for any of the colonists who found occasion for skills that most of them, to the general detriment of the colony, lack'. He thought the books especially useful to agriculturalists and craftsmen, both which arts, he said, 'stand at quite a low stage' on the Coast. He thus identified the main difficulty on such a frontier: if one did not know how to do something oneself, there was simply nowhere to turn. 'The tropical colonial plants, like coffee, indigo, cotton, sugar, cacao, rice, and others', he wrote, 'require, as is well known, a particular and to us Europeans altogether unfamiliar process', and they were not grown here because 'one can obtain no information about how one most economically and most advantageously goes

by Bernhard Schlesinger, 1820]); *Dansk Biografisk Leksikon*, 2nd ed., s.v. "Schmidt-Phiselde(c)k, [Conrad] v."

94   Kai L. Henriksen, "Oversigt over dansk entomologis historie," *Entomologiske Meddelelser* 15 (1921–37): 201; Nørregård, *Danish Settlements in West Africa*, 199–200.

95   *Fonden ad Usus Publicos*, vol. 3, 1827–42, ed. Henny Glarbo (Copenhagen: Ejnar Munksgaards Forlag, 1947), 63–4; Trentepohl, 7 July 1828, GJ 60/1828; Reinhardt [Copenhagen], 5 February 1828, and Schouw, Copenhagen, 5 February 1828 (both *ad* GJ 60/1828), filed among Guineiske Resolutioner, 1821–26, GTK.

96   Royal resolution, 7 October 1828, GJ 148/1828.

97   Peter Thonning's notes, Materialier, Diverse, d. Guineiske Kommission af 9. Januar 1833 (hereinafter abbreviated GK), box III, GTK.

98   The government on the Coast, 22 March 1829, GJS 332/1829.

99   Trentepohl [Christiansborg], 23 July 1829, GJS 414/1829.

about cultivating same'. Since the Africans among whom they lived knew nothing of these crops, either, he presumed that the two works he recommended would be very valuable in this regard, and he hoped that others unknown to him might also be purchased for the fort's little library.

The colonial office thought it wisest to consult scientific authorities in the capital and forwarded Trentepohl's request to Jens Wilken Hornemann, professor of botany and editor of *Flora Danica*, and to Ole Rawert, a leading technocrat in the ministry of commerce and publisher of the *Handels- og Industrie-Tidende* (Trade and Industry News), a government organ.[100] Hornemann recommended Sloane's *Voyage* to the West Indies, by then a hundred years old, and Patrick Browne's *Civil and Natural History of Jamaica*.[101] He further recommended F.R. de Tussac's (1808–1827) *Flore des Antilles, ou Histoire générale botanique, rurale, et économique des végétaux indigènes des Antilles, et des exotiques qu'on est parvenu à y naturaliser*. Such books would be difficult to obtain and expensive, but Philip Miller's (1724) *Gardeners Dictionary*, which the director of the Botanic Garden called 'an essential work in the art of gardening' and which was available in many editions and languages, could surely be found in Copenhagen at less cost.[102] 'Regarding the colonial plants' [wrote Hornemann], 'Miller has very thorough accounts, taken from the best sources and collected from the colonists' observations'. Hornemann also recommended Abbé Raynal's *Histoire philosophique et politique des etablissements et du commerce des deux Indes*, published in Amsterdam in 1770, a Danish translation of which had been available for many years.[103] This work contained 'very reliable accounts of the cultivation of the tropical commercial productions', and Hornemann cited specific pages dealing with indigo, cocoa, cotton, sugar cane, and what he called the cochineal cactus.

---

100   Thomas Hansen Erslev, comp., *Almindeligt forfatter-lexicon for Kongeriget Danmark med tilhørende bilande fra før 1814 til efter 1858*, 1814–1840, reprint (Copenhagen: Rosenkilde og Bagger, 1962–63 [1843–53]), s.v. "Hornemann (Jens Wilken)"; *Dansk Biografisk Leksikon*, 2nd ed., s.v. "Rawert, Ole Jørgen".

101   Hornemann, at the Botanic Garden, 19 January 1830, GJ 462/1830.

102   According to Hornemann, the book had first come out in 1724. A second edition of Miller's *The Gardeners and Florists dictionary, or A Complete System of Horticulture* was published in London in 1724, and there were many other editions and translations. A later title, which expresses the usefulness of the book, was *The Gardeners Dictionary, Containing the Methods of Cultivating and Improving all Sorts of Trees, Plants, and Flowers, for the Kitchen, Fruit, and Pleasure Gardens; as also those which are used in Medicine. With Directions for the Culture of Vineyards, and Making of Wine in England. In which likewise are included the Practical Parts of Husbandry*, 4th edition (London: printed for the author, 1754).

103   Guillaume Thomas Raynal, *Raynals philosophiske og politiske Historie om Europæernes Handel og Besiddelser i Ost- og Vest-Indien*, trans. Matthias Nascou (Copenhagen: published by the translator, 1804–08).

To these suggestions Ole Rawert added a French work on the cultivation of cotton, even enclosing a copy of it for the fort's collection with his letter.[104] He pointed out that he had the year before published an account of sugar cultivation on the Danish West Indian Island of St. Croix in the *Handels- og Industrie-tidende*; and that two Danish books on the islands contained material on sugar cane; these also should be sent to the Coast.[105] Rawert also recommended the fifth (newly-published part) of Humboldt's *Reise in die Aeqvinoctial-gegende*. He doubted, moreover, that the books Trentepohl had requested would actually serve his purposes well and suggested instead recent books on mechanics,[106] a new German translation of a French handbook of chemistry, and Cuvier's *Geschichte der Fortschritte in den Naturwissenschaften seit 1789 bis auf den heutigen Tag*.[107] He also drew the colonial office's attention to the 'Annales maritimes et coloniales' published monthly by the French Marine and Colonial Department. 'It is a publication as interesting as it is useful', although he admitted it would be 'more suitable' for the central government's library in Copenhagen 'than for the fort's book collection'.

Imperial libraries, like all others, are subject to budgetary constraints. From Rawert's list, the colonial office selected Humboldt, the two books on mechanics, and Cuvier; and from Hornemann's list only Raynal's history of the Indies. These, and the book on cotton supplied by Rawert, were bound in Russia leather (the tanning process imparts resistance to insects), stamped *Ft Christiansborg*, and sent out in the East Indiaman *Alexander* in 1830. The government on the Coast was ordered to add them to the inventory of the fort's books.[108]

---

104   Rawert, Copenhagen, 12 March 1830, GJS 481/1830: C[harles] de Lasteyrie, *Du cotonnier et de sa culture, ou, Traité sur les diverses espèces de cotonniers, sur la possibilité et les moyens d'acclimater cet arbuste en France, sur sa culture dans différens pays* (Paris: Chez Arthur-Bertrand, 1808).

105   These were Peter Lotharius Oxholm, *De Danske Vestindiske Øers Tilstand i Henseende til Population, Cultur og Finance-Forfatning* (Copenhagen: Johan Frederik Schultz, 1797) and Hans West, *Bidrag til Beskrivelse over Ste Croix med en kort Udsigt over St. Thomas, St. Jean, Tortola, Spanishtown og Crabeneiland* (Copenhagen: trykt hos Friderik Wilhelm Thiele, 1793).

106   Georg Frederik Ursin, *Haandbog i den mecaniske Deel af Naturen* (based on John Millington's *Epitome of Natural and Experimental Philosophy*) (Copenhagen: [published by Ursin], 1826); Olinthus Gregory, *Theoretische, praktische und beschreibende Darstellung der mechanischen Wissenschaften*, trans. J.F.W. Dietlein (Halle: bei Hemmerde und Schwetschke, 1828).

107   George Cuvier, *Geschichte der Fortschritte in den Naturwissenschaften seit 1789 bis auf den heutigen Tag* (*Histoire des progrès des sciences naturelles depuis 1789 jusqu'à ce jour* [1826–28]), trans. F.A. Weise (Leipzig: Baumgärtner, 1828–29).

108   Pakhusforvalteren, 20 July 1830, GJ 558/1830; Chamber of Customs to the government on the Coast, 24 July 1830 (no. 566), Guin. kopibog.

Trentepohl was dead before the books arrived on the coast of Africa.[109] The government wrote, in announcing the news of his death, that 'his conscientiousness in his office and ardor for plantation operations make his loss felt'. It was indeed this dreadful mortality, which so interrupted continuity of effort and enthusiasm, that most hampered Denmark's colonial undertaking on the Coast. Trentepohl's books remained in circulation, however. Balthazar Christensen, a colonial clerk who had also come out in *Alexander*, recorded in his diary that Trentepohl's effects, when auctioned off, 'went very high, with the exception of some scientific works, which, however, I persuaded the governor to buy for our library'.[110]

Christensen amused himself immensely with his diary in the year he spent on the Coast, and it is an important source for the period.[111] In the first entry, written en route to Africa, he recorded that he was reading Goethe, Schiller and Las Casas. He was also studying accounting. The ship called at Monrovia, the capital of Liberia, the newly-founded colony of African-American freedmen, and Christensen compares the town to frontier scenes called up in his imagination by James Fenimore Cooper's novel *The Pioneers, or the Sources of the Susquehanna*. Christensen's English was poor, but a German translation of *The Pioneers* had been published in the 1820s.[112] Elsewhere, Christensen compares one of his fellow officers to Hawkeye, in Cooper's *The Last of the Mohicans*.[113] The strange spectacle of an educated Danish traveller in Africa referring, however ironically, to the American romances of Cooper is the very stuff of early nineteenth-century African colonialism: notions drawn from historical or purely literary constructions of the colonial experience of other times in other continents were being projected back across the Atlantic to Africa. Both the economic precedent and the huge potential of the Americas figured centrally in African colonial thinking at this period.

Christensen met the editor of the local Monrovia monthly, the *Liberia Herald*, and we know that this paper subsequently reached him down the coast at Accra.[114] Christensen wrote of Liberia, 'It is interesting to see how this North American colony progresses, and it is obvious that it will soon mark a new era in the history

---

109   The government on the Coast, 16 March 1830, GJ 615/1830.

110   Entry for 20 November 1830, Balthazar Christensen's Diary.

111   He shipped sections of his diary to friends in Copenhagen at every opportunity (for example, the entry for 15 May 1831). It remains unpublished. Some of his letters home made their way into print in a Copenhagen periodical, with the editor's apologies for taking such a liberty with his friend's correspondence; the diary (28 March 1831) indicates that the letters were in fact intended for publication: [Christensen], "Breve fra og om Guinea," *Valkyrien* 2 (1831): 262–9 and 269–78, and 3 (1831): 56–63.

112   James Fenimore Cooper, *Die Ansiedler, oder Die Quellen des Susquehannah* (Frankfurt am Main: J.D. Sauerländer, 1826–27). There is, to be sure, no direct evidence that Christensen actually took the book with him to Africa.

113   Entry for 15 May 1831, Balthazar Christensen's Diary.

114   [Balthazar Christensen], "Nogle Bemærkninger om fremmede Colonier paa Vestkysten af Afrika," *Valkyrien* 1 (January-March 1832): 264.

of the colonization of the West Coast, and contributes much more than everything that has been done hitherto to cultivate this portion of the world'. He also noted that an American missionary in Monrovia had presented him with a copy of 'lost Paradise', which he looked forward to reading in the original English. He later recorded that he spent several hours a day in the study of French and English, as well as of a couple of African languages.[115]

At Fort Christiansborg, Christensen sneered at the government's collection of books, of which Schmidt-Phiseldek, 'with much unctuousness', had led him to expect so much before he had left Copenhagen.[116] He urged his friends at home to take up a subscription for a really good library, which the fort's officers sorely felt the lack of. He was confident that formal relations could be established with the inland kingdom of Asante 'if we at this moment had books that could bring me to the point of being able to write a single letter in comprehensible Arabic'. Christensen found himself on a friendly social footing with the wealthy and influential Accra merchant Henrich Richter, a man of mixed African and Danish parentage, who, like other such individuals on the Coast, occupied a social position between two worlds. The African side of their lives is mainly implicit in their political and economic transactions and is difficult to reconstruct, but something of their perspective on the colonial Atlantic can be read on Richter's bookshelves. Christensen recorded that he was reading everything he could lay his hands on about Africa and indeed about the whole colonial world, and that Richter owned a great number of such works.[117] The young officer confided in his diary that he and Richter were working together on a history of relations between Asante and the European establishments on the Coast with the intention to publish it in Danish and English.[118] Christensen scoffed at everything else that had yet seen print, but at the same time presumed that anyone reading his journal was already familiar with Monrad's book, which he allowed was of 'great worth'.[119] He thought Isert's account, however, remained the best of all, at least until he read Henry Meredith's

---

115 Entries for 17 July 1831 and 1 October 1830, Balthazar Christensen's Diary, at which last entry a pencil note in the margin indicates that in 1898 this same copy of *Paradise Lost* was back in Denmark), see also entry for 14 December 1830.

116 Ibid., entry for 14 November 1830.

117 A well-travelled copy of Meredith's book now at the Royal Library in Copenhagen is inscribed "Richter", with the name "Thonning" superimposed in another ink. Richter was on good terms with Peter Thonning and his younger brother Matthias, who also served on the Coast. See Richter, 25 October 1817, Guineiske Resolutioner 1816–20, GTK, and Richter, Danish Accra, 13 June 1841, to Thonning, Breve fra alle om Guinea, Kommissionens Korrespondance, GK, box I. Meredith bequeathed to Richter what Christensen said was an excellent English encyclopedia, which he often consulted at the great merchant's home in Accra; see Justesen, "Henrich Richter," 187.

118 Entry for 17 July [1831], Balthazar Christensen's Diary. No such work was forthcoming.

119 Ibid., entry for 1 January 1831.

book, which he then declared to be the most important West African account ever published.[120]

European newspapers made their way down the Guinea coast. The whole 1791 run of *Morgenposten*, a Copenhagen weekly, was sold with Governor Olrick's estate in 1793. The merchant Peder Meyer read the *Politisk og Physisk Magazin*, a monthly published in Copenhagen between 1793 and 1800. In July 1831, Christensen recorded that Henrik Richter had recently received the *Times* of London up to 5 May, and that he himself was working his way through March and April.[121] He exhorted his friends in Denmark to collect and send him the Copenhagen papers (and several provincial papers), but it does not appear that they ever did so.[122] In the 1840s, the government on the Coast urged the colonial office, as a matter of formal policy, to send the most important Copenhagen papers, even if it had to be at the officers' own expense. The administration yielded so far as to subscribe to one paper for the fort, *Fædrelandet* (The Fatherland), an important voice in Denmark's incipient democratization in the 1830s.[123] A late inventory of the library at Fort Christiansborg also lists the official Danish ministerial journal, *Collegial-Tidende* (the Collegial Times, so named for the structure and functioning of the government) from 1836 until 1848.[124]

There is little indication that many Africans read very far at all in any of these books. It can be speculated that atlases were laid on tables between Europeans and their African interlocutors, and that illustrated botanical and zoological works eased communication between natural historians and their field collectors. But the archives of this period put the printed word only in the hands of the Danes and of their Danish-African offspring. King Frederik VI was an enthusiast for Joseph Lancaster's influential system of mutual or monitorial instruction, which allowed the reduction of outlays for teachers through reliance on simple standardized rote instruction by older pupils working with large printed tables of letters, words and figures hung on school-room walls. The king was persuaded that the system would benefit his Danish-African subjects on the Coast and, in 1821, ordered four copies of a new Danish work on the monitorial

---

120   Ibid., entry for 17 July [1831]. Another young officer, Wulff Joseph Wulff, also commented on Monrad's and Isert's books: *Da Guinea var dansk*, 52–3, 125.

121   Entry for 13 July 1831, Balthazar Christensen's Diary. Wulff learned of news in the English papers over dinner with visitors from Cape Coast Castle.

122   Entry for 20 November 1830, Balthazar Christensen's Diary.

123   The government on the Coast, 10 September 1843, GJ 499/1843; Jette D. Søllinge and Niels Thomsen, *De danske aviser 1634–1991*, vol. 1 (Odense: Dagspressens Fond i Kommission hos Odense Universitetsforlag, [1988]), 156–7.

124   "Fortegnelse over de fra Guinea hjemsendte Sager," GJS 161/1851. A bound volume of the *Roeskilde Stændertidende* (vol. 2, 1842), the published proceedings of one of the estates assemblies, elected regional advisory bodies, that marked the liberalization of the Danish political system beginning in the 1830s, is also listed.

system sent to the Coast.[125] In a 1789 inventory taken a couple of years before the establishments reverted to direct royal control, the schools in the four forts were each recorded to possess several ABCs.[126] The forts were also full of books for the spiritual edification of the officers, troops and their African families: psalm books, catechisms, epitomes of biblical history for children, collections of sermons and mid eighteenth-century Christian apologies are listed in such inventories.[127] At Fort Christiansborg, there were thirty or forty copies of some of these books, which were recorded in 1789 to have been 'somewhat damaged by cockroaches and bookworm'.

As men perish and books decay, so colonies also fade from currency and memory. Denmark abandoned its African colonial ambitions much earlier than the other colonial powers. In 1850, Denmark liquidated its establishments and transferred its forts and territorial claims to Britain. What remained of the government library at Fort Christiansborg was inventoried and shipped home to Copenhagen, where it was lost to view, presumably having been dispersed into the working libraries of various government agencies and public and private collections.[128] Many of the titles one would expect to see on this last inventory are missing, but this is perhaps in the nature of all libraries, and especially of libraries in such wild tropical outposts of literacy as the Danish forts on the Guinea coast.[129] Books are lost or moulder away in damp storerooms. They are read and as quickly forgotten: even the greatest works of literature become dated and inaccessible. Nevertheless, books' influence in any literate society is

---

125   The king, 6 October 1821, GJ 109/1821; P.H. Münster and J. Abrahamson, *Om den indbyrdes Underviisnings Væsen og Værd* (Copenhagen: trykt hos A. Seidelin, 1821–28). A copy of Lancaster's own book on his system, *Improvements in Education as it Respects the Industrious Classes of the Community* (London: Darton and Harvey, 1803) was sent to the Coast: Chamber of Customs to the government on the Coast, 25 October 1821, no. 470, Guin. kopibog.

126   "Udskrift af Afleverings-Forretningen over de Danske Forteresser [sic] og Loger med tilhørende Inventarium paa Kysten Guinea," 1 November 1789, Finanskollegiets correspondence journal files, *ad* 661/1791, Finanscollegiet; see also "Overleverings forretning," 1 January 1800, under church and school inventory, *ad* GJS 32/1800.

127   Hans Jørg Birch, *Nye Evangelie Bog eller kort Udtog af den Bibelske Historie for Børn, især paa Landet* (Copenhagen, 1788); a book of Herman Treschow's sermons; Hans Jørg Birch's *Morgen og Aftenbønner og Psalmer for Børn* (Copenhagen, 1803); Erik Pontoppidan, *Sandhed til Gudfrygtighed udi en eenfoldig og efter Mulighed kort, dog tilstrækkelig Forklaring over Sal. Mort. Luthers Liden Catechismo*, which was published in many editions beginning in 1737. Ramus's and Glad's libraries both included one or another of Christian Bastholm's books of "spiritual orations".

128   "Fortegnelse over de fra Guinea hjemsendte Sager," GJS 161/1851.

129   In 1825, Governor Richelieu, in response to a query from the colonial office, reported that the inventory at Fort Christiansborg included a hundred "books in various languages (completely unusable)", a "Fortegnelse" with Richelieu, Christiansborg, 19 August 1825, GJS 477/1825.

incalculable. Without writing – without books – the European colonialism that helped define the modern era is scarcely imaginable. Books do not merely record colonial history: they powerfully shaped it.

Acknowledgements

The author is particularly grateful to Hanne Balslev and Ole Justesen for their advice. The archival research in Copenhagen was supported by the Carlsberg Foundation and the University of Missouri Research Board.

Chapter 9

Volney's *Tableau*, Medical Geography and Books on the Frontier

Michael L. Dorn

Can a book change how a society sees itself and is seen by others? Historians and literary critics have argued in the affirmative for *Notes on the State of Virginia* (1787), Thomas Jefferson's response to questions posed in 1781 by M. Barbé de Marbois, secretary of the French legation in Philadelphia. Jefferson's *Notes* offered a comparative table of flora and fauna in the Old World and the New World – even the mastodon made its appearance – as Jefferson refuted Buffon's depictions of North America as a place where Old World plants and animals reappeared in degenerate forms.[1]

Constantin Volney's *Tableau du climat et du sol des États-Unis d'Amérique* (1803) was another key contributor to transatlantic debates over natural philosophy and the prospects for American self-government. Responses to the *Tableau* drew Americans into a print discourse linking savants in European and American capitals – Paris, New York, Philadelphia – and frontier intellectuals in Pittsburgh, Cincinnati and Lexington. The creation of knowledge and the exchange of ideas between European centres of calculation and colonial and post-colonial peripheries was not a one-way process. The need for information from the hinterlands, whether of people, plants or climate, created complex relationships between metropolitan expertise and local knowledge. Volney's ideas stimulated further observation, collection and publication in regions whose intellectual and social leaders were not content to have others write about them without rejoinder. Because Volney made remarks about soil, climate and the health of their country, his work made a key contribution to the articulation of an American medical geography. But Volney's book has additional interest given recent work not just in the history of medical geography, but in Enlightenment studies and in book history. Understanding the making and, more importantly, the reception of this book on the frontier can shed light on the ways in which the new American Republic was understood and understood itself.

---

1 Ralph H. Brown, "Jefferson's Notes on Virginia," *Geographical Review* 33 (1943): 467–73; Paul Semonin, *American Monster: How the Nation's First Prehistoric Creature Became a Symbol of National Identity* (New York: NYU Press, 2000).

Historians of medicine in recent years have explored the wellsprings of the geographical-medical impulse in its theoretical, practical and textual manifestations.[2] Such work has often been couched in terms of national narratives, but it can learn from recent work in Enlightenment studies which stresses the multi-dimensional geographies of the Enlightenment, and work which has investigated the role of books on the American frontier.[3] America's western landscape was being cast through Euro-American orders of meaning, through the division and sale of land, the creation of literary societies and the establishment of early newspapers and book presses. This provides an important backdrop to Constantin Volney's study of health and society, and the trans-Atlantic production of, and cis-Atlantic reception for, his *Tableau*. The *Tableau* drew on observations Volney gathered during residence and travel in the United States between 1795 and 1798 to provide an exemplar of late Enlightenment environmental medical thought. The book then engendered a wide variety of responses, notably after the publication of an American edition in 1804.[4] Examination of these responses, and other geographical guides and grammars published on the frontier between 1804 and 1815, reveals what we may think of as a rich inter-textual conversation responding to Volney's concerns about the state of civilization in the West and its prospects for health and political unity. In these works the Ohio Valley served not only as a terrain for systematic observations – books about it were a vessel for discussion of the future of a newly-emergent America.

Enlightenment and Print on the Frontier

Judging upon probable grounds, the Mississippi was never designed as the western boundary of the American empire. The God of nature never intended

---

2   For one overview, see Nicolaas A. Rupke, ed. *Medical Geography in Historical Perspective* (London: Wellcome Trust, 2000).

3   Charles W.J. Withers, *Placing the Enlightenment: Thinking Geographically about The Age of Reason* (Chicago: University of Chicago Press, 2007); Hugh Amory and David D. Hall, eds, *The Colonial Book in the Atlantic World: A History of the Book in America* (Cambridge: Cambridge University Press, 2000); Richard W. Clement, *Books on the Frontier: Print Culture in the American West, 1763–1875* (Hanover: University Press of New England, 2003); Rosalind Remer, "Preachers, Peddlers, and Publishers: Philadelphia's Backcountry Book Trade, 1800–1830," *Journal of the Early Republic* 14:4 (1994): 497–522.

4   C.-F. Volney, *Tableau du climat et du sol des États-Unis d'Amérique*. 2 vols. (Paris: Courcier, 1803). Unless stated otherwise, this paper draws on the first American edition, *A View of the Soil and Climate of the United States of America: with supplementary remarks upon Florida; of the French colony on the Sciota, and in Canada; and the aboriginal tribes of America*, trans. C.B. Brown (Philadelphia: J. Conrad & Co., 1804) [hereafter *View of the Soil and Climate*].

that some of the best of his earth should be inhabited by subjects of a monarch, 4000 miles from them.[5]

Early American pedagogues like Jedidiah Morse took seriously their responsibility in narrating the nation. Published to coincide with George Washington's inauguration to the presidency and viewed from the perspective of New England, Morse's *American Geography* is, for Brückner, indicative of an expanding northeastern and mid Atlantic geographical authority.[6] But different stories emerge, of creole voices speaking back to Eastern centres of calculation, if one looks at the geographies of Enlightenment and of print being produced on the American frontier. Withers's *Placing the Enlightenment* expands our understanding of the reach of Enlightenment societies and books, from the reading practices of New England villagers to the collecting and surveying activities of French naturalists and engineers along the Mississippi frontier.[7] Lexington, identified as the lone western outpost of scientific society, is 7 degrees west of Richmond, the nearest enlightened capital city in North America. Lexington and New Orleans were also the first communities in the American West to establish printing presses and to publish books. They were university towns while much of the territory drained by the Mississippi and Ohio Rivers remained Native American territory.

The Ohio Valley region, celebrated in Europe as a cornucopia, featured prominently in the colonial ambitions of the British, the French and the Spanish, as well as the American Republic. Marginal by location, it was to assume centre stage as a site for the construction of knowledge claims concerning American ingenuity, perseverance and potential for improvement. Claims to authority and knowledge in the trans-Appalachian West were crafted through practices of printing and land surveying. Over the course of the 1780s the region north of the Ohio River was formally ceded by Virginia into the Federal public domain but remained largely closed to settlement. Over the same decade, the frontier crossroads at Lexington became the region's market, established a newspaper and a college, and supported six different stores offering books for sale.[8] Within months of its first number, *The Kentucke Gazette* [sic] noted the formation of the Lexington Society for the Promotion of Useful Knowledge, modeled on the Virginia society of the same name.[9]

---

5   Jedidiah Morse, *The American Geography; or, a View of the Present Situation of the United States of America* (Elizabethtown: Shepard Kollock, 1789), 469.

6   Martin Brückner, *The Geographic Revolution in Early America: Maps, Literacy, and National Identity* (Chapel Hill: University of North Carolina Press, 2006).

7   Withers, *Placing the Enlightenment*, 34.

8   Howard H. Peckham, "Books and Reading on the Ohio Valley Frontier," *The Mississippi Valley Historical Review* 44 (1958): 652.

9   The Virginia Society for the Promotion of Useful Knowledge [VSPUK] was founded in 1773, the Lexington Society in 1787, the same year the Transylvania Seminary moved to Lexington from Danville. Huntley Dupre, "The *Kentucky Gazette* reports on

Yet the public sphere that emerged in Lexington during this period was highly circumscribed. The revolutionary enthusiasm of leading figures such as John Brown, George Nicholas and John Breckinridge was fueled by their political and legal education at the College of William and Mary at Williamsburg.[10] But their participation in the public sphere also depended upon oppressive practices that restricted rights to democratic self-determination: slavery and land speculation. Many Virginia planters brought all of their possessions including their libraries and their slaves on their passage through the Cumberland Gap.[11] Because of the ability of Virginia landowners to commission land jobbers and use informal metes-and-bounds surveying methods to claim land sight unseen in the backcountry, most of the valuable land was quickly locked away from the long hunter and the yeoman farmer.[12] With an Ordinance of 1785, rules were laid down that would extend the power of the executive branch over vast Western lands that Eastern states had ceded claim to under the Articles of Confederation.[13] The federal responsibility of implementing a new system of land division and distribution was assigned to a new office, the United States Surveyor General. Land division, like early print, became a mode for the exercise of political authority.

Captain Thomas Hutchins initiated the first rectangular survey, the Seven Ranges, from a point near where the Ohio River crossed the western boundary of Pennsylvania. Although historians of the public lands have noted that this effort was not an unmitigated success, for it slowed in the face of continued Indian resistance and competing priorities amongst members of the surveying team, it was an important 'point of beginning'.[14] The unwavering lines of rectangular survey symbolized a new relationship between settler and landscape from that practiced amongst the Indian nations. It was also a purposeful response to the individualistic land practices of the contentious backcountry Virginians, who, in

---

the French Revolution," *The Mississippi Valley Historical Review* 26 (1939): 163–80. On the VSPUK, see Lyon G. Tyler, "Virginia's Contribution to Science," *William and Mary College Quarterly Historical Magazine* 24 (1916): 222.

10   Marion Nelson Winship, "Kentucky in the New Republic: A Study of Distance and Connection," in *The Buzzel About Kentuck: Settling the Promised Land*, ed. Craig Thompson Friend (Lexington: University Press of Kentucky, 1999), 109.

11   Lowell Harrison, "A Virginian Moves to Kentucky, 1793," *The William and Mary Quarterly* 15 (1958): 201–13.

12   Stephen Aron, "Pioneers and Profiteers: Land Speculation and the Homestead Ethic in Frontier Kentucky," *The Western Historical Quarterly* 23 (1992): 179–98.

13   Andrew R.L. Cayton, "'Separate Interests' and the Nation-State: The Washington Administration and the Origins of Regionalism in the Trans-Appalachian West," *Journal of American History* 79 (1992): 39–67.

14   Andro Linklater, *Measuring America: How an Untamed Wilderness Shaped the United States and Fulfilled the Promise of Democracy* (New York: Walker & Co., 2002), 74–88; Silvio Bedini, "Captain Thomas Hutchins – Part 2: Geographer of the United States," *Professional Surveyor* 18 (1998): 76–8.

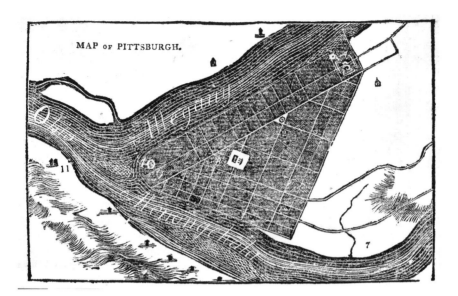

Figure 9.1   Plan of Pittsburgh, from Zadok Cramer's *Navigator*

Source: Reproduced with permission of the Library Company of Philadelphia.

the winter of 1784–5 convened to discuss the need to create a state of their own.[15] Since they had no printing press with which to publicize their convention, local resident John Bradford was recruited for the project. Before he could publish the first number of the *Kentucke Gazette*, his brother Fielding Bradford visited Fort Pitt at the confluence of the Allegheny and Monongahela Rivers to purchase types and to learn the craft from John Scull and Joseph Hall, the editors of the *Pittsburgh Gazette*, the first newspaper in the West.[16] For a time, John Bradford was the only book publisher between Pittsburgh and New Orleans.

When the state capital moved to Frankfort, another book press was established there. Hunter and Beaumont of Frankfort would publish the first edition of the key traveler's guide, the *Ohio Navigator*.[17] While the earliest Kentucky editions merely offered an account of places and their distances down the Ohio River from Pittsburgh, the *Navigator* editions published after 1802 by Zadok Cramer served a broader purpose, including geographical information on the cities, states and

---

15   Thomas P. Abernethy, "Journal of the First Kentucky Convention Dec. 27, 1784 – Jan. 5, 1785," *The Journal of Southern History* 1 (1935): 67–78.

16   Douglas C. McMurtrie, "A History of Printing in the United States," in *Douglas C. McMurtrie, Bibliographer and Historian of Printing*, ed. Scott Bruntjen and Melissa L. Young (Metuchen: Scarecrow Press, 1979), 73–81.

17   Karl Yost, "Introduction," in *The Ohio and Mississippi Navigator of Zadok Cramer: Third and Fourth Editions*, ed. Zadok Cramer (Morrison: Karl Yost, 1987), xi–xxxiv.

territories that lay on either side of the river, and beyond the mouth of the Ohio River to New Orleans (Figure 9.1).

While navigation of the Ohio River was imperiled by ongoing frontier bloodshed, the Lower Mississippi River was controlled by Spain. President Washington recognized that a show of federal military authority would do much to allay worries and build national sentiment. Ohio Valley fortifications would give rise to active market centers to compete with Pittsburgh. The first military outpost north of the Ohio River, Fort Washington, generated trade and attracted migrants to the village of Losantiville. Soon renamed Cincinnati after the Society of the Cincinnati, a military fraternal organization, this village of 200 supported the first newspaper north of the Ohio River, *The Centinel of the North-Western Territory*, in 1793.[18] In time Cincinnati would grow to support its own paper mill, and two newspapers, each with its own book press.[19] Along with legal proceedings and the ever popular almanacs, Reverend John W. Brown published the first geographic account of Southwestern Ohio to be written in the West, Daniel Drake's *Notices Concerning Cincinnati* (1810).[20] In short, America's frontier history and geography was laid out in print and on the map as it was being laid out in the political realm. Drake's work in particular will be returned to. Let me turn now, however, to Constantin Volney's view of this part of frontier America.

Volney's Philosophy and Method

Michael Heffernan and David Livingstone have noted that the secular idea of progress grew out of mid eighteenth-century efforts to construct a natural history of man incorporating speculative histories of his development and possible futures.[21] Assessments of the 'state of society' in travelers' accounts frequently included discussions of health and disease. The more philosophically-minded sought to explain them through the systematic description of the physical situation of the community and moral development of its inhabitants, often citing the Hippocratic treatise *Airs, Waters, Places*.[22] For the provincial scientist and physician 'on the

---

18   C.B. Galbreath, "The First Newspaper of the Northwest Territory: The Editor and His Wife," *Ohio Archaeological and Historical Society Publications* 13 (1904): 332–49.

19   Walter Sutton, *The Western Book Trade* (Columbus: Ohio State University Press, 1961).

20   Daniel Drake, *Notices Concerning Cincinnati* (Cincinnati: John W. Browne & Co., 1810).

21   Michael Heffernan, "On Geography and Progress: Turgot's *Plan d'un ouvrage sur la geographie politique* (1751) and the Origins of Modern Progressive Thought," *Political Geography* 13 (1994): 328–43; David N. Livingstone, "Putting Progress in its Place," *Progress in Human Geography* 30 (2006): 559–87.

22   Conevery Bolton Valencius, "Approaches to the History of Medical Geography," in *Medical Geography in Historical Perspective*, ed. Rupke, 3–28.

spot', the medical topographic genre afforded the opportunity to assert a form of local knowledge and authority.[23]

Programmatic descriptions in the Enlightenment of the term 'medical geography' (and 'medical topography') can be found in the writings of Jean-Nöel Hallé and Ludwig Leonhard Finke.[24] Hallé offered his vision of medical geography in 1792 while describing the approach that was being used to organize the 'hygiene' section of the *Encyclopédie Méthodique*.[25] Finke offered his definition that same year in the introduction to his *Versuch einer allgemeinen medicinisch-praktischen Geographie*. For both authors, 'medical geography' was a global survey of patterns of health as they related to the conditions of the environment and of human society. Work in this genre drew upon, and implicitly critiqued, Montesquieu's *The Spirit of the Laws* (1748) which offered broadly deterministic reasoning in explaining the connections between physical geography, particularly climate, and the development of human societies. Volney's *Leçons d'histoire* (first delivered in 1795), along with Cabanis' *Rapports du physique et du moral de l'homme* (1802), further developed the philosophical and methodological foundations for the natural history of man and disease.

Volney traveled to North America intending to make observations on the current state and prospects of the American republican experiment. He was particularly keen to test his philosophical theories against the observed experience and character of the indigenous American populations, and to assess the role of climate in shaping the immigrant populations that were quickly replacing them. By the time Volney visited the Ohio Valley in the mid 1790s, he had developed a comparative approach to the physical and moral geography of states and nations based in the methods of natural history. Following education in law at Angers, Volney's principal course of study in Paris after 1775 was medicine and he quickly established a close friendship with the vitalist medical philosopher Cabanis.[26] Inspired also by the works of Montesquieu, Volney devoted himself to oriental languages and in 1782, at the age of 25, undertook an exploration of Egypt, Libya and the Levant. The resulting publication, *Voyage en Syrie et en Egypte* (1787) was an immediate success and made his reputation. This publication set a pattern for later work by linking population health and wellbeing to physical situation

---

23   Mark Harrison, *Climates and Constitutions: Health, Race, Environment and British Imperialism in India, 1600–1850* (New Delhi & New York: Oxford University Press, 1999).

24   Frank A. Barrett, ed., *Disease and Geography: The History of an Idea* (Toronto, York University Press, 2000).

25   Jean-Noël Hallé, "Exposition du plan du'un traité complet d'Hygiene," *Le Médicine eclairée pars les sciences physiques* 4 (1792): 225–35.

26   On Volney and the *École Normale* see Charles Coulston Gillispie, *Science and Polity in France: The Revolutionary and Napoleonic Years* (Princeton: Princeton University Press, 2004), 514–18.

(principally climate and soil) while also considering the importance of social institutions in catalyzing or impeding progress.

As a liberal concerned with social reform, Volney welcomed the French Revolution. After publishing the visionary *Ruins, ou Méditations sur les Revolutions des Empires* (1791), Volney settled briefly in Corsica, where he undertook experiments in orange cultivation. This was a failure, however, and he returned to Paris in 1793 with the idea of looking to the New World for his next comparative study. His departure was delayed, however, by the radicalization of the Revolution, and he spent ten months imprisoned for supporting the Girondin faction. With the reintegration of the Girondins to the Convention, Volney assumed the position of history chair at *L'Ecole Normale* in 1794.[27]

As Michel Foucault has shown, the revolution of society was linked to a revolution in the received systems of knowledge. It is telling that the one Royalist institution not shut down by the Jacobins was the King's Garden, the *Jardin des Plantes*. With his *Histoire Naturelle*, George-Louis Leclerc, comte de Buffon (1707–1788) provided a new way of conceiving of natural organization that incorporated those findings in natural knowledge flooding into Paris, and for devising strategies to reorganize systems of knowledge in other fields. An opponent of mechanist natural philosophy, Buffon emphasized the primacy of living over inanimate matter and sought to explain an order of things 'that elevated dynamic relationships and qualitative change over time'. Buffon also directed the plantings at the *Jardin* to present the visitor with a visual example of this new system of natural history.[28]

Constantin Volney's lectures at the *Muséum de Histoire Naturelle* in the winter of 1794–5 give a sense of his philosophical approach to historical scholarship on the eve of his departure for the United States. Volney, in keeping with many contemporaries, found the authors of classical antiquity to be insufficiently rigorous, and instead held the writing of history to standards of evidence dominant in natural philosophy. The last of Volney's lectures considered various models for historical scholarship: the chronological arrangement; the dramatic or systematic arrangement; the textbook approach where each topic is treated separately; and the analytic or philosophic approach, which began with a survey of the physical and moral characteristics of a region and its inhabitants following a schedule similar to the *Questions de statistique à l'usage des voyageurs* which Volney had presented to the External Relations Commission in 1795. 'After

---

27 The *Ecole Normale* lectures in history, geography and political economy have been published by Daniel Nordman, *L'Ecole Normale de l'an III: leçons d'histoire, de géographie, d'économie politique* (Paris: Dunod, 1994).

28 Michel Foucault, *Les mots et les choses: une archéologie des sciences humaines* (Paris: Gallimard, 1966); Emma C. Spary, *Utopia's Garden: French Natural History from Old Regime to Revolution* (Chicago: University of Chicago Press, 2000); Peter Hanns Reill, *Vitalizing Nature in the Enlightenment* (Berkeley: University of California Press, 2005) 69.

investigating the effects of the 'physical state' – climate, soil, and natural resources – on the habits, customs, and character of a nation, and its government and kind of laws, then the physical appearance and health of the population, occupations, agricultural practices, industrial arts and methods, commerce, type of government and legal system, family authority and education would be examined'.[29] As in his *Travels Through Egypt and Syria*, this method involved discussion of the physical and moral foundations of society before considering the strengths, weaknesses and resultant character and prospects of systems of government. This approach closely allied with medical geographic practice as described by Hallé in the *Encyclopédie Méthodique*. Volney's method would find fertile intellectual soil when he turned his gaze to the American frontier. And since the French Republic regarded the reacquisition of Louisiana through the fortification and colonization of the Mississippi Valley as key to containing Anglo-American expansion, Volney's travels in America were to serve geopolitical as well as scientific purposes.[30]

Volney in America

Upon his arrival in Philadelphia in October 1795, Volney was welcomed into the French liberal émigré community that focused on the bookstore at the corner of Front and Walnut Streets owned by Moreau de St. Méry. From this setting, Volney had access to news accounts, tracts and treatises from across the French Atlantic world. Volney joined Moreau as a resident member of the American Philosophical Society, where he forged friendships with American scientists Thomas Jefferson and Benjamin Smith Barton. He began to be drawn into the Republican camp in a city dominated by Federalists.

Volney's arrival in Philadelphia coincided with that of another geographer, General Victor Collot. Until recently Governor of Guadeloupe, Collot had accepted a commission from French minister Adet to conduct a military survey of the Ohio and Mississippi Rivers during the summer of 1796.[31] That same summer Volney would leave Philadelphia to conduct his own survey of the western states and territories. Collot left Philadelphia with an assistant, Adjutant-General Joseph Warin, on 21 March 1796. After reconnoitering the situation of Pittsburgh, on 6 June they began

---

29   Martin S. Staum, *Minerva's Message: Stabilizing the French Revolution* (Montreal and Kingston: McGill-Queen's University Press, 1996), 157.

30   Before departing, Volney secured a commission as naturalist from the Conseil Executif. See "Extrait des Registres des deliberations du Conseil Exécutif provisoire du 3ème jour du 2è mois de la 2 année de la République une et Indivisible" (3 Brumaire an 2), Archives Nationales, Paris.

31   Jack D.L. Holmes, "Some French Engineers in Spanish Lousiana," in *The French in the Mississippi Valley*, ed. John Francis McDermott (Urbana: University of Illinois Press, 1965), 133.

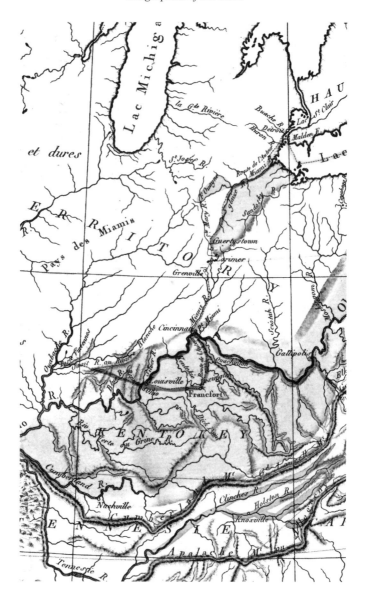

Figure 9.2   Detail from the Map of the United States of North America
prepared for Volney's *Tableau*. The shading (tinted green in
the original) indicates a calcareous region underlain with
limestone. Volney's route from Gallipolis to Poste Vincennes
and from Cincinnati to Detroit is indicated with a dashed line

Source: Reproduced with permission of The Academy of Natural Sciences, Ewart Sale
Stewart Library and the Albert M. Greenfield Digital Imaging Center for Collections.

their descent of the Ohio River. They plumbed the river's depth, surveyed its route and noted evidence of human habitation. Volney began his tour by heading south to Charlottesville, where he conferred with Thomas Jefferson regarding what he should expect to encounter further west (see Figure 9.2).[32] His route from Staunton to Greenbriar traversed the Virginia Springs region, where Volney made a detour to the bath-house at Hot Spring – facilities that helped the weary traveler to reclaim his vitality before traversing the 'rugged desart'.[33]

The arrival in the West of French military topographers Collot and Warin and scientific traveler Volney could not fail to attract the attention of American military authorities. In preparation for his own departure to the West, General Anthony Wayne received instructions from U.S. Secretary of War James McHenry requesting that he gather evidence on the missions of three emissaries: 'Powers, Warin and DeCallot', employed to 'gain knowledge of our military posts in the Western union, and form a political and separate connexion with a foreign power'.[34] On his arrival in Pittsburgh in late July, Wayne learned that Collot had told American friends there that the secret provisions of a peace treaty between the French Republic and Spain meant Louisiana would revert to French hands. With command over the Lower Mississippi, the French spies expected to take possession of the entire Mississippi Valley including four key Appalachian mountain passes, with the assistance of sympathetic Kentuckians.[35]

Volney made a point of visiting settlements in the U.S. Northwest Territory and Spanish Louisiana with substantial French populations, noting their history, demographic composition and future prospects. The first of these, Gallipolis, lay a few miles down the Ohio River from Point Pleasant. This had been settled by French emigrants of middle class background who had been convinced by the glowing reports of Joel Barlow and William Playfair, the American land agents for the *Compagnie du Scioto*, who spoke of the Ohio Country as having 'a climate wholesome and delightful, frost, even in the winter, almost entirely unknown, and a river called, by way of eminence, the *beautiful*, and abounding with excellent fish, of a vast size'.[36] Volney thought the settlement forlorn. In contrast, Poste Vincennes offered a pleasing prospect.

32   Gilbert Chinard, *Volney et l'Amérique d'après des Documents Inédits et sa Correspondance avec Jefferson* (Baltimore: The Johns Hopkins Press, 1923).

33   Volney, *View of the Soil and Climate*, 324.

34   James McHenry to Anthony Wayne, 25 May 1796, in Richard C. Knopf, ed., *Anthony Wayne, a Name in Arms: Soldier, Diplomat, Defender of Expansion Westward of a Nation; The Wayne-Knox-Pickering-McHenry Correspondence* (Pittsburgh: University of Pittsburgh Press, 1959), 481–4.

35   Anthony Wayne to James McHenry, 27 July 1796, in *Anthony Wayne*, ed. Knopf, 489–90.

36   Robert F. Durden, "Joel Barlow in the French Revolution," *The William and Mary Quarterly* 8 (1951): 327–54; Volney, *View of the Soil and Climate*, 322–30.

The eye is first presented with an irregular savannah, eight miles in length by three in breadth, skirted by eternal forests, and sprinkled with a few trees, and abundance of umbelliferous plants, three or four feet high. Maize, tobacco, wheat, barley, squashes, and even cotton, grow in the fields around the village, which contains about fifty houses, whose cheerful white relieves the eye, after the tedious dusk and green of the woods.[37]

Established by the French as a fort in 1731, Vincennes had enjoyed forty years of peaceful interactions between the French trappers, fur traders, ex-soldiers and their families and a mixed Indian population before the arrival of English troops from Kentucky 'who plundered and insulted them, and killed or drove off the cattle which formed their chief wealth'. By the time of Volney's visit, Vincennes had been an English and then American possession for less than twenty years. The population in 1796 was still substantially French, but considerable enmity existed between the Americans and the Indian, Métis and French populations.[38]

Volney observed the behavior of the Indians who gathered for the late summer trading at Vincennes: 'The men and women roamed all day about the town, merely to get rum, for which they eagerly exchanged their peltry, their toys, their clothes … never ceasing to drink till they had lost their senses'. The lack of fluent translators frustrated his efforts to communicate with key leaders. From what he gathered, the years 1788 to 1794 had been ones of constant warfare in the Ohio Valley. Before then, the region's residents had been 'united and peaceful' and the Indians were learning to raise corn like the Whites. The Weeaw chief mourned, 'but now we are poor hunted deer, scattered around without house or home, and, unless somebody come to our assistance, no trace will be left of us'. Disheartened by the vulgar drunkenness and domestic violence, Volney abandoned the original plan of residing a few months among the Native Americans, as he had with the Bedouin of Arabia: 'all was anarchy and disorder. … As they scarcely made provision for themselves, a stranger would run the risk of being starved'. Volney's original intention had been to continue to New Madrid across the Mississippi River. Instead he retraced his steps to the Falls of the Ohio. Traveling overland through Frankfort and Lexington in Kentucky, and then to Cincinnati, Volney was able to view the stark contrasts between the Bluegrass market center and its Ohio River counterpart. 'In 1782, not a house was to be seen' at Lexington, but it now 'contained near five hundred habitations, well built in brick'.[39] The more

---

37   Volney, *View of the Soil and Climate*, 332.

38   Ibid., 331–51, quote from p. 334. On intercultural relations at Vincennes, see Florence G. Watts, "Fort Knox: Frontier Outpost on the Wabash, 1787–1816," *Indiana Magazine of History* 62 (1966): 51–78; Denise Marie Wilson, "Vincennes: From French colonial village to American frontier town, 1730–1820." Ph.D. dissertation, West Virginia University, 1997.

39   Volney, *View of the Soil and Climate*, 354–5, 382, 339 and 356 respectively.

Figure 9.3   Transect of the Ohio River as measured by Volney. Cincinnati is located on the second levee, north of the river

Source: Reproduced with permission of The Academy of Natural Sciences, Ewart Sale Stewart Library and the Albert M. Greenfield Digital Imaging Center for Collections.

impressive landscape to Volney was that of Cincinnati, the upstart military town on a bend of the Ohio (Figure 9.3).

In a settlement where everyone had connections to the military, it is perhaps surprising that Anthony Wayne missed the opportunity to detain the suspected spy Powers and the surveyors Collot and Warin as they passed through that June. Collot and Warin would continue their explorations of the Mississippi Valley all the way to New Orleans, attracting unwelcome attention from American, British and Spanish military authorities.[40] These same informants could not, however, overlook Volney as he collected specimens and took measurements. In order to take a different route back to the settled East, Volney left on 8 September under the protection of a U.S. military convoy that was heading north to Fort Detroit. This was an arduous two-week journey through 250 miles of forest under military jurisdiction. The road between Cincinnati and Detroit passed through a series of five pallisaded fortifications occupying the high ground between the watersheds of the Miami and Maumee Rivers. These forts were kept free from Indian attack, but as Volney noted, there was no protection from the pestilential environment:

> We did not encamp one night without one at least of the party being seized with a periodical fever. [At Fort Green Ville] … three hundred persons, from among three hundred and seventy, were sick with fevers. On arriving at Detroit only three of our party [of 25] were in good health, and on the ensuing day, our commander major Swan and myself were both seized with a malignant fever. This fever annually visits the garrison at Miami Fort, where it has more than once assumed the form of yellow fever.[41]

Even as he traveled thus guarded, Volney was regarded with caution: 'I feel myself [noted Wayne] in a delicate position, with respect to this man, having been

---

40 Georges-Henri-Victor Collot, *A Journey in North America, Containing a Survey of the Countries Watered by Mississippi, Ohio, Missouri, and other Affluing Rivers* (Paris: Arthus Bertrand, 1826), Vol. 1, 132.

41 Volney, *View of the Soil and Climate*, 229.

particularly introduced to him at the House of Mr. George Clymer, in the City of Philadelphia, in the presence of a Numerous and select Company of friends, assembled to dine with (as a Gentleman of the first scientific knowledge, & a practical Philosopher) & having frequently met with him afterwards, in many of the first Circles of that City'. For Wayne, Volney was another 'link in the Chain' of French espionage.[42]

Volney's route to Philadelphia took him past Niagara Falls and through the State of New York along the route of the eventual Erie Canal, thus enabling him to return to Philadelphia before winter. Collot and Warin were detained and transported to New Orleans, where they became exposed to an epidemic of yellow fever. Back on the Eastern seaboard, Volney began organizing his notes and artifacts, and continued his research into Indian languages, American geography and diseases. He was helped in this by the publication in New York of *The Medical Repository*, a new source of 'original essays and intelligence', its mission in concert with ideas regarding hygiene and environmental medicine then current in British and French scientific literature. In introducing the first volume, the editors stated their intention to illustrate 'the connection subsisting between Climate, Soil, Temperature, Diet, &c. and Health'. In order to lay broad foundations to their analysis, the editors mailed a circular to scientists and physicians across the country requesting the transmission on a regular basis of information on 'histories of disease in humans and domestic animals, accounts of insects, histories of the progress and condition of vegetation, and the state of the atmosphere'.[43]

As an exemplar of late Enlightenment medical epistemology and of knowledge through circulated queries, *The Medical Repository* kept Volney informed of contemporary scientific debates and helped vindicate his own enquiries. The contributors, which included Joseph Priestley, Noah Webster, Benjamin Waterhouse, Philip Syng Physick, Benjamin Rush and Valentine Seaman offered material vital to Volney's analysis of patterns in human and physical geography and insight into the effect of climate on health. Many of the contributions came as letters from the prolific co-editor Dr Samuel Latham Mitchill; Volney also maintained a scientific correspondence with Mitchill regarding the geological structure of Eastern North America.[44]

---

42   Anthony Wayne to James McHenry, 30 September 1796, in Knopf, *Anthony Wayne*, 530–2.

43   Edward Miller, Samuel Latham Mitchill, Elihu Hubbard Smith, "Introduction," *Medical Repository* 1 (1797): 1–2; Idem., "Circular address," *Medical Repository* 1 (1797): v–xii; Myrl Ebert, "The Rise and Development of the American Medical Periodical, 1797–1850," *Bulletin of the Medical Library Association* 40 (1952): 247.

44   Volney drew on Mitchill's classification of American rocks for the description and maps featured in his *Tableau*: John West Wells, "Notes on the Earliest Geological Maps of the United States, 1756–1832," *Journal of the Washington Academy of Sciences* 49 (1959): 199; Samuel L. Mitchill, "Sketch of the Mineralogical History of New York," *The Medical Repository of Original Essays and Intelligence* 1 (1797/8): 279–303; 431–9.

Over the next eighteen months, Volney consulted with University of Pennsylvania professor Benjamin Smith Barton on the natural history, cultural traditions and languages of the aboriginal populations of North America (Volney would later lend credence to Barton's theory that Native Americans' oral accounts for violent earthquakes could help explain the geological structure of the maritime region).[45] He dined with local intellectuals and literary men. But the political current was running against him. In his writings, Volney had drawn attention to himself as a Parisian ideologue, someone who had written against the influence of religious systems. His works were singled out for criticism even by fellow radicals such as Joseph Priestley, and a translation of *The Ruins* by Charles Brockden Brown was used to impart a sinister mood to novels exploring the dark side of Enlightenment utopianism.[46] Public policy was swinging away from the French, and, through campaigns organized by Federalist politicians, newspaper editors, theologians and scientists to rally support for president John Adams and silence dissent, the situation of émigrés in America was rapidly deteriorating.[47]

Further munitions for the ideological warfare against French Enlightenment thought arrived in the form of the first American editions of counterrevolutionary manifestos: John Robison's *Proofs of a Conspiracy against all the Religions and Governments of Europe* and Abbé Barruel's *History of Jacobism*. These works were welcomed by New England ministers Jedidiah Morse and Timothy Dwight, president of Yale College, and particularly by the scientist and linguist Noah Webster, who like his counterparts stoked the flames of anti-French sentiment from pulpit and public platform.[48] The unfavorable intellectual climate was exacerbated by the return in August 1798 of yellow fever to the coastal cities. By then, newspapers there had been reprinting petitions for signature at local meetings in demonstration of support for Federalist principles.[49]

---

45   Volney, *View of the Soil and Climate*, 97–101; Benjamin Smith Barton, *A Discourse on Some of the Principal Desiderata in Natural History, and on the best means of promoting the study of this science, in the United-States: Read before the Philadelphia Linnean Society, on the tenth of June, 1807* (Philadelphia: Denham & Town, 1807). Barton discusses Volney's *Tableau* under 'geology' on pages 51–65.

46   Brown's translation, appearing as C.-F. Volney, *The Ruins, or, A Survey of the Revolutions of Empires* (Philadelphia: James Lyon, 1799), is discussed in Peter Kafer, *Charles Brockden Brown's Revolution and the Birth of American Gothic* (Philadelphia: University of Pennsylvania Press, 2004), 204.

47   Stanley Elkins and Erik McKitrick, *The Age of Federalism: The Early American Republic, 1788–1800* (Oxford: Oxford University Press, 1994); Seth Cotlar, "The Federalists' Transatlantic Cultural Offensive of 1798 and the Moderation of American Democratic Discourse," in *Beyond the Founders: New Approaches to the Political History of the Early American Republic*, ed. Jeffrey L. Pasley, Andrew W. Robertson and David Waldstreicher (Chapel Hill: University of North Carolina Press, 2004), 276–99.

48   Ibid.

49   Thomas M. Ray, "'Not One Cent For Tribute': The Public Addresses and American Popular Reaction to the XYZ Affair, 1798–1799," *Journal of the Early Republic* 3 (1983):

Leaving the epidemics of yellow fever and francophobia behind in Philadelphia, Constantin Volney boarded ship for Europe on 23 August 1798. In time, most of Moreau de St. Méry's circle would leave as well, returning to France or to Saint-Domingue. Victor Collot, stranded by the quasi-war between American and French ships in the Atlantic, did not leave Philadelphia for Paris until 1800.[50] Volney returned to France in despair, his overall impression of American society one of rapid decline following political revolution. In his eyes, the new Republic suffered from a weak constitution.

Volney in Print

As I have suggested, knowledge of the geographies of health and disease was fundamental to scientific and medical leaders in a country experiencing its own neoclassical revival. To many it appeared that invasive diseases were propagated by the very act of clearing land and turning over the soil. Yet, in spite of their apparent inevitability, diseases still attracted strong moral associations. According to the widely-shared neo-Hippocratic view, when the soil and climate exhibited a generalized epidemic constitution, the selection of who would be visited and who passed over by the disease would be based on the individual's physical makeup and moral behavior. If 'inflammatory diseases of an exalted grade' were to be expected in a robust and free people, then certain constitutional protections needed to be in place.[51] Much early American geographical writing was concerned with the influence of the physical and moral environment, and the role that environmental improvements might play in ensuring regional and national growth and vitality. In doing so, it followed the obvious models of Jefferson's *Notes on the State of Virginia*, and Constantin Volney's *Tableau*.

Volney's Tableau *and the State of America*

Volney only ever completed the first 'physical' volume of his report, which was published in 1803 as the *Tableau*. He included there extended commentary on the culture, manners and character of the 'savages' of North America, whose stunted development served, he believed, as a cautionary moral example to the young nation. Volney had copies of the work sent to his scientific collaborators, including Jefferson, Barton and Mitchill. Given their shared interest in the earth sciences,

---

389–412.

    50   George W. Kyte, "A Spy on the Western Waters: The Military Intelligence Mission of General Collot in 1796," *The Mississippi Valley Historical Review* 34 (1947): 141.

    51   Charles Caldwell, *An Oration on the Causes of the Difference, in Point of Frequency and Force, Between the Endemic Diseases of the United States of America, and Those of the Countries of Europe, Delivered, by Appointment, to the "Philadelphia Medical Society," on the Fifth Day of February, 1802* (Philadelphia: T. and William Bradford, 1802), 31.

much of Mitchill's initial review of *Tableau* for *The Medical Repository* was devoted to Volney's considerations of geology and use of colored tints on a map of North America to indicate the underlying structures.[52] Its more significant impact had to await publication in English.

In order to understand the reception and impact of Volney's book, it is important to realize that, despite the move toward American geographical publishing signaled by Morse's works, geographical and natural historical works of the size and scope of Volney's were a challenge for American publishers. Even publishing British imprints was a financial risk: Mathew Carey was driven to the brink of bankruptcy in publishing a two-volume American edition of William Guthrie's *Geographical Grammar* in 1792 with accompanying atlas.[53] By the early nineteenth century, printers and booksellers were looking to find new ways of sharing their risks. Associations of printers met from 1802 to 1806 to regulate prices and printing. Literary fairs were organized to provide an opportunity for publishers in a particular region to exchange recent publications. These were typically held in large Eastern cities – Boston, New York and Philadelphia – but one was held at Cincinnati.[54] Combinations of printers would share publication costs and arrange to sell popular French works such as Barthélemy's *Travels of Anacharsis the Younger in Greece* and Saint-Pierre's *Studies of Nature* (with additional notes and illustrations by Dr. Barton) at a discounted price.[55] The Conrad family of printers in Philadelphia formed a powerful combination during the first decade of the nineteenth century, and it was they who published the first American edition of Volney's *Tableau* in 1804.

Offered to American readers in the more portable octavo format, *View of the Soil and Climate* was widely read and cited. American physicians took great interest in Volney's observations regarding the geographical basis for patterns of wellbeing, health and illness in the new Republic. While they found the discussion of mineralogy, disease and the history of the Native Americans instructive, Mitchill found fault with the way that Volney mixed observation with opinion in

---

52   John C. Greene, *American Science in the Age of Jefferson* (Ames: The Iowa State University Press, 1984); Samuel L. Mitchill, "[Review] *Tableau du Climat et du Sol des Etats Unis D'Amerique*," *The Medical Repository of Original Essays and Intelligence* 2 (1804): 172–96.

53   J. Brian Harley, "Atlas Maker for Independent America," *The Geographical Magazine* 49 (1979): 766–71; on reprint publishing and the extension thereby of the Enlightenment, see Richard B. Sher, *The Enlightenment & the Book: Scottish Authors & their Publishers in Eighteenth-Century Britain, Ireland & America* (Chicago: University of Chicago Press, 2006). Sher discusses Guthrie on pages 155–6, 487–93, 573–82.

54   Charles W. Nichols, "The Literary Fair in the United States," in *A Tribute to Wilberforce Eames*, ed. George Parker Warship (Cambridge: Harvard University Press, 1924), 85–92; Richard C. Wade, *The Urban Frontier: The Rise of Western Cities, 1790–1830* (Urbana: University of Illinois Press, 1996), 140.

55   Rosalind Remer, *Printers and Men of Capital: Philadelphia Book Publishers in the New Republic* (Philadelphia: University of Pennsylvania Press, 1996), 88–9.

his description of Native Americans' potential for education and industry, and in his questioning of the basis for political union amongst the American states. As Mitchill put it, 'He mingles in his philosophy angry and disrespectful remarks against our government, fastidious criticisms on the diet and regimen of the inhabitants, and the most unfavorable judgment of the climate'. Mitchill put this down to Volney having suffered ill health during his time in America, including 'two very severe attacks of malignant fever, five or six violent colds, and rheumatic affections which have proved incurable'.[56]

Yet these unfavourable views were not so easily dismissed. An outline in Volney's preface shows that they were a key part of the work:

> After laying a suitable foundation in an examination of the climate and soil, I proposed, agreeably to the most natural and instructive method, to consider the numbers of the people, their diffusion over the surface of their territory, their distribution into classes and professions, their manners as influenced by their actual situation, and by the habits and prejudices derived from their ancestors. By simply tracing their history, laws, and language, I proposed to detect the error of those who represent, as a sort of new-born race, as an *infant nation*, a mere medley of adventurers from all parts of Europe, but more especially from the three British kingdoms.[57]

Volney stated that his methodology for the second volume on the statistical and moral geography of the United States of America would have been to treat each colony, its founders, early principles and institutions separately in order to discover 'the sources of that diversity, which grows daily more conspicuous, in the character and conduct of the different parts of the union'. In traveling the East Coast and the American interior, Volney followed key New England Federalists and geographers such as Morse in finding that the behavioral legacies of colonial history and government still operated to divide state from state in the young Republic. The mercantilist interests of the northeast had particularly benefited from the U.S. government's policy of neutrality toward the English and French combatants. At the same time, southern agricultural interests had suffered. In his western travels, he had not observed a nation capable of conforming its diversity to a unified purpose: 'I should have explained the disjunction of interests and views, and contrariety of habits, which already separate the eastern from the southern, the Atlantic from the Mississippi states'. Without the institutions necessary to promote stronger nation-state identification, he argued, the topographical and climatological diversity of the different regions conspired to 'dismember this body into a few powerful parts'.[58]

---

56   Mitchill, "Review of *Tableau*," 73.
57   Volney, *View of the Soil and Climate*, ix–x.
58   Ibid., xii, xiv.

American geographers differed on the best approach to marrying regional difference and national unity. In his *Notes on the State of Virginia*, Thomas Jefferson had reorganized Marbois's questionnaire so that his response took the form of a natural history, 'preferring to respond to a question about the exact limits and boundaries of the state of Virginia first and deferring his answer to the first query – about the constitution of the state and its charters – until the thirteenth section'. As with Montesquieu, Jefferson believed that systems of government needed to respond to the physical setting and climate of the region as well as its population. Jedidiah Morse's *American Geography* of 1789, on the other hand, cultivated the sense that distinctive state and regional types had emerged from sets of artificial, human-derived, boundaries, and that a strong federal constitution was necessary to regularize the 'language, manners, customs, political and religious sentiments of the mixed mass of people who inhabit the United States'.[59] These questions of how geographical works on climate and health could speak to matters of political unity on the frontier are also apparent in the work that the Cincinnati physician Daniel Drake produced in response to Volney's *Tableau*.

Daniel Drake and Volney on the Frontier

Daniel Drake was raised in the Kentucky backcountry after being transported by his parents down the Ohio with other migrants from New Jersey. Apprenticed to a frontier physician at Cincinnati before sitting the winter 1806 session of medical lectures in Philadelphia, Drake self-consciously positioned himself as the leader of a new generation of creole Western intellectuals upon return to the Ohio Valley.[60] He was first secretary to the Cincinnati Lyceum debating society, and contributed an essay on the 'duties of periodical essayists' to Charles Brockden Brown's *Literary Magazine* while composing his first medical topographies and reading and critically reflecting on Volney's *View of the Climate and Soil of the United States*.[61]

Like other Ohio Valley natives, Drake saw significant limitations in Volney's natural history of the Western Country, believing that it misrepresented the climate

59   Eric Slauter, "The Dividing Line of American Federalism: Partitioning Sovereignty in the Early Republic," in *American Literary Geographies: Spatial Practice and Cultural Production, 1500–1900*, ed. Martin Brückner and Hsuan L. Hsu (Newark: University of Delaware Press, 2007), 72, 76.

60   Henry D. Shapiro, "Daniel Drake: The Scientist as Citizen," in *Physician to the West: Selected Writings of Daniel Drake on Science & Society*, ed. Henry D. Shapiro and Zane L. Miller (Lexington: University Press of Kentucky, 1970), xi–xxii.

61   Daniel Drake, "Duty of Periodical Essayists," *The Literary Magazine and American Register* 6 (1806): 265–6; Idem., "Observations on Debating Societies & the Duties of Their Members: Read before the Cincinnati Lyceum, 1807," Torrence Papers, Box 5, Cincinnati Historical Society, Cincinnati Museum Center, Cincinnati, Ohio; Idem., "Some account of the epidemic diseases which prevail at Mays-Lick, in Kentucky," *The Philadelphia Medical and Physical Journal* 3 (1808): 85–90.

and culture of the American West for potential emigrants. Utilizing meteorological data he had gathered with Jared Mansfield, the US Surveyor-General, Drake set out to correct misconceptions about the West, many of which derived from Volney's reliance upon authorities from the eastern side of the Appalachians, such as Jefferson and Morse. Drake objected principally to the depiction of the climate of the West. Volney had observed that 'the temperature of the vallies [sic] of the Ohio and the Mississippi is warmer, in the proportion of three degrees of longitude, than of the maritime districts'. This temperature difference, Volney maintained, held true from the Gulf of Mexico north to the Great Lakes. North of the southern shore of Lake Erie, however, the cold 'incessantly and prodigiously' increased, bringing eastern and western temperatures into accord. Volney offered all manner of circumstantial observations on the timing of harvest and the level of snowfall in support of this law. He even repeated Jefferson's claim in *Notes on the State of Virginia* that 'paroquets [parakeets] winter on the Sciota [River in Ohio]'.[62]

East coast intellectuals had taken these assertions on the temperature and precipitation patterns of the Ohio Valley seriously, coming as they did from a renowned philosopher who could offer a scientific explanation to the assertions of Morse and Jefferson. The same year that the Philadelphia translation of Volney's *Tableau* appeared in print, Charles Brockden Brown placed the following notice in his *Literary Magazine*: 'A remark has been often made, that the climate of the United States becomes more temperate as we recede from the sea-shore, westward. The difference in temperature, in the same parallels of latitude, has been reckoned equal to ten degrees, in winter, between the sea and the Ohio and Mississippi'. Philadelphia physician James Mease in his *Geological Account* merely reiterated Volney's division of the climate and winds of the United States, and his arguments about the flora and fauna of the West.[63]

Daniel Drake responded to these assertions in the first scientific publication composed and published in the West, *Notices Concerning Cincinnati* (1810). Drake was conspicuously aware that the existence of the book itself would be taken as evidence of civilization on the frontier and had great ambitions for his work. He followed the example of Jefferson's *Notes on the State of Virginia*, beginning with a description of the physical environment. This meant both using Volney's geological observations and questioning his climatic conclusions. The tropical southwest wind, according to Volney's account, originated in the Caribbean and traveled up the Mississippi Valley before arriving at Cincinnati (see Figure 9.4). Comparing records kept at locations at the same latitude outside Cincinnati and Philadelphia showed differences in the Fahrenheit range across the year for Western locations, but not nearly as much difference when compared for the year as a whole. As Drake

---

62   Volney, *View of the Soil and Climate*, 119, 128–9, 120.

63   Charles Brockden Brown, "Remarkable Occurrences," *The Literary Magazine, and American Register* 3 (1804): 147–8; James Mease, *A Geological Account of the United States; Comprehending a Short Description of their Animal, Vegetable, and Mineral Productions, Antiquities, and Curiosities* (Philadelphia: Birch & Small, 1807).

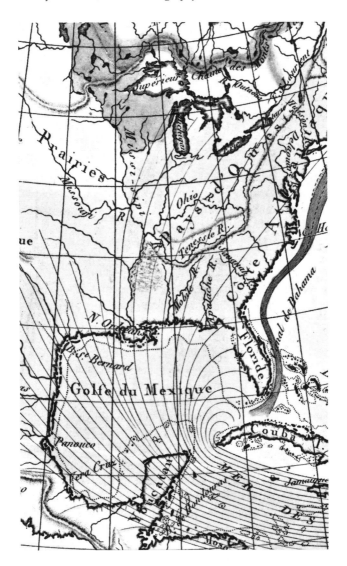

Figure 9.4 **Detail from the Map of the Continent of North America prepared for Volney's *Tableau*. The region encompassed by the tropical winds (tinted yellow in the original) is shown extending from the 'Gulfe du Mexique' across the 'Missi-sipi' Valley all the way to Michigan and across the Ohio Valley all the way to Lake Champlain. The Gulf Stream is also shown, traveling from west to east along the Florida Keys before flowing north along the Eastern Seaboard**

Source: Reproduced with permission of The Academy of Natural Sciences, Ewart Sale Stewart Library and the Albert M. Greenfield Digital Imaging Center for Collections.

put it, 'It is not denied that this country is possibly warmer than the same parallels in the eastern states, but not any means in as much a degree as has been supposed; Mr. Volney's tropical summers, during which the thermometer rises to 90 deg. and upwards, for 60 successive days, have never yet occurred here'.[64] Unchallenged, views like Volney's threatened the agricultural and commercial development of the Ohio Valley. Such depictions of a tropical Western climate reprised longstanding Eastern and European suspicions that Western pioneers degenerated, taking on the physical and moral features of aboriginal populations.[65] The flow of immigrants to Ohio could well turn away from what Drake saw as its natural geographical sources in the New England and the mid Atlantic states towards the Southern states and the Spanish possessions. This would have moral and political implications for the well-being of the frontier, and of America.

The importance that Drake attached to this issue is clear in a 'Climate' note which appeared with his signature in *The Western Spy*. Volney had claimed that 'cotton, which flourishes at Cincinnati and Vincennes, in latitude 39°, cannot be raised, in Carolina, further north than 35° or 36°'. 'Thus a Carolinian [immigrant] must wander as far as the Ohio river', retorted Drake. 'A Virginian must fix himself on the borders of lake Erie; a Pennsylvanian beyond Detroit, and a New-Englander on the waters of lake Superior, to avoid the scorching fervour of our summer sun'.[66] Such advice to potential immigrants, if left uncountered, would introduce the patriarchal character of the southern plantation culture into Cincinnati's demographic mix, thus threatening the viability of Ohio's democratic institutions. Drake was concerned with the variety of peoples and politics being introduced into the Ohio Valley, from which the character and constitution of its creole population would be formed. Like many other Ohioans, he saw rising numbers of free blacks entering the state. According to the 1816 census there were only 247 blacks and mulattos living in Cincinnati, less than 2 per cent of the total population of 6,493 – yet the fear of black immigration from the South was already strong. The Ohio General Assembly passed the restrictive Black Laws of 1804 and 1807 to assuage the fears of the leaders of Virginia and Kentucky that Ohio would become the home of fugitives and abolitionists. The 1804 law was, in essence, a fugitive slave law, requiring that free blacks register with local authorities as they entered the state. Underlying Drake's satirical instructions for immigrants were his fears that immigration to Ohio from Southern states would warp the political culture of the state even further.[67]

---

64   Drake, *Notices Concerning Cincinatti*, 16–17.

65   Kariann Yokota, "'To Pursue the Stream to Its Fountain:' Race, Inequality, and the Post-Colonial Exchange of Knowledge across the Atlantic," *Explorations in Early American History* 5 (2001): 173–229.

66   Daniel Drake, "Climate," *The Western Spy*, 15 June 1811, 3; Volney, *View of the Climate and Soil*, 122–3.

67   Carter G. Woodson, "The Negroes of Cincinnati prior to the Civil War," *The Journal of Negro History* 1 (1916): 1–22; Richard C. Wade, "The Negro in Cincinnati,

Drake also saw the need to challenge New England emigrants' depictions of the southern influence upon the character of the West. Diary entries and letters sent to Eastern friends and relatives suggest that New Englanders weathered a difficult period of acclimatization after settlement. Drake, like his many colleagues who studied with Dr Benjamin Rush at the University of Pennsylvania, favored a depletive regimen including bloodletting and calomel in assaulting 'Western' diseases. The young New Englander Bellamy Storer, locating to the West in order to begin his career as a lawyer, shared his concerns in a letter to Reverend Dr Appleton, the president of Bowdoin College:

> The climate here, I should not consider as healthy as that of Maine notwithstanding many opinions to the contrary: I notice every morning and evening a thick vapour arise from every part of the town. It checks perspiration and induces sleep. So convinced are many of the inhabitants of its pernicious effects, that they seldom venture in the open air at these periods – The neighboring forests supply a vast body of vegetable matter, which the frequent rains tend to putrify, and the annual inundation of the low lands, without proper precaution being taken to remove the alluvial, is another prolific source of disease – In the months of September & October, the fever and ague generally prevail to a great extent on the banks of the Ohio, particularly in the towns between this place and the Mississippi – The constant trade with New Orleans & the facilities now afforded by the numerous steamboats, for uninterrupted commerce, frequently introduce the contagious disorders of the Southern climate and what is eventually more pernicious, all the vices and crimes of a city proverbial for its refined voluptuousness.[68]

Daniel Drake seems, even, to have felt personal responsibility for seeing that the city itself weathered its awkward youth and entered adolescence in good health. His views to this effect appear in an 1814 anniversary address to the Cincinnati School of Literature and the Arts, and in his second book, *Natural and Statistical View; or, Picture of Cincinnati* (1815). Drake presents a resonant image of the Ohio Valley as a vessel of amalgamation for European peoples.[69] He believed that the foundation for the city's success could be found in its magnificent physical situation, located on a high, flat levee of the Ohio, like ancient Rome sheltered

1800–1830," *Journal of Negro History* 39 (1954): 43–55; Thomas D. Matijasic, "The Foundations of Colonization: The Peculiar Nature of Race Relations in Ohio during the Early Ante-Bellum Period," *Queen City Heritage* (1991): 23–30.

68   Bellamy Storer, letter to Reverend Dr Appleton, president of Bowdoin College, Brunswick, Maine, 18 August 1817, Bellamy Storer Papers, Cincinnati Historical Society, Cincinnati Museum Center.

69   Daniel Drake, *Natural and Statistical View: or, Picture of Cincinnati and the Miami Country* (Cincinnati: Looker and Wallace, 1815); idem., "Anniversary Address to the School of Literature and the Arts (1814)," in *Physician to the West*, ed. Shapiro and Miller, 57–65.

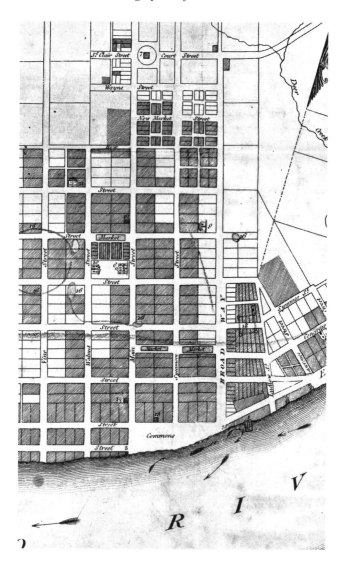

Figure 9.5 **Detail from the 'Plan of Cincinnati, Including All the Latest Additions & Subdivisions. Engraved for Drake's Statistical View. 1815'. Third Street, the first street north of the river levee, is shown bisecting the remains of Fort Washington (18). The curvilinear remains of 'Ancient works' (16) lay to the south and east of the Central Square, framed by Fourth and Fifth (with its market stalls), and Walnut and Main**

Source: Reproduced with permission of The Academy of Natural Sciences, Ewart Sale Stewart Library and the Albert M. Greenfield Digital Imaging Center for Collections.

by seven hills, and with easy access to the fertile hinterlands of the Miami River Valley. Drake took exception to Volney's characterization of the American people as a 'mere medley of adventurers' irreparably divided by sectional and moral divisions. He subscribed instead to the romantic Arcadian notions that Volney strongly decried: Drake's 'Buckeyes', named after a hearty tree native to the Ohio Valley, were a newly-compounded people who created political constitutions to meet the challenge of their environment.

Immigrants from every corner of Europe and North America found their way into the Western country: 'In no country of the same age and numbers, do the immigrants exhibit more diversity. The sister states, from Georgia to Maine – the Canadas and West-Indies – the united kingdom of Great Britain and Ireland, and the empires of Europe, from the shores of the Baltic to the Meditterranean [sic], have contributed to increase and variegate our population'. Various national characters, displaying the 'principles on which they were designed and executed', were proving themselves useful in all of the occupations necessary for a productive society. When menaced with war, as in 1812, they rose to defend their rights. And after proving themselves the rightful overseers of the Western territory, Drake argued that the time had come for Cincinnatians to build institutions befitting their position as the cultural center of the Ohio Valley. 'A society, thus compounded, has *within itself* no indifferent substitute for travelling; and exhibits, in the lapse of time, what belongs naturally to change of place'.[70] A frontispiece 'Plan of Cincinnati' served as a permanent spatial record of the stages of the town's advance by incorporating the locations of ancient works and the original military fortifications established by Washington (Figure 9.5).

But the marks of the past were also evidence that societies did not last. Drake devoted twenty pages of his *Picture of Cincinnati* to descriptions of the ruins of ancient forts and burial tumuli that lay scattered around the Ohio Valley.[71] Following Benjamin Smith Barton's exhortation to 'open the tombs of the ancient Americans', Drake returned to an ancient mound originally surveyed by Col. Winthrop Sargent.[72] To the items already sent by Sargent to the American Philosophical Society, Drake added the following: a number of beads; the teeth of a carnivorous animal; several marine shells 'cut in such a manner as to serve for domestic utensils'; several copper articles; and human bones. Utilizing the taxonomic approach of Linnaeus and the comparative anthropology techniques of Petrus Camper and Johann Blumenbach, Drake examined two human skulls, from

---

70   Drake, "Anniversary Address," 61.

71   Drake, *Picture of Cincinnati*, 199–218.

72   Benjamin Smith Barton, "Observations and conjectures concerning certain articles which were taken out of an ancient tumulus, or grave, at Cincinnati, in the county of Hamilton, and Territory of the United-States, north-west of the River Ohio: In a letter from Benjamin Smith Barton, M.D. to the Reverend Joseph Priestley, LL.D.F.R.S. &c. (read May 20th, 1796)," *Transactions of the American Philosophical Society* 4 (1799): 214.

a middle-aged 'ancient' American male, and a Wyandot warrior.[73] Daniel Drake failed to find a rational basis for classing the ancient and modern Indian skull as separate varieties or races.

Like his brother Benjamin, biographer of the Shawnee chief Tecumseh, Daniel Drake believed that Indians were amenable to improvement, particularly under the influence of Christian missionary educators.[74] He concluded his discussion of Ohio Valley antiquities in *Picture of Cincinnati* by citing the authority of Barton and of the Scottish historian William Robertson that the aboriginal population of North America was once much more numerous. The mound-building ancients of the Ohio Valley had trading ties that extended into Middle America. The first European to penetrate the interior found scattered remnants of this once-powerful people. Drake argued that they had since degenerated further 'into the present savage hordes'.[75] Unlike Volney, Drake saw in the native population the necessary seeds of improvement, improvement that would only come with cultivation of the land and training in European agricultural methods.

Print Geographies and Geography Books on the American Frontier

If we are to take seriously the geographies of print on the Western frontier and understand their significance, we need to note more than the location of printers, lists of imprints and the presence of authors in localities on the margins. Douglas McMurtrie of the American Imprints Inventory long ago popularized the story of the diffusion of printing to the Ohio Valley.[76] Many of the earliest books on the frontier, besides compilations of laws, almanacs and school primers, were works of geography and health, designed with travelers and philosophers in mind. These works convey a complicated story of intercultural exchange in the West, in a borderland crossed by individuals and groups with multiple identities.

Much scholarship on early American geography and nation-building has focused on Jedidiah Morse's texts for the home and the school. They have been recognized as the hardly impartial works of a New England Federalist who, while envisioning an American empire that was continental in scope, frequently found opportunities to emphasize sectional disparities in industry and ethos – as

---

73   Drake, *Picture of Cincinnati*, 206–7.

74   Benjamin Drake, "Our aborigines," *The Cincinnati Literary Gazette* 2 (1824): 206–7; idem., "Civilization of the Indians," *The Cincinnati Literary Gazette* 2 (1824): 194–6; Daniel Drake, "Travelling letters from the senior editor, No. IV (Kanzas River, 'Indian Territory,' August, 1844)," *Western Journal of Medicine and Surgery* 2 (1844): 270–79.

75   Drake, *Picture of Cincinnati*, 218.

76   Douglas C. McMurtrie, "The Westward Migration of the Printing Press in the United States, 1786–1836," *Gutenberg-Jahrbuch 1930*, 269–88.

when he condemned the infidelity of southern planters and educators.[77] But other geography texts were also in circulation. Partly because of the debatable authority of its original authorship, and the fact that American editions modified the original text, less attention has been given to William Guthrie's *Geographical Grammar,* first published in Edinburgh in 1770.[78] As Mayhew notes, this was a key work in the dissemination of a Scottish version of the Enlightenment stadial model of societal evolution.[79] The quarto American edition was first published by Mathew Carey in 1794 with an accompanying atlas.[80] Carey's agents in the Southern states and Western territories marketed the work aggressively. Zadok Cramer ordered several copies from Carey for the large bookstore and circulating library he had established at Pittsburgh, and if Daniel Drake's memoirs are to be trusted, a copy of Guthrie played an important role in his Kentucky backcountry education.[81]

As Westerners, publishers and readers began to engage in their own Enlightenment projects. They also reclaimed Guthrie as a product of western intellect. The Philadelphia firm of Johnson and Warner assumed the Guthrie franchise from Mathew Carey, publishing their substanially revised 'first edition' in 1809.[82] They employed several traveling salesmen to push their catalogue of books in the West, but even more significantly, 'traveled themselves to establish large wholesale accounts with country merchants, printers, and booksellers'.[83] While drumming up business, they also solicited western authors to compose descriptions of their states for the new edition. In the same letter where he ordered medical books on credit from Johnson and Warner, Daniel Drake reiterated his plan to compose a description of Ohio along with his partner in science Jared Mansfield.[84] The article, eventually appearing in the 1815 edition of Guthrie,

77   Ralph Hall Brown, "The American Geographies of Jedidiah Morse," *Annals of the Association of American Geographers* 31 (1941): 145–217; David N. Livingstone, "'Risen Into Empire': Moral Geographies of the American Republic," in *Geography and Revolution*, ed. David N. Livingstone and Charles W.J. Withers (Chicago: University of Chicago Press, 2005), 304–35; Brückner, *Geographic Revolution in Early America.*

78   W.G. East, "An Eighteenth-Century Geographer: William Guthrie of Brechin," *Scottish Geographical Magazine* 72 (1956): 32–7. On Guthrie's mediation in later Irish and American editions, see Sher, *Enlightenment & the Book.*

79   Robert Mayhew, "William Guthrie's *Geographical Grammar*, the Scottish Enlightenment and the Politics of British Geography," *Scottish Geographical Magazine* 115 (1999): 19–34.

80   Harley, "Atlas Maker for Independent America."

81   Remer, "Preachers, Peddlers, and Publishers"; Daniel Drake, *Pioneer Life in Kentucky, 1785–1800* (New York: H. Schuman, 1948).

82   William Guthrie, *A New Geographical, Historical, and Commercial Grammar; and Present State of the Several Kingdoms of the World.* 2 vols (Philadelphia: Johnson & Warner, 1809).

83   Remer, "Preachers, Peddlers, and Publishers," 516.

84   Daniel Drake, letter to Mssrs Johnson and Warner, Philadelphia, Pennsylvania, from Cincinnati, 10 December 1811, Emmet Horine Collection, Drake Papers – Box

highlighted the development of Cincinnati and the predictability of its climate. The description of the Missouri Territory in the 1815 edition, on the other hand, is drawn entirely from Henry Marie Brackenridge's *Views of Louisiana*, published by Cramer and partners in Pittsburgh the year before.[85] This addition to Guthrie's *Grammar* contributes a more sympathetic account of the French colonial legacy than Morse's *American Universal Geography*. By allowing Westerners to become producers as well as consumers of geographical accounts, the publishers Johnson and Warner reinforced the republicanism evident in the original work.

With the publication of his *Tableau*, Volney offered a panoramic vista and a challenge to American intellectuals. If they were put off by Volney's tone, Eastern reviewers were still largely positive, and frequently reprised his views on soil and climate. As shown, however, cultural leaders in frontier localities like Pittsburgh, Marietta, Lexington and Cincinnati that Volney passed through in the summer of 1796 read his work more as a personal affront and took advantage of opportunities to compose their own natural and statistical views in response. Ironically, Western readers like Daniel Drake, sympathetic to Volney's holist vision, would come around to adopting his systematic geographical approach as the most appropriate model for their critique. In the end Drake devoted much of his career to founding a distinctly Western geographical tradition, constructing new tableaux based upon a first-hand knowledge of the physical and moral character of the Ohio Valley and, eventually, of the Mississippi Valley as a whole.[86] James Fleming and James Cassedy have discussed how pioneer climatologists and ecologists like Drake and Samuel Forry drew on extensive networks of correspondents as well as the topographical maps, journals and logs produced by officers serving on government-sponsored expeditions, and stationed at Western military posts.[87] But largely neglected in their writings has been the role of Volney's physical and moral *Tableau* as a key early model and guide to the geographic craft.

13, Department of Special Collections and Archives, University of Kentucky Libraries, Lexington.

85   William Guthrie, *A New Geographical, Historical, and Commercial Grammar* 2nd American edition. 2 vols (Philadelphia: Johnson & Warner, 1815); Henry Marie Brackenridge, *Views of Louisiana; Together with a Journal of a Voyage up the Missouri River, in 1811* (Pittsburgh: Cramer, Spear and Eichbaum, 1814).

86   Emmet Field Horine, "Daniel Drake and His Medical Classic," *Bulletin of the Kentucky State Medical Association* 50 (1952): 68–79; Frank A. Barrett, "Daniel Drake's Medical Geography," *Social Science and Medicine* 42 (1996): 791–800; Michael L. Dorn, "(In)temperate Zones: Daniel Drake's Medico-Moral Geographies of Urban Life in the Trans-Appalachian American West," *Journal of the History of Medicine and Allied Sciences* 55 (2000): 256–91; idem., "Climate, Alcohol and the American Body Politic: The Medical and Moral Geographies of Daniel Drake (1785–1852)" (Ph.D. dissertation, University of Kentucky, 2003).

87   James R. Fleming, *Meteorology in America, 1800–1870* (Baltimore: Johns Hopkins University Press, 1990); James H. Cassedy, *Medicine and American Growth* (Madison: The University of Wisconsin Press, 1986).

Constantin Volney occupies an important position in the history of European travelers' accounts of the North American environment and human population. Arriving for his American sojourn in the middle of the turbulent 1790s, a decade in which French authors were increasingly viewed with suspicion and read selectively, Volney and the English editions of his *Tableau* would find a more engaged reading public in the first decade of the new century. Western states were being carved out of territories with deep French colonial pasts, as, at the same time, American medical students were finding unparalleled opportunities for professional improvement in the reorganized clinics of post-revolutionary Paris.[88] Critical though they were, physicians, natural philosophers and politicians found much to learn from Volney's *Tableau*. He provided young scientist-physicians with a new way of seeing their country, a concrete example of what a systematic geographical account attuned to questions of ecology and population health would look like, based not on the 'artificial' divisions of nation-states, but on the physical topography and human settlement patterns of America's great 'Interior Valley' all too hesitantly emerging into view.

---

88   Edward Watts, *In This Remote Country: French Colonial Culture in the Anglo-American Imagination, 1780–1860* (Chapel Hill: University of North Carolina Press, 2006); John Harley Warner, *Against the Spirit of System: The French Impulse in Nineteenth-Century American Medicine* (Princeton: Princeton University Press, 1998).

Chapter 10
Reading the Messy Reception of *Influences of Geographic Environment* (1911)

Innes M. Keighren

Introduction: Miss Semple's *Influences*

For much of the early twentieth century, geographical inquiry in the United States was characterized by questions of environmental influence.[1] Scientific efforts to describe and explain the ways in which environmental circumstances conditioned and constrained human societies – an approach interchangeably termed anthropogeography and environmentalism – unified the discipline's intellectual focus and thus facilitated its academic institutionalization. The principal spur to geography's engagement with what would later be termed, pejoratively, "environmental determinism", was the work of American geographer Ellen Churchill Semple (1863–1932), particularly her 1911 volume *Influences of Geographic Environment*.[2] In the opinion of one of Semple's contemporaries, her book shaped 'the whole trend and content of geographic thought in America'.[3] The history of geography in the United States is, in this respect, intimately connected with Semple's work. For a brief time, in an American context at least, *Influences was* geography and was implicated in the discipline's 'scramble for intellectual turf'.[4]

In retrospect it is clear that Semple's book facilitated a (re)turn to environmentalism in geographical scholarship, and that it contributed to the discipline's 'entry into modern science' – two issues bound up in a positive reception of the text and the ideas it sought to communicate. The response to *Influences* varied considerably through time, however, as well as between and within the different contexts of its reception: institutional, disciplinary, metropolitan, among

---

1   Geoffrey J. Martin, *All Possible Worlds: A History of Geographical Ideas*, 4th edition (Oxford: Oxford University Press, 2005).

2   Ellen C. Semple, *Influences of Geographic Environment on the Basis of Ratzel's System of Anthropogeography* (New York: Henry Holt and Company, 1911).

3   Wallace W. Atwood, "An Appreciation of Ellen Churchill Semple, 1863–1932," *The Journal of Geography* 31:6 (1932): 267.

4   Neil Smith, "Geography as Museum: Private History and Conservative Idealism in *The Nature of Geography*," in *Reflections on Richard Hartshorne's* The Nature of Geography, ed. J. Nicholas Entrikin and Stanley D. Brunn (Washington: Association of American Geographers, 1989), 93.

others.[5] The dissimilarity in the responses that Semple's text provoked at the various sites of its reading are striking. These differences in reception and what they reveal about the circulation and consumption of environmentalist thought in American geography are the central concern of this chapter.

In what follows, I interrogate *Influences'* reception by reconstructing Semple's different academic audiences and the intellectual concerns and preconceptions – whether shared or private – which they brought to their reading. In attempting to understand how Semple's text conditioned geography as a discipline, I consider the ways in which the response to *Influences* was itself shaped by the social and spatial geographies of its reading. By reference to geography's engagement with environmentalism, and to the role of Semple's book in that process, I also describe a partial prosopographical history of the discipline which attends to the negotiation and circulation of ideas in printed form.

By exploring the particular qualities of *Influences'* diffusion, this chapter considers more broadly the processes by which scientific and geographical knowledge circulates in, and beyond, the guise of the book. Informed by recent work in the history of science on the reception of scientific texts and by emergent scholarship on the geography of reading, I examine the locational particularities of *Influences'* reception revealed by its incorporation into, and excision from, the geographical curricula of various academic institutions in the United States. I am not here suggesting that geographical location straightforwardly determined the reading of Semple's book. Rather, I question the notion of a taken-for-granted "American" or "institutional" response to *Influences*, by showing that its reading and reception were messier and more complicated than these apparently neat categories allow.

The geography of reception, at least as far as it concerns the reception of texts, necessitates attention to individual and collective reading practices. Examinations of the reception of Nicolas Copernicus' *De Revolutionibus* (1543), Robert Chambers's *Vestiges of the Natural History of Creation* (1844), and Charles Darwin's *On the Origin of Species* (1859) have demonstrated how readers' understandings of those works varied on account of their political, religious, scholarly, and other interests.[6] By attending as such studies have done to the material traces of reading – published reviews; jottings in diaries and correspondence; impromptu or carefully considered

---

5   Richard Peet, "The Social Origins of Environmental Determinism," *Annals of the Association of American Geographers* 75:3 (1985): 310.

6   Owen Gingerich, *An Annotated Census of Copernicus'* De Revolutionibus *(Nuremberg, 1543 and Basel, 1566)* (Leiden: Brill, 2002); James A. Secord, *Victorian Sensation: The Extraordinary Publication, Reception, and Secret Authorship of* Vestiges of the Natural History of Creation (Chicago: University of Chicago Press, 2000); Thomas F. Glick, ed., *The Comparative Reception of Darwinism* (Austin: University of Texas Press, 1974).

marginal annotations – it is possible to 'move from [non-spatial] reception histories to the historical geography of reception'.[7]

Any attempt to reconstruct the historical geography of a text's reception is also an attempt to reconstruct its various audiences. Although *Influences* secured an audience that transcended disciplinary divisions and encompassed scholarly and lay communities, Semple's intended readers were university students of geography. She envisioned a clear pedagogical role for her book, the principal function of which was as an aid to education in anthropogeography. She used lectures at the University of Chicago, where she taught from 1906, to adapt her emergent text 'to students' needs'.[8] Whilst the response of Semple's various audiences – defined, variously, by discipline, political orientation, or nation – tell different stories of the book's reception, what follows considers how *Influences* was read and understood by its academic and geographical audience in the United States. After tracing the intellectual origins and expression of Semple's geography, I examine the different ways in which her book was employed pedagogically. Drawing upon evidence of individual reading experiences at different academic institutions, I attempt to situate these uses and readings within the context of then-contemporary geographical debates. In so doing I hope to explain something of the motivating factors which underpinned the teaching of geography at these different institutions, and the use made of Semple's work in that teaching.

By attending to the pedagogical function of *Influences* and to its disciplinary implications, my broader intention is to outline a geography of the book which is concerned with more than the act of reading. In tracing the circulation of Semple's ideas – in their textual guise and in their other representational forms – I hope to demonstrate the importance of social networks and to show that the hermeneutic groups which comprised Semple's audience were not defined straightforwardly by scale – metropolitan, regional, or national, for example. This is to offer, then, some thoughts on how the geography of the book as a question of reception might attend both to the spatialized practices of reading and to the spatially-transcendent nature of interpretative communities. In so doing, I reflect upon the methodological and epistemic consequences of envisioning the printed book as something defined not only by its physical form but also by its more-than-material social and intellectual function.

---

7   Robert J. Mayhew, "Denaturalising Print, Historicising Text: Historical Geography and the History of the Book," in *Practising the Archive: Reflections on Methods and Practice in Historical Geography*, ed. Elizabeth A. Gagen, Hayden Lorimer and Alex Vasudevan (London: Royal Geographical Society, 2007), 26.

8   Ellen C. Semple to John S. Keltie, 6 March 1910, Correspondence Block 1881–1919, Manuscript Archive, Royal Geographical Society [RGS].

Situating Semple's Environmentalism

Semple was introduced to environmentalist thought in the libraries and parlours of postbellum Louisville, Kentucky.[9] Having completed a Bachelor of Arts degree in history at Vassar College, she spent much of the 1880s in Louisville pursing an independent programme of teaching and study, supplemented by discussions with 'widely read and cultivated lawyers and a brilliant Jewish Rabbi'.[10] Conversations with these men, complemented by access to their libraries, had an important influence on Semple's intellectual development. She became interested in questions of environmental influence, but found little beyond 'the purely pseudo-scientific writings of Henry Buckle' with which to engage.[11] As she later put it, 'I began to scent the importance of geographic influences, tho' at that time ... I struck no trail of a previous investigator that was reliable enough to follow'.[12] In 1887 she encountered, somewhat by chance, the work of German geographer Friedrich Ratzel and found in him an investigator she felt worthy of pursuit.

Ratzel's understanding of the relationship between society and environment emerged from a materialist and Darwinian reinterpretation of early-nineteenth-century German geography. From the work of, among other geographers, Alexander von Humboldt and Carl Ritter, Ratzel inherited an understanding of the 'reciprocal and evolutionary relation of environment and society'.[13] Supplemented by the Neo-Lamarckian principles of Charles Darwin and Herbert Spencer – and inspired by the systematic approach to physical geography advocated by his near-contemporaries Oscar Peschel and Ferdinand von Richtofen – Ratzel sought to outline a scientific approach to human geography which would take as its focus the role of environmental conditions in determining the distribution and comparative success of human populations.

Ratzel's principles were most fully articulated in his two-volume *Anthropogeographie* (1882, 1891). Although each volume treated the distribution and comparative success of human populations as an approximate function of their environmental conditions, the first was coloured by an obvious 'deterministic tint'.[14] It was this volume which Semple encountered first, and with which she felt the closest intellectual affiliation. Following a period of study under Ratzel in

---

9   Allen D. Bushong, "Ellen Churchill Semple 1863–1932," in *Geographers: Biobibliographical Studies*, vol. 8, ed. Thomas W. Freeman (London: Mansell, 1984), 8.

10   Charles C. Colby, "Ellen Churchill Semple," *Annals of the Association of American Geographers* 23:4 (1933): 231.

11   *The Evening Post* (New York City), 9 November 1912.

12   Ellen C. Semple to John S. Keltie, 30 October 1912, Correspondence Block 1911–1920, RGS.

13   Carl O. Sauer, "Recent Developments in Cultural Geography," in *Recent Developments in the Social Sciences*, ed. Edward C. Hayes (Philadelphia: J. B. Lippincott, 1927), 166.

14   George Tatham, "Geography in the Nineteenth Century," in *Geography in the Twentieth Century*, ed. Griffith Taylor (London: Methuen, 1957), 64.

Leipzig in the 1890s, Semple undertook to communicate his ideas to the English-speaking world, 'but clarified and reorganized'.[15] Her first opportunity to address her concerns to a geographical audience came in 1897 with a paper she contributed to the inaugural volume of the *Journal of School Geography*. Semple's text, 'The influence of the Appalachian barrier upon colonial history', was an attempt to apply aspects of Ratzel's method to the historical study of North America.[16]

Semple's use of environmentalist principles to explain the historical development of the United States was not unprecedented. At a meeting of the American Historical Association in Chicago four years previously, the historian Frederick Jackson Turner had presented a 'penetrating essay' on 'The significance of the frontier in American history'.[17] Turner's paper motivated an intellectual reassessment of the nation's frontier experience in which the American West was seen as a metaphor and an explanation for the distinctive historical development of the United States. For Turner, the physical and cultural distance which separated the frontier from the eastern seats of power promoted an individualism and ad hoc democracy among the frontier's pioneers. The frontier was seen to be responsible, in part, for facilitating a national character – and consequently national institutions – in which individual liberty was prized and emphasized.[18]

Turner's formulation was similar to the anthropogeographical principles outlined by Ratzel. In applying to the historical study of the United States the Lamarckian metaphor of the social organism, Turner's perspective corresponded with the 'new science of evolutionary human geography'.[19] As such, Turner has been credited as the 'cofounder, along with Ellen Churchill Semple' of the subfield of American geography concerned with environmental influence.[20] Whilst Turner's and Semple's contributions to the post-Darwinian project exerted a novel influence upon the disciplinary focus of American geography, aspects of their intellectual interests were evident earlier still in the work of, among others, George Perkins Marsh and Nathaniel Southgate Shaler.[21] Semple's work on the

---

15   Preston E. James and Geoffrey J. Martin, *All Possible Worlds: A History of Geographical Ideas*, 2nd edition (New York: John Wiley & Sons, 1981), 304–5.

16   Ellen C. Semple, "The Influence of the Appalachian Barrier upon Colonial History," *Journal of School Geography* 1 (1897): 33–41.

17   Lawrence E. Gelfand, "Ellen Churchill Semple: Her Geographical Approach to American History," *The Journal of Geography* 53:1 (1954): 30.

18   John A. Campbell and David N. Livingstone, "Neo-Lamarckism and the Development of Geography in the United States and Great Britain," *Transactions of the Institute of British Geographers* 8:3 (1983): 267–94.

19   William Coleman, "Science and Symbol in the Turner Frontier Hypothesis," *The American Historical Review* 72:1 (1966): 24.

20   Robert H. Block, "Frederick Jackson Turner and American Geography," *Annals of the Association of American Geographers* 70:1 (1980): 31.

21   David Lowenthal, *George Perkins Marsh: Prophet of Conservation* (Seattle: University of Washington Press, 2000); Wilford A. Bladen, "Nathaniel Southgate Shaler and Early American Geography," in *The Evolution of Geographic Thought in America: A*

Appalachian barrier and Turner's frontier thesis were, then, representative of a relatively venerable intellectual trend in American scholarship.

Between 1897 and 1900 Semple contributed five further papers on various aspects of anthropogeography to the *Journal of School Geography*, and one on that subject to the *Journal of the American Geographical Society of New York*. These papers were based largely upon secondary sources and were restatements of ideas absorbed during her time in Leipzig rather than genuinely original pronouncements. Her contributions to the *Journal of School Geography* were considered by its editor, Richard Dodge, to be important in communicating the principles of anthropogeography to 'common school teachers'.[22] For the journal, geography was 'the science of man's relation to his earth environment', and its object of inquiry 'the mutual dependence of man and nature upon one another'.[23] Given this formulation, Semple's anthropogeography could be seen not only to parallel the initial contours of academic geography in the United States, but also to represent a model for the teaching of school geography.

Keen to test her anthropogeographical perspective in the field, Semple completed, in 1899, a 350-mile horseback journey through isolated stretches of the Kentucky Mountains. Her observations of the social, economic, and cultural characteristics of the mountain inhabitants formed the basis of her most personal contribution to the anthropogeographical literature: 'The Anglo-Saxons of the Kentucky Mountains: a study in anthropogeography', a paper said to take 'high rank among the geographical articles in the English language'.[24] Semple's demonstration that anthropogeography could be studied in the field, and that environmental influence was an apparently legitimate and demonstrable causal explanation, was significant for those geographers who believed that the promotion of the discipline depended upon an ability to adhere to a scientific and nomothetic approach. In appearing thus to satisfy a desire for rational and deductive geographical research, Semple's paper drew positive attention; it 'fired more American students to interest in geography than any other article ever written'.[25]

Concurrent with her fieldwork in the Kentucky Mountains, Semple was encouraged by Dodge to collate her earlier articles on North America in a single volume. Her resultant book – *American History and its Geographic Conditions* (1903) – attended to the environmental factors which she understood to have conditioned war, migration, commercial development, the location of cities,

---

Kentucky Root, ed. Wilford A. Bladen and Pradyumna P. Karan (Dubuque: Kendall/Hunt, 1983), 13–27.

22   Richard E. Dodge, "The Social Function of Geography," *Journal of School Geography* 2:9 (1898): 335.

23   James A. Merrill, "A Suggestive Course in Geography," *Journal of School Geography* 2:9 (1898): 321.

24   Ellen C. Semple, "The Anglo-Saxons of the Kentucky Mountains: A Study in Anthropogeography," *The Geographical Journal* 17:6 (1901): 588–623;

25   Colby, "Ellen Churchill Semple," 232.

the provision of transportation, and international trade.[26] Together with her examination of the Kentucky mountaineers, Semple demonstrated that an anthropogeographical approach might be applied with equal success to the study of historical and contemporary society. Her work was read by geographers as a contribution to then-current debates regarding the infant discipline's epistemic and methodological foundation.[27]

Ratzel's death in 1904 was a prompt to Semple's long-standing ambition to communicate the full scope (as she saw it) of his anthropogeography. When not teaching in Chicago, she divided her time between Louisville and the Catskill Mountains in New York State, where she lived in a tent and worked on her book without interruption. Semple's project was not one, however, of straightforward translation. It was, rather, adaptation. She intended to relocate Ratzel's book linguistically; to reframe its contents, revise its arguments, and supplement its sources. She sought to 'make the research and induction as broad as possible, to draw conclusions that should be elastic and not rigid or dogmatic ... to be Hellenic in form but Darwinian in method'.[28] For Semple, it was important that *Influences* should be 'adapted to the Anglo-Celtic and especially to the Anglo-American mind'.[29] The purpose of this cultural reframing was to place Ratzel's work more obviously on a scientific foundation and, by so doing, to 'throw it into the concrete form of expression demanded by the Anglo-Saxon mind'.[30] Semple's concern was, as it had been in her earlier work, to reform Ratzel's conclusions, which she regarded as 'not always exhaustive or final', and to present them in a manner more clearly supported by real-world examples.[31]

In the two years following its publication in summer 1911, *Influences* was reviewed in more than forty periodicals, including local and national newspapers; geographical and non-geographical journals; and popular and literary magazines. The reaction to Semple's scholarship – to her extensive citation of authorities – was almost unanimous approbation. In nearly every case, her effort to situate her work within an established literature, and to support her claims by reference to contemporary research, was seen as a warrant of credibility. Securing authority in this way mattered particularly to geographical reviewers of *Influences* – it was seen to relate to the then-current project of defining geography as an independent and scientific discipline. That geography could be considered a science was, however, a 'strange assertion', particularly for those non-geographers for whom

---

26   Ellen C. Semple, *American History and its Geographic Conditions* (Boston: Houghton Mifflin, 1903).

27   Andrew J. Herbertson, "Geography in the University," *Scottish Geographical Magazine* 18:3 (1902): 124–32.

28   Ellen C. Semple to John S. Keltie, 2 April 1911, Correspondence Block 1911–1920, RGS.

29   Semple, *Influences of Geographic Environment*, v.

30   Ibid., vi.

31   Ibid., v.

the subject inevitably recalled 'certain grammar-school exercises in locating rivers, mountains, [and] political boundaries'.[32] *Influences* was seen, then, to have a unique and particular importance in helping to place geography on a nomothetic footing, by showing that it was more than simply 'descriptive and mnemonic'.[33]

The critical response to Semple's text seemed to confirm the success of her intended representation of Ratzel's ideas. One newspaper spoke of *Influences* as a 'German dose sweetened'.[34] For the *Boston Herald*, so impressive was Semple's reformulation of Ratzel's work – characterized by a 'gain in clearness of statement and concreteness of expression' – that *Influences* could, conceivably, 'be advantageously re-translated for the use of Germans themselves'.[35] It was the breadth of Semple's scholarship, however, which drew the most distinct praise. Her 'extended reference to books and personal authorities', which numbered more than one thousand separate citations, was seen as fundamental to extending the value and credibility of her conclusions.[36] Semple's book was, in this respect, seen to have more than simply pedagogical value. It had national importance as a 'distinct credit to American scholarship'.[37]

The link between Semple's scholarship and her contribution to the nation's intellectual standing was further emphasized by *The Nation* – a weekly New York City magazine. Semple's book was seen by that publication to be not only a significant scholarly accomplishment, but also, and more importantly, a national triumph. *Influences* was, in the view of *The Nation*, 'a remarkable book, one of the few products of American contemporary science which may safely challenge the best that has been put forth in this field by any foreign scientist whatsoever'.[38] Semple's achievement was seen, moreover, to have subverted the erroneous conflation of femininity and unreason.[39] As *The Nation* made clear, 'Let us add, without any condescension, that it [*Influences*] places Miss Semple among the handful of women in the world over who are the peers of the foremost men of science'.[40]

The response of academic periodicals to Semple's book echoed the popular press's commendation of her scholarship and national intellectual contribution. *The Journal of Geography*, for example, described Semple's text as 'unquestionably

---

32   Albert G. Keller, "Review of *Influences of Geographic Environment*, by Ellen C. Semple," *The Yale Review* 1 (1912): 333.

33   Ibid.

34   *Providence Daily Journal*, 20 August 1911.

35   *Boston Herald*, 2 September 1911.

36   *Boston Evening Transcript*, 5 July 1911; John K. Wright, "Miss Semple's 'Influences of Geographic Environment': Notes Towards a Bibliobiography," *Geographical Review* 52:3 (1962): 346–61.

37   *Boston Evening Transcript*, 5 July 1911.

38   *The Nation* (New York City), 21 December 1911.

39   Genevieve Lloyd, *The Man of Reason: "Male" and "Female" in Western Philosophy* (London: Methuen, 1984).

40   *The Nation* (New York City), 21 December 1911.

the most scholarly contribution to the literature of geography that has yet been produced in America'.[41] The fact that Semple had bolstered the intellectual position of the nation mattered almost as much as did her geographical achievement. As the review's author noted, 'that a volume of such evident and unquestionable merit has been produced by an American geographer, is a matter of just pride to us'.[42] By setting out a systematic approach to the study of human geography, *Influences* also provided a disciplinary model for which 'present geographers cannot but feel a deep sense of gratitude'.[43]

The laudatory assessments which accompanied *Influences*' publication seemed to predict for it no less than revolutionary importance – both to geography's position as a scientific discipline, and to the international standing of American scholarship. Although Semple's text was deemed to afford a 'new vantage ground for the study of man', its perceived significance was not a question so much of its novelty (environmentalism having, after all, been a concern for geographers, historians, and biologists, among others, for much of the previous two decades), but rather its apparent comprehensiveness and rigour.[44] *Influences* was seen to have particular significance as the apotheosis of Semple's anthropogeographical perspective, and as the printed totem of the discipline's emergent scientific manifesto. Perhaps inevitably, the book's predicted potential was realized differently in the different sites of its reading. This was particularly true in respect of the University of Chicago and the University of California at Berkeley – institutions which, to some extent, are representative of geography's different engagements with Semple's text and with her environmentalist ideas.

The University of Chicago and the Emergence of Environmentalist Geography

A concern with environmentalism can be traced at Chicago to the last decade of the nineteenth century when courses offered there in botany and zoology addressed aspects of environmental influence.[45] Part of the reason for the establishment of the department of geography – first proposed in 1902 – was to provide a professional and unified focus to the study of these environmental factors. The new department's potential scope was set out, in part, by the geologist-geographer John Paul Goode,

---

41   Anonymous, "Review of *Influences of Geographic Environment*, by Ellen C. Semple," *The Journal of Geography* 10:1 (1911): 33.

42   Ibid.

43   Ibid., 34.

44   Orin G. Libby, "Review of *Influences of Geographic Environment*, by Ellen C. Semple," *The American Historical Review* 17:2 (1912): 357.

45   William D. Pattison, "Rollin Salisbury and the Establishment of Geography at the University of Chicago," in *The Origins of Academic Geography in the United States*, ed. Brian W. Blouet (Hamden: Archon Books, 1981), 151–63.

who proposed three courses on broadly environmental themes: 'Geographic Ethnology', 'Geographic Factors in Social Development', and 'Racial Problems in America'.[46] Although Goode's approach to human-environment relations was largely deterministic, he understood that whilst environmental influences were generally persistent, the response of societies to them was not. As a consequence most probably of the moderating authority of the department's head, Rollin Salisbury (who saw the physical environment as only one factor influencing the development of societies), Goode's proposed classes did not materialize. Even so, an environmentalist rhetoric – apparent in the 'discourse signalizing' vocabulary of '"geographical influence", "geographic factor", and "geographic condition"' – was central to the department's early curriculum.[47]

In the years before the First World War, the principal function of Chicago's department of geography was 'to train men and women for posts in other universities and colleges'.[48] Historical geography formed an important component of this education, and Semple's paper on the Anglo-Saxons of the Kentucky Mountains and her *American History* were required reading – notably for Harlan Barrows's course on 'Influences of Geography on American History', which ran from 1904. The intellectual stimulus for Barrows's course came from two principal sources: Semple's anthropogeography and Frederick Jackson Turner's historical geography.[49] Barrows's lectures 'introduced historical geography to the curriculum of the American university' and articulated a position for environmentalist themes in geographical study.[50] Of those students who graduated with a Ph.D. in geography in the United States during the first half of the twentieth century, more than one third were Barrows's academic descendants.[51]

Semple joined the department on a part-time basis in 1906. Her appointment, the first of her professional career, reflected both the topicality of her geographical interests, and also the perceived authority of her scholarship. The course which Semple developed – 'Some Principles of Anthropo-geography' – was intended as a general introduction to her intellectual position and to her existing body of work.

---

46   William D. Pattison, "Goode's Proposal of 1902: An Interpretation," *The Professional Geographer* 30:1 (1978): 3–8.

47   Pattison, "Rollin Salisbury and the Establishment of Geography at the University of Chicago," 157.

48   Charles C. Colby to John K. Wright, 23 May 1961, Replies to a Questionnaire Relating to Ellen C. Semple's *Influences of Geographic Environment*, American Geographical Society Library [hereafter cited as Replies to *Influences* Questionnaire, AGSL].

49   Harlan H. Barrows, *Lectures on the Historical Geography of the United States as Given in 1933*, ed. William A. Koelsch (Chicago: Department of Geography, University of Chicago, 1962).

50   Harry Roy Merrens, "Historical Geography and Early American History," *The William and Mary Quarterly* 22:4 (1965): 429–48.

51   Allen D. Bushong, "Geographers and Their Mentors: A Genealogical View of American Academic Geography," in *The Origins of Academic Geography in the United States*, ed. Blouet, 193–219.

Her lectures were considered to be the department's 'most stimulating & inspiring'.[52] The opportunity for Semple to present her work to an audience of enthusiastic students, the first in the United States to receive an explicitly geographical education at graduate level, proved valuable in shaping the subsequent content of her writing, especially *Influences*. Semple's students included 'many who went on to play important roles in the development of professional geography'.[53]

For most graduate students of geography at Chicago during the first three decades of the twentieth century, environmentalism was fundamental to their education; it was present in the teaching of, among other faculty, Barrows, Goode, and Semple. The publication of *Influences* in 1911 captured the 'enthusiasm of the moment' and 'completely illuminated' the potential scope of anthropogeography.[54] Although students' exposure to environmentalism at Chicago was thus near universal, their reaction to it varied considerably. This disparity is illustrated in the differing responses of two students – near contemporaries who went on to exert important but distinct influences on the discipline's development: Carl Ortwin Sauer and Stephen Sargent Visher.

Visher was raised in a remote agricultural community in South Dakota; his boyhood shaped by 'direct contact with the rigorous regime of the upper mid latitude continental climate'.[55] Exposure to the 'day to day vicissitudes of the South Dakota natural environment' provided an important background to Visher's later environmentalist concerns.[56] His emergent interest in the role of climate in shaping nature and society was cemented in the first decade of the twentieth century by study at Chicago under the geographer-botanist Henry Chandler Cowles. Cowles, a pioneer of plant ecology, conducted research which drew upon environmentalist precepts to define 'a causal relation between plant[s] and environment'.[57] Inspired by Cowles's perspective, and by subsequent geological training under Salisbury, Visher undertook research on the biogeography and regional ecology of South Dakota, paying particular attention to the ways in which 'settlers, cowboys, and trappers' had historically adapted to life on the steppe.[58] This research formed the basis to 'The Geography of South Dakota' – a course Visher offered at the

---

52   Robert S. Platt, "Changes in Geographic Thought," 19 March 1950, Folder 16, Box 10, Robert S. Platt Papers, University of Chicago Special Collections Research Center [UCSCRC].

53   Preston E. James, Wilford A. Bladen, and Pradyumna P. Karan, "Ellen Churchill Semple and the Development of a Research Paradigm," in *The Evolution of Geographic Thought in America*, ed. Bladen and Karan, 33.

54   Charles C. Colby, "Changing Currents of Geographic Thought in America," *Annals of the Association of American Geographers* 26:1 (1936): 21.

55   John K. Rose, "Stephen Sargent Visher, 1887–1967," *Annals of the Association of American Geographers* 61:2 (1971): 394.

56   Ibid., 395.

57   Hugh M. Raup, "Trends in the Development of Geographic Botany," *Annals of the Association of American Geographers* 32:4 (1942): 331.

58   Rose, "Stephen Sargent Visher," 395.

University of South Dakota between 1911 and 1913. His lectures examined the 'industrial development of South Dakota as dependent upon ... geographic conditions, especially location, topography, climate and resources'.[59] Visher returned to Chicago for doctoral work in geography in 1913, where his ecological training, research in the field, and teaching experience proved useful preparation, particularly for Semple's course in anthropogeography.

At the time Visher took Semple's course, environmental influence was a predominant geographical concern in the United States and United Kingdom, not least as a consequence of the publication of *Influences*. A survey conducted in 1914 found that three quarters of those geographers questioned considered the determination of the influence of geographical environment to be 'one of the chief problems facing the modern geographer'.[60] In this context, it is perhaps unsurprising that Visher responded to Semple's lectures, and to her text, with alacrity. He later introduced *Influences* to 'a succession of ... advanced students' at Indiana University, where he taught between 1919 and 1957.[61] One of Visher's students at Indiana, Clifford Maynard Zierer, went on in the mid 1920s to propagate Semple's environmentalism at the University of California at Los Angeles in a graduate course on the 'Development of Geographic Thought' and, later, as the department's chair.[62] Under Zierer's direction, the 'departmental philosophy' was derived 'largely from Ellen C. Semple ... and others of similar bent'.[63] As a result of the frequent reference to *Influences* in Zierer's course, 'the copies placed on reserve at U.C.L.A. were well worn' – a material trace of Semple's intellectual significance at that institution.[64] Environmentalism remained a component of the department's curriculum until at least 1956 – a consequence of the presence on the faculty of one of Semple's most devoted students, Ruth Emily Baugh.

Although not entirely uncritical of Semple's work, most of the courses Baugh offered at UCLA were 'on regions and topics that had been of interest to Miss Semple – Europe, Historical Geography of the Mediterranean Region, the Geographic Basis of Human Society'.[65] Whether directly by reference to *Influences*,

---

59   Anonymous, *Thirtieth Annual Catalogue of the University of South Dakota 1911–1912* (Vermillion: University of South Dakota, 1911).

60   George B. Roorbach, "The Trend of Modern Geography," *Bulletin of the American Geographical Society* 46:11 (1914): 803.

61   Stephen S. Visher to John K. Wright, 25 March 1961, Replies to *Influences* Questionnaire, AGSL.

62   James O. Wheeler and Stanley D. Brunn, "An Urban Geographer Before his Time: C. Warren Thornthwaite's 1930 Doctoral Dissertation," *Progress in Human Geography* 26:4 (2002): 463–86.

63   Joseph E. Spencer, "A Geographer West of the Sierra Nevada," *Annals of the Association of American Geographers* 69:1 (1979): 46.

64   Ruth E. Baugh to John K. Wright, 24 May 1961, Replies to *Influences* Questionnaire, AGSL.

65   Henry J. Bruman, Archine Fetty and Joseph E. Spencer, "Ruth Emily Baugh, Geography: Los Angeles," *In Memoriam* (1976): 11.

or indirectly through her own teaching, Baugh facilitated the communication and dissemination of Semple's geographical work during much of the first half of the twentieth century. In a career which spanned five decades at UCLA, Baugh influenced the undergraduate experience of a number of future geographers, including Robert Cooper West, Evelyn Lord Pruitt, and Peter Hugh Nash. Although, as Nash recalled, 'Baugh almost worshiped Semple, and much of this admiration rubbed off on me', Semple's anthropogeography was not necessarily inherited by Baugh's students.[66] Baugh's personal affection for Semple did not translate to an evangelical espousal of anthropogeography, but it was manifest in her inclination to direct her students to Semple's work (either through her own teaching, or by direct reference to *Influences* or *American History*).

The environmentalist orientation of the geography curriculum at UCLA was challenged, however, most notably by Joseph Earle Spencer, an undergraduate between 1925 and 1929. For Spencer, the 'simplistic and one-sided views' embodied in the environmentalist position 'caused me considerable difficulty, and I was summarily ejected from class on several occasions for arguing with instructors'.[67] Spencer was supported in his dubiety by Jonathan Garst, an Iowa-born, Edinburgh-educated geographer, who joined the faculty in 1927. As Spencer recalled, Garst's 'views were very different from those of the American-trained faculty', and he introduced Spencer to the work of European geographers, particularly those of the French school of regional geographers, who advocated a non-deterministic understanding of the relationship between society and its environment.[68]

Geography as practiced in the 1920s at UCLA, Indiana, and Chicago was linked by the common sensibility of three scholars – Zierer, Visher, and Semple – but was not defined solely by them. The circulation of Semple's anthropogeography in both its pedagogical and textual guises was facilitated by the literal movements of Visher and Zierer, and by Chicago's initial remit to train students for teaching careers. It was also challenged by the emergence of alternative, non-deterministic perspectives. Scholastic lineages such as that which connected the otherwise distant cities of Chicago, Bloomington, Indiana, and Los Angeles – and conditioned the reading of *Influences* in those places – were nevertheless central to the successful diffusion of environmentalist thought, particularly in the form of Semple's anthropogeography. Semple's students – her intellectual descendants – were not uniformly evangelical. In addition to the disciples who emerged from her classroom, there was an equally vociferous lineage of sceptics, of whom Carl Sauer became the foremost representative.

---

66   Peter Hugh Nash, Leonard Guelke and Richard E. Preston, *Abstract Thoughts, Concrete Solutions: Essays in Honour of Peter Nash* (Waterloo: Department of Geography, University of Waterloo, 1987), 280.

67   Spencer, "A Geographer West of the Sierra Nevada," 47.

68   Ibid.

Carl Sauer and the 'rebellion against determinism'

Sauer came to graduate geography in 1909, having previously trained in geology. His entrée to geographical research was shaped by the environmentalist perspectives of Semple and Barrows, but he came later to be considered synonymous with 'criticism of environmental determinism'.[69] His exposure to what he subsequently termed the 'rather simple mechanical theory of behaviour' advanced by Barrows and Semple, whilst seeming not to trouble him unduly at first, was an idea against which he later rebelled.[70] By the 1920s, Sauer viewed environmentalism as an unduly dominant component of American geography and advanced a 'detailed and devastating refutation' of it which 'singled out for critical consideration' Semple and likeminded contemporaries.[71] His disillusion with environmentalism can be traced to his time 'as apprentice and journeyman geographer', during which his observations in the field appeared increasingly to contradict the causal mechanisms which Semple's anthropogeography described.[72]

In 1910, having completed a year of study at Chicago, but before having read *Influences*, Sauer was sent under the titular supervision of Salisbury to complete a geological/geographical examination of the upper valley of the Illinois River. With little direction from Salisbury as to his research focus, Sauer chose to examine the physical origin of the grassland environment and the historical influence of the plains upon pioneer settlers. As Sauer subsequently noted, his study was, in this later respect, 'an attempt to apply the orientation then prevailing of human adaptation to physical environment'.[73] Even at this stage, however, Sauer harboured 'some early doubts that this direction [environmental influence] was adequate or proper'.[74] Despite his concerns as to the explanatory validity of anthropogeography, it is apparent that Sauer undertook this and subsequent fieldwork with certain environmentalist presuppositions in place. His field notebook from an expedition to the Missouri section of the Ozarks in 1914 is, for example, scattered with comments which reflect this student training in geographical influence: 'The people are the typical Missouri hillfolk'; 'typical backwoods – log cabins – log everything'; 'Fertile population still largely of French descent & decided Frenchtypes. They

---

69   Joanna E. Beck, "Environmental Determinism in Twentieth Century American Geography: Reflections in the Professional Journals" (PhD diss., University of California, 1985), 17.

70   Anonymous, "Annual General Meeting," *The Geographical Journal* 141:3 (1975): 521.

71   William W. Speth, *How it Came to Be: Carl O. Sauer, Franz Boas and the Meanings of Anthropogeography* (Ellensburg: Ephemera Press, 1999), 182.

72   Anonymous, "Annual General Meeting," 521.

73   Carl O. Sauer, *Seventeenth-Century North America* (Berkeley: Turtle Island Foundation, 1980), 9.

74   Ibid.

speak a very broken lingo & are said not to be able to read the printed French'; 'Good example of influence of isolation'.[75]

Sauer came to realize, however, that for each observation which appeared to confirm the role of environment in shaping social organization, another seemed contradictory. The most significant of these observations was that the architectural, agricultural, and social traditions of different immigrant groups to the Ozarks persisted, despite the fact that they were in (culturally speaking) a "new" environment. If geographical conditions were truly the predominant mechanism for determining these cultural expressions, then it might be assumed that descendents of immigrant Germans, French, New Englanders, and Kentuckians, among others, would work the land in similar ways, and adapt their architectural practices to reflect the requirements of the environment, rather than maintaining the traditions of their cultural heritage. The fact that the Ozarks were geographically relatively homogeneous, yet culturally heterogeneous, gave Sauer pause. As previously suggested, Sauer's concerns emerged only gradually; it took extended methodological debate and revision before he rejected environmentalism in its entirety. As Sauer later noted, 'Most of the things I was taught … as a geographer I had either to forget or unlearn at the cost of considerable effort and time'.[76]

Part of the process of unlearning occurred at Chicago in the years between Sauer's field seasons in the Illinois River valley and the Ozarks, when 'a vigorous group of graduate students' began to discuss alternatives to the then-dominant environmentalism.[77] It was in this context that Sauer read *Influences* for the first time, and subjected its ideas to sceptical appraisal, but not, at least initially, to rejection. The fact that the principles of anthropogeography were tested at Chicago so vigorously was a consequence, in part, of Semple's approach to its teaching. As one student recalled, 'she often said she hoped her work would prove or disprove the values of anthropogeography' and 'that if better theory came along she hoped she would have strength of mind to embrace it'.[78] In this respect, Semple permitted and facilitated an environment which was productive of critical and independent thought. It was at Chicago, then, that the 'rebellion against environmentalism' found its earliest expression in the United States.[79]

---

75   "Field Notes, 1914," Folder 24, Carton 4, BANC MSS 77/170 c, Carl O. Sauer Papers, The Bancroft Library, University of California at Berkeley [UCB].

76   John Leighly, "Carl Ortwin Sauer 1889–1975," in *Geographers: Biobibliographical Studies*, vol. 2, ed. Thomas W. Freeman and Philippe Pinchemel (London: Mansell, 1978), 100.

77   Robert S. Platt to Walter M. Kollmorgen, 12 May 1956, Folder 6, Box 3, Robert S. Platt Papers, UCSCRC.

78   Charles C. Colby to Whittlesey, 4 February 1954, HUG 4877.417, Derwent S. Whittlesey Papers, Harvard University Archives.

79   Robert S. Platt to Walter M. Kollmorgen, 12 May 1956, Folder 6, Box 3, Robert S. Platt Papers, UCSCRC.

The rebellious mood which Sauer and likeminded dissenters promoted was not, as Visher's experience has shown, shared by all the department's graduate community. Bernard Henry Schockel, Mary Lanier, Almon Ernest Parkins, and Mary Dopp – whose periods as graduate students coincided with that of Sauer – were all important in promoting environmentalism in their later teaching careers: Schockel at the Indiana State Normal School; Lanier at Wellesley College in Massachusetts; Parkins at the George Peabody College for Teachers in Nashville, Tennessee; and Dopp at various high schools in Chicago. It is worth noting, however, that Sauer's dissatisfaction with anthropogeography did not in its initial expression equate to a straightforward rejection of *Influences*. When he attained his first academic appointment – at the University of Michigan in 1915 – the general introductory course he developed was 'built about Ellen Semple's ideas as expressed in her *Influences of Geographic Environment*'.[80] The fact that Sauer employed Semple's text in this way, even whilst harbouring misgivings about the anthropogeographical project, illustrates both something of his ambivalence, and also the fact that the pedagogical value of Semple's text could be judged in isolation from Sauer's more general concerns about environmentalism.

Sauer's time at the University of Michigan provided the opportunity to pursue new research concerns and to develop his pedagogic skills. Fieldwork in Michigan, Kentucky, and New England in the early 1920s – mapping '"natural" and "cultural" landscapes' – was important in reinforcing his non-deterministic perspective on human-environment relations.[81] In 1923, Sauer accepted the positions of professor of geography and chairman of the department at the University of California at Berkeley – a move which brought him into contact with the anthropologist Robert Henry Lowie. Lowie encouraged Sauer to reappraise the work of Ratzel – specifically the second volume of *Anthropogeographie*. As Sauer recalled, 'Lowie got me to understand Ratzel against whom I had been prejudiced by Miss Semple's enthusiasm for her great master environmentalist'.[82] From Ratzel, Sauer inherited an epistemic concern – evident in the second volume of *Anthropogeographie* – which took culture (rather than environment) as an organizing factor. Concurrent with his refamiliarization with Ratzel's work, Sauer completed a methodological monograph which he saw as part of a process of 'emancipating myself from the dictum then ruling at Chicago'.[83] Sauer's methodological reappraisal – 'The morphology of landscape' – outlined an approach to geography that placed empirical focus upon cultural landscapes.[84]

---

80   John Leighly, "Carl Ortwin Sauer," 338.

81   Ibid.

82   Carl O. Sauer to William W. Speth, 3 March 1972, Box 18, BANC MSS 77/170 c, Carl O. Sauer Papers, UCB.

83   Carl O. Sauer to Richard Hartshorne, 22 June 1946, Box 11, BANC MSS 77/170 c, Carl O. Sauer Papers, UCB.

84   Carl O. Sauer, "The Morphology of Landscape," *University of California Publications in Geography* 2:2 (1925): 19–54.

Sauer's paper – described later as 'the famous piece that blasted determinism' – sought to disrupt what he saw as the mechanistic and deterministic bases of environmentalism.[85] In this respect, his work represented an overt criticism of Semple, but seems not to have affected their 'friendship of long standing'.[86] Whilst Sauer admitted to Semple that his work was in 'quite a different direction from that in which you have worked', he believed that she would be 'sympathetic toward what we are trying to do in the study of the succession of natural and cultural landscapes'.[87] Whilst Semple had been subject to similar critical appraisal in France – notably in Lucien Febvre's *La terre et l'évolution humaine* (1922) – Sauer's paper was the most explicit condemnation of her work in the North American literature.[88]

Sauer's accession to chairmanship of the Berkeley geography department effectively eliminated the teaching of environmentalism there. It had, however, occupied an important place in the curriculum during the previous decade. The principal exponent of environmental influence at Berkeley was Sauer's predecessor, Ruliff Stephen Holway.[89] Holway had trained as a geologist at Stanford University before joining the geography faculty at Berkeley in 1904. There, Holway offered a number of courses which 'mirrored the prevailing geographic opinion of the time' – dealing, in various ways, with human-environment relations.[90] His course on 'General Physical Geography' attended, for example, to 'Land forms, climatology, oceanography, and planetary relations, and their effect upon human affairs', whilst his course on 'Geographical Influences in the Western United States' dealt with 'The geographic conditions which have influenced the exploration and early settlement of the west and the present effect of physical factors on the life of the people'.[91]

Although there was an implicit move away from the environmentalist imperative in geographical research at Berkeley from the mid 1920s, this did not equate to a straightforward excision of Semple's work: *Influences* was used for 'many years' in 'Geography 151, American Geographic Thought'.[92] The book

85   Beck, "Environmental Determinism in Twentieth Century American Geography," 123.

86   Carl O. Sauer to Ellen C. Semple, 20 June 1926, B4-18-11, Wallace W. Atwood Collection, Clark University Archives and Special Collections.

87   Ibid.

88   Lucien Febvre, *La terre et l'évolution humaine: introduction géographique à l'histoire* (Paris: La Renaissance du Livre, 1922).

89   Carl O. Sauer, "Memorial of Ruliff S. Holway," *Annals of the Association of American Geographers* 19:1 (1929): 64–5.

90   William W. Speth, "Berkeley Geography, 1923–33," in *The Origins of Academic Geography in the United States*, ed. Blouet, 224.

91   Anonymous, *Annual Announcement of Courses of Instruction for the Academic Year 1910–11* (Berkeley: University of California Press, 1910), 113–5.

92   Clarence J. Glacken to John K. Wright, 11 May 1961, Replies to *Influences* Questionnaire, AGSL.

was also employed in other parts of the university, notably in the Department of Social Instructions. It was there that Clarence James Glacken, who went on to write a celebrated history of environmentalism, *Traces on the Rhodian Shore*, was introduced to the book by Frederick John Teggart.[93] Glacken undertook Teggart's year-long course 'The Idea of Progress' in 1928, and as preparation for that class read Semple's book. Teggart recommended *Influences* as 'a significant book in the general field of the history of ideas', and Glacken read it alongside 'the *Kleine Schriften* of Ratzel, and the writings of the French possibilist school'. [94]

Glacken's encounter with Semple's book – juxtaposed as it was with the work of Ratzel and of the French school – was distinct from those earlier students for whom *Influences* was presented principally as a source of instruction, rather than as a point of comparison. Glacken's approach to the text was, then, somewhat more critical and considered than that of certain of his predecessors. Teggart presented Semple's work 'as an example of environmental explanation of cultural differences … not as a necessarily valid exposition of the problem'. For this reason, Glacken did not feel a particular need at this point to express an opinion as to the validity of Semple's conclusions. He was inclined, rather, to view *Influences* in its intellectual context as 'an important landmark in the history of ideas'.[95]

During the 1950s, Glacken returned to Berkeley as a faculty member. He inherited responsibility for 'Geography 151' and devoted 'at least an hour, often more, to a discussion of selected chapters' of *Influences*. Glacken also made use of Semple's book in his course on 'Relations Between Nature and Culture'. As he later recalled, 'Students are always interested in some of Miss Semple's more detailed analyses and of course are critical'.[96] This period was marked by the increasing dominance of quantitative methods in geography, and the cultural geography that Sauer had developed in response to environmentalism was itself being challenged. The geography faculty at Berkeley was 'divided into factions either defending the "Berkeley School" … or trying to turn geography's course into a more "modern" direction'.[97] In much the same way that concerns emerged in the 1920s over the validity of Semple's method, so too the authority and value of Sauer's geography was questioned in the 1960s.

In different ways, and for different reasons, Semple's book had an important role in the teaching of environmentalist thought at Berkeley. Although the principles upon which the book depended had been effectively refuted by Sauer

---

93   Clarence J. Glacken, *Traces on the Rhodian Shore: Nature and Culture in Western Thought from Ancient Times to the End of the Eighteenth Century* (Berkeley and Los Angeles: University of California Press, 1967).

94   Clarence J. Glacken to John K. Wright, 11 May 1961, Replies to *Influences* Questionnaire, AGSL.

95   Ibid.

96   Ibid.

97   Anne Macpherson, "Clarence James Glacken 1909–1989," in *Geographers: Biobibliographical Studies*, vol. 14, ed. Geoffrey J. Martin (London: Mansell, 1993), 33.

in the 1920s, Semple's book fulfilled two distinct roles: an instructional tool for Holway during the second decade of the twentieth century, and an illustrative example of environmentalist thought for Teggart and Glacken from the 1930s. In these ways it continued to function, albeit in an altered capacity, after its thesis had been gainsaid by Sauer and others at Berkeley. The rejection of *Influences* at Berkeley, and one might suppose also at other institutions, cannot be regarded straightforwardly, then, as coterminous with the end of its "career". Semple's book had an afterlife – its usefulness transcended its ability to shape and to direct the course of geographical research.

Although Chicago and Berkeley are often regarded as synonymous with certain approaches to geographical research – Chicago with Semple's anthropogeography; Berkeley with Sauer's cultural geography – the differing responses to environmentalism at (and within) these institutions belie the notion of a unanimity of geographical perspective.[98] As communities of individual scholars and students, it is unsurprising that opinion as to the validity of Semple's anthropogeography varied within or between different institutions. When the concerns and intellectual interests of individual geographers are juxtaposed with those of their institutional or disciplinary context, however, some of the complicating factors which determined the geography of *Influences'* reception become clear. Moreover, the circulation and reception of Semple's anthropogeography was not simply a matter of reading; it had to do with, among other things, the connectedness and mobility of geographers and the personal networks which linked Semple's proponents and opponents.

Conclusion: Towards a Geography of Textual Reception

In the forty years before Semple's death in 1932, the nature and position of American geography changed significantly. Semple's career as a geographer paralleled the discipline's institutionalization and professionalization; her own contribution to its methodological focus was significant and, for a time, dominant. Semple was not, of course, the only proponent of environmentalism during this period. Yet *Influences* proved unusually important in communicating the principles of anthropogeography (in part because there was 'practically nothing in English' on the topic).[99] Given that Semple's book 'was published when few English geographers read German', her text served also as an important means of presenting to students and scholars of geography a selected part of Ratzel's

---

98   Chauncy D. Harris, "Geography at Chicago in the 1930s and 1940s," *Annals of the Association of American Geographers* 69:1 (1979); Speth, "Berkeley Geography, 1923–33".

99   John N.L. Baker to John K. Wright, 30 April 1961, Replies to *Influences* Questionnaire, AGSL.

work.[100] Semple's presentation of anthropogeography coincided with and helped to define a period of methodological realignment in disciplinary geography in the United States. Her book succeeded in fulfilling a pedagogical role associated with this realignment – that of providing 'a firm interpretation of the influence of the environment' written 'in such a way that the [student] reader understands Semple's meaning'.[101]

Influences built upon a number of earlier anthropogeographical pronouncements on Semple's part. Its contents were not, strictly speaking, novel. The book's intended function was to provide a complete and coherent statement of Semple's perspective on environmental influence – not a definitive statement on the remit and methods of anthropogeography, but an indication of potential scope and possible approaches. It was, then, in its educational role that Semple's text had its most direct influence upon the teaching of geography during the second decade of the twentieth century, and upon the discipline's subsequent research focus.

Semple's book was, at turns, a textbook for students of anthropogeography; a convenient proxy of Ratzel's thought; an emblem of an outmoded and erroneous deterministic environmentalism; and an exemplar of one particular expression of the environmentalist tradition. The reading, reception, repudiation, and reappraisal of *Influences* varied with time and across space, and varied between people in (sometimes different) institutional contexts. The institutional uses of Semple's text and their engagement with environmentalism more generally cannot be separated, then, from the individuals of whom they were comprised. The individual readings of *Influences*, together with the different and particular uses to which it was put in educational settings, show that the reception of Semple's text was not a fixed and singular event – an either/or judgement of acceptance or repudiation – but was an ongoing process, changing either in terms of location or time as a consequence of shifting attitudes, novel experiences in the field, or the vagaries of scholarly topicality.

In attending to individual engagements with Semple's work, this chapter has been concerned to show that the social networks upon which the circulation, dissemination, and reception of *Influences* depended were not inscribed or constrained by particular spatial scales. That is, whilst certain scales facilitated particular types of engagement, the interpretative networks and hermeneutic communities in which Semple's ideas circulated existed both within and beyond the institutional scale. These networks depended, however, upon a series of nodes – people, lecture theatres, field sites, periodicals – whose physical location meant that they were not "beyond" the influence of their spatial situation. Whilst certain of these networks can be seen, broadly, to correspond to a particular scale (the metropolitan in the case of the writers and readers of newspapers reviews, for

---

100   George Tatham to John K. Wright, n.d., Replies to *Influences* Questionnaire, AGSL.

101   Albert S. Carlson to John K. Wright, n.d., Replies to *Influences* Questionnaire, AGSL.

example), a number (members of a disciplinary community, for instance) cannot be rendered so neatly as cartographical abstractions. It is necessary, when considering the reception of *Influences*, to understand the connections between scales, not to treat them as discrete entities.

The plural and disparate readings of *Influences* show that the reception of Semple's ideas was not a binary defined by acceptance or rejection *in toto* (as the example of Sauer makes particularly clear), but was a complicated process in which her anthropogeography was remade, and remade differently, by individual readers' encounters with it. What was being received, then, was always new and was always the result of negotiation between the text and the reader. This interpretative complexity was not a function of the veracity of *Influences* as a representative of Semple's ideas, but a reflection of the fact that anthropogeography did not have singular and fixed meaning. It existed as an effectively infinite number of potential meanings. Although the book's intended purpose was to convey Semple's ideas, it might more properly be understood to have acted as the prompt to the creation of new knowledge.

The reception of *Influences* was not a matter simply of its reading: the circulation of Semple's anthropogeography depended upon its representation and reproduction in a number of distinct forms and different spaces. The lecture theatre and classroom were, for example, often as important as the text itself in the communication of Semple's ideas. With time, different interpretations of Semple's text became codified in lecture courses and examinations. These mediated representations often mattered as much in informing audiences' opinion as to the particular qualities of anthropogeography than did the actual reading of *Influences*. It is evident, then, that in thinking about the communication of Semple's knowledge, it is important to acknowledge that its reading and its reception were not the same thing.

The parallel between geography's engagement with environmentalism and geographers' reading of *Influences* illustrates the disciplining influence of Semple's book and its function in the communication of knowledge and practice. As a number of recent studies have shown, the history of geography is also a history of geography's printed texts.[102] Attending to this common history – an approach which might be through of as bibliographical historiography – unites biographical and prosopographical concerns in addressing the reasons why knowledge and ideas are conceived of and received differently in different places. The spatial variation in the reception of Semple's text exposes differences in the individual and collective practices of reading. These interpretative dissimilarities belie the notion of uniform institutional engagements with environmentalism, and show how the experiences and preconceptions of individual geographers, and the networks

---

102   Robert J. Mayhew, "The Character of English Geography *c.* 1660–1800: A Textual Approach," *Journal of Historical Geography* 24:4 (1998): 385–412; Trevor J. Barnes, "Performing Economic Geography: Two Men, Two Books, and a Cast of Thousands," *Environment and Planning A* 34:3 (2002): 487–512.

which they formed, conditioned how Semple's text functioned pedagogically in different academic venues. More particularly, however, it is apparent that what environmentalism was taken to be was as much the consequence of conflict and disputation as it was of concurrence and unity.

Acknowledgements

I acknowledge with thanks the Arts and Humanities Research Council, whose funding permitted the research upon which this chapter is based. For permission to quote from archives in their care, I thank the American Geographical Society Library, the Bancroft Library of the University of California at Berkeley, the University of Chicago Special Collections Research Center, Clark University Archives and Special Collections, Harvard University Archives, and the Royal Geographical Society. For guidance and assistance, I am indebted to the editors of this volume.

Index